217

Topics in Current Chemistry

Springer

Berlin
Heidelberg
New York
Barcelona
Hong Kong
London
Milan
Paris
Tokyo

Dendrimers IV

Metal Coordination, Self Assembly, Catalysis

Volume Editors: Fritz Vögtle,
Christoph A. Schalley

With contributions by
H.-F. Chow, A. Hirsch, A. W. Kleij,
R. J. M. Klein Gebbink, R. Kreiter, L. J. Lawless,
C.-F. Leung, T. K. Lindhorst, D. N. Reinhoudt,
N. Röckendorf, G. van Koten, H.-J. van Manen,
F. C. J. M. van Veggel, O. Vostrowsky, G.-X. Wang,
J. Zhang, S. C. Zimmermann

Springer

The series *Topics in Current Chemistry* presents critical reviews of the present and future trends in modern chemical research. The scope of coverage includes all areas of chemical science including the interfaces with related disciplines such as biology, medicine and materials science. The goal of each thematic volume is to give the non-specialist reader, whether at the university or in industry, a comprehensive overview of an area where new insights are emerging that are of interest to a larger scientific audience.

As a rule, contributions are specially commissioned. The editors and publishers will, however, always be pleased to receive suggestions and supplementary information. Papers are accepted for *Topics in Current Chemistry* in English.

In references *Topics in Current Chemistry* is abbreviated Top. Curr. Chem. and is cited as a journal.

Springer WWW home page: http://www.springer.de
Visit the TCC home page at http:/link.springer.de/series/tcc/
or http://link.springer-ny.com/series/tcc/

ISSN 0340-1022
ISBN 3-540-42095-9
Springer-Verlag Berlin Heidelberg New York

Library of Congress Catalog Card Number 74-644622

Springer-Verlag Berlin Heidelberg New York
a member of BertelsmannSpringer Science+Business Media GmbH

Printed in Germany

Cover design: Friedhelm Steinen-Broo, Barcelona; MEDIO, Berlin
Typesetting: Fotosatz-Service Köhler GmbH, 97084 Würzburg

SPIN: 10753435 02/3020 ra – 5 4 3 2 1 0 – Printed on acid-free paper

Volume Editor

Prof. Dr. Fritz Vögtle
Dr. Christoph A. Schalley
Kekulé-Institut für Organische Chemie
und Biochemie der Universität Bonn
Gerhard-Domagk-Straße 1
53121 Bonn, Germany
E-mail: voegtle@uni-bonn.de

Editorial Board

Preface

Dendrimers stand within the focus of quite an interdisciplinary area of research: Metallodendrimers bring inorganic chemistry into play. Organic synthesis contributes much to the preparation of dendrimers, which are then studied by various physicochemical methods such as small angle neutron scattering, photochemistry, and many others. The relation to macromolecules is straightforward, but their routine use in biochemistry, e.g., as gene transfection vectors may be less obvious. All these different aspects have been combined in the Topics tetralogy in order to provide an overview as broad as possible in this fascinating field of chemistry.

The fourth and final issue in the series starts with a chapter by Chow on the synthesis of dendritic oligoethers, which represent polypodands soluble in many solvents. Two contributions deal with dendrimers based on the "less-than-covalent" bond. While metal coordination as described in the review by Reinhoudt still employs rather strong bonds with bond energies close to covalent bonds, Zimmerman's overview comprises dendrimers that self-assemble via weak forces such as hydrogen bonding. Biologic activity is one of the major topics in Lindhorst's overview of glycodendrimers, which have become a useful tool for the study of carbohydrate-protein interactions and multivalency. The article by Hirsch on fullerenes containing dendrimers provides extensive information on their properties as new materials. Finally, function again is a major topic, when catalysis (van Koten) is achieved using dendrimers.

With this volume of Topics in Current Chemistry, the tetralogy ends, but definitely not the development in dendrimer chemistry. Let us therefore make a few concluding remarks. The history of dendrimers teaches us again a well-known lesson: It is sometimes quite a long way from fundamental science to applications. When the research on regular multi-branched molecules started in 1978 with what was originally coined "cascade" molecules, the analysis was difficult and hampered development for several years, until modern mass spectrometric methods provided a more solid basis for detection of defects and impurities. In the mid 1980s, dendrimer research started to flourish. By now, many synthetic and analytical problems have been solved and the properties of the new compounds are studied with respect to their many potential functions putting dendrimers on the verge of commercial applications. Some of the advantages of multibranched molecules result from their monodispersity which is a valuable feature also for a more profound analysis and understanding of the features of dendritic molecules. However, once fully understood, polydispersity might no longer be considered a problem, but used deliberately to fine tune the desired properties by generating tailor-made blends of dendrimers.

With respect to the future, we see several prospects for further research in dendrimer chemistry: (i) Efficient methods for the controlled high-yield functionalization of only several branches still have to be devised. This would open the door towards more complex structures that are also capable of fulfilling more complicated functions. (ii) Although much work has been devoted to the host-guest chemistry of dendrimers (which has been under dispute for quite a lengthy while), it is not fully understood. (iii) We expect further improvement in the synthetic protocols leading to structurally almost perfect dendrimers with even higher generations and larger numbers of branches than available hitherto. (iv) Chirality in such highly mobile and dynamic multicenter molecules will provoke new questions and answers and might also have an important impact on our knowledge of chirality. These few examples should suffice to make clear it that dendrimer chemistry still is a lively field of challenging research from which we can expect a whole bunch of intriguing new results and additional distinct applications.

Bonn, July 2001

Fritz Vögtle, Christoph A. Schalley
Kekulé-Institut für Organische Chemie und Biochemie
Universität Bonn

Contents

Contents of Volume 197

Dendrimers I

Volume Editor: Fritz Vögtle

ISBN 3-540-64412-1

Contents of Volume 210

Dendrimers II

Architecture, Nanostructure and Supramolecular Chemistry

Volume Editor: Fritz Vögtle

ISBN 3-540-67097-1

Contents of Volume 211

Dendrimers III

Design, Dimension, Function

Volume Editor: Fritz Vögtle

ISBN 3-540-67828-X

Dendritic Oligoethers

Hak-Fun Chow, Cham-Fai Leung, Guo-Xin Wang, Jie Zhang

Department of Chemistry, The Chinese University of Hong Kong, Shatin, NT, Hong Kong, HKSAR
E-mail: hfchow@cuhk.edu.hk

This manuscript summarizes the recent research progress on the synthesis of oligoether-based dendrons. Methods for preparing the individual dendrons and branching agents are outlined, along with a survey of their uses in the preparation of functional and structural dendrimers that possess a wide range of chiroptical, physical, photochemical, electrochemical, or catalytic properties.

Keywords. Dendrimers, Oligoethers, Synthesis

1 Introduction

Dendrimer chemistry has begun to merge with different research avenues due to the recent surge of interest in its applications in biological and medicinal chemistry, catalysis and material sciences. This is partly due to the availability of various synthetic armories that greatly facilitate the construction of a wide variety of dendritic structures with pre-defined architecture. Another reason is attributed to the fact that dendritic macromolecules do sometimes possess unusual structural features and properties that are not often seen in the study of conventional polymer molecules. A number of review articles have already been devoted to the synthesis and property investigations of dendrimer molecules [1]. The purpose of this chapter is not to provide an exhaustive account of the latest development in this vast area of chemistry. Instead we wish to focus our attention on a particularly small subclass of dendritic molecules, namely oligoether dendrimers, and to show how complex supramolecular systems can be built from this simple class of dendritic fragments. In the next section, we will begin to review the chemistry of the most commonly encountered dendritic oligoethers in the literature.

dendrimer [Gn]-$(f)_s$ dendron [Gn]-f

Fig. 1. Schematic diagram of a divergently synthesized dendrimer and a convergently synthesized dendron

Strictly speaking, most dendrimers reported in the literature are copolymers and may contain non-ether-based functional groups in part of their structure. We believe it is necessary to include these compounds in our discussions and therefore our attention will be focused on those dendritic fragments having repeating units made up solely of aryl ether or alkyl ether functionalities, disregarding the nature of their core and surface functionalities. Particular emphasis will be paid to their synthetic efficiency (i.e., % yield), molecular properties (i.e., molecular weight and diameter), and structural purities (i.e., polydispersity index), which are of crucial importance for choosing the appropriate dendritic fragments as 'Lego' sets towards the construction of complex dendritic structures. We will also provide a brief survey on the use of various subclasses of oligoether dendritic fragments towards the construction of functional dendrimers. However, elaborated discussions will be concentrated on those derived from the 3,5-dihydroxybenzyl ether repeating unit, which is by far the most popular and commonly used dendritic oligoether building block in the literature.

In this manuscript, dendrimers synthesized by a divergent approach are designated as [Gn]-$(f)_s$, where n is the generation number, i.e., the number of layers of repeating branching units, $(f)_s$ is the functional group on the dendrimer surface. For the structural diagrams, this type of dendrimer will be represented by a circle with the surface functionality drawn inside a small bracket (Fig. 1). On the other hand, the notation [Gn]-f (without the small brackets and the subscript 's') originally proposed by Fréchet is adopted to represent dendrons synthesized by the convergent approach [2], where f denotes the reactive functional group located at the local point. The corresponding structural diagram is shown as a pie with the focal functional group labeled at the focal point.

2 Synthesis and Properties of Dendritic Oligoethers

2.1 Dendritic Oligoethers Based on a 2,2,2-Tris(hydroxymethyl)-1-ethoxy Repeating Unit

This series was probably the earliest example of dendritic oligoethers reported in the literature. In 1987, Hall and coworkers described the use of pentaerythritol as a repeating unit for the construction of a series of alkyl ether dendrimers (e.g., **1**) [3]. The synthesis was based on a four-step divergent iterative cycle (Scheme 1) [4]. A dendritic oligoalcohol **2** [Gn]-$(OH)_s$ was converted into the corresponding bromide **3** [Gn]-$(Br)_s$ via the oligo-tosylate in two steps. The oli-

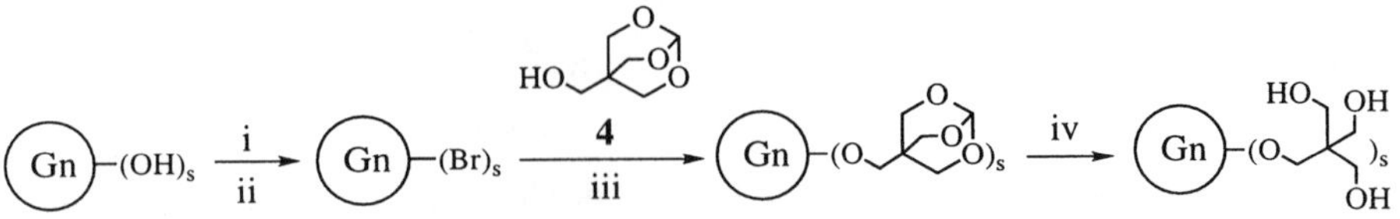

2 [Gn]-$(OH)_s$ **3** [Gn]-$(Br)_s$ **1** [G(n+1)]-$(orthoester)_s$ **5** [G(n+1)]-$(OH)_s$

Scheme 1. i) pyridine, TsCl; ii) dimethylacetamide, NaBr, 150 °C; iii) KH, diglyme or diethylene glycol, 162 °C; iv) MeOH, HCl

goether **1** [G(n + 1)]-$(orthoester)_s$ was prepared by the Williamson reaction between the bromide **3** and the potassium salt of a branching agent **4**. Subsequent acid-catalyzed hydrolysis of the orthoester then afforded the oligo-alcohol **5** [G(n + 1)]-$(OH)_s$ of the next generation. Starting from pentaerythritol as the core, the [G2]- and [G3]-dendrimers were prepared (Table 1). However, substantial structural defect was detected for the [G3]-series of compounds as revealed by size exclusion chromatography (SEC).

This series of oligoether dendrimers was used by Ford to construct catalytically active dendrimers **6** bearing multiple quaternary ammonium surface groups (Fig. 2) [5]. The higher generation catalyst was shown to exhibit higher reactivity than the lower generation ones on a per catalyst center basis.

Table 1. Selected data for 2,2,2-tris-(hydroxymethyl)-1-ethyl ether-based dendrimers[a]

n	Yield (%) [Gn]-$(OH)_s$ → [G(n + 1)]-$(OH)_s$	Calcd MW of [Gn]-$(orthoester)_s$	Molecular diameter of [Gn]-$(OH)_s$ (Å)[b]
0	38	–	–
1	33	649	9.8
2	12	2146	18.6
3	–	6639	25.0

[a] Core = [G0]-$(OH)_s$ = pentaerythritol.
[b] Determined by SEC, see [3].

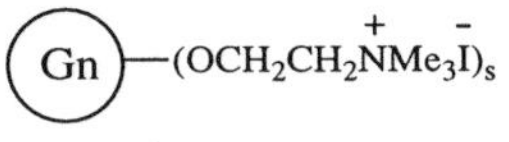

6 (n = 1, 2)

Fig. 2. Functional dendrimers constructed from the 2,2,2-tris(hydroxymethyl)-1-ethyl ether-based dendritic skeleton

2.2 Dendritic Oligoethers Based on a 1,3-Dihydroxy-2-propoxy Repeating Unit

Yamamoto reported the convergent preparation of oligoether dendrons using glycerol as the branching unit [6]. The key reaction was the Williamson ether synthesis between a dendritic alcohol [Gn]-OH **7** (2 mol equiv) and epichlo-

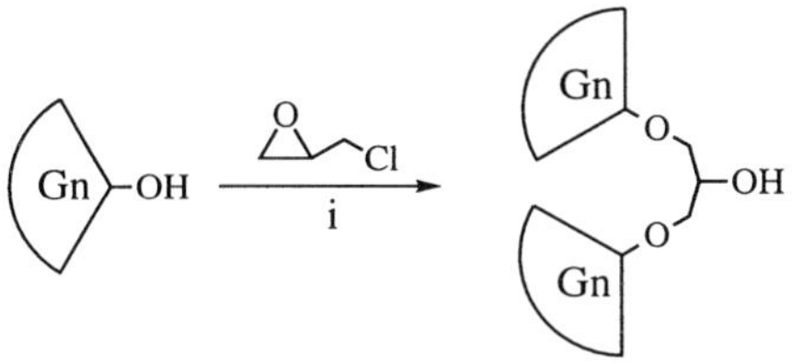

7 [Gn]-OH **8** [G(n+1)]-OH

Scheme 2. i) Bu_4NI, KOH, H_2O

rohydrin in the presence of an aqueous base to produce [G(n + 1)]-OH **8** (Scheme 2). One advantage of this method was that no protecting group was needed. However, the reaction scheme had not been applied towards the synthesis of dendrons higher than [G3] (Table 2). This series of oligoether dendrons had been used to ligate to a carborane unit, followed by cleavage of the benzyl surface groups to produce water soluble carborane bound dendrimers **9** for use in boron neutron capture therapy (Fig. 3) [6].

Table 2. Selected data for 1,3-dihydroxy-2-propyl ether-based dendrons[a]

n	Yield (%) [Gn]-OH→[G(n + 1)]-OH	Calcd MW of [Gn]-OH
0	88	–
1	88	272
2	–	601

[a] [G0]-OH = benzyl alcohol.

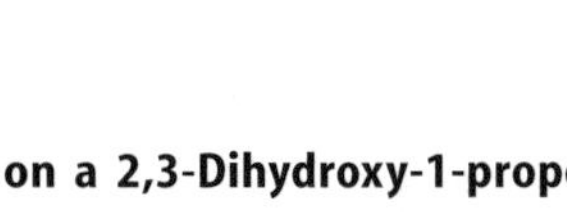

9 (n = 1, 2)

Fig. 3. Functional dendrimers based on the 1,3-dihydroxy-2-propyl ether dendritic skeleton

2.3 Dendritic Oligoethers Based on a 2,3-Dihydroxy-1-propoxy Repeating Unit

Recently Haag reported a new series of oligoether dendrimers that was isomeric to the series reported in Sect. 2.2 (Scheme 3) [7]. The branching unit used was still glycerol but the C-2 and C-3 hydroxyl groups, instead of the ones at C-1 and C-3 positions, were used to create further branching. The divergent iterative cycle involved an exhaustive allylation of an oligoalcohol [Gn]-$(OH)_s$ **10** with allyl chloride in the presence of aqueous NaOH. The resulting dendritic olefin [Gn]-

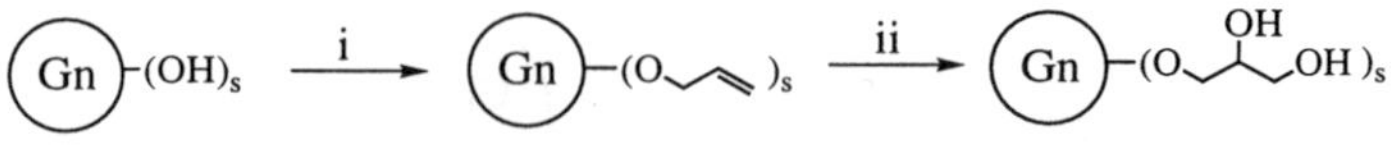

10 [Gn]-(OH)$_s$ **11** [Gn]-(ene)$_s$ **12** [G(n+1)]-(OH)$_s$

Scheme 3. i) NaOH, allyl chloride, Bu_4NBr, H_2O, 45 °C; ii) NMO, acetone, H_2O, *t*-BuOH, OsO_4

(ene)$_s$ **11** was then subjected to catalytic dihydroxylation in the presence of osmium tetraoxide and *N*-methyl morpholine oxide (NMO) to furnish the dendritic alcohol [G(n+1)]-(OH)$_s$ **12**. Starting from 2,2-bis(hydroxymethyl)-1-butanol as the core, the oligoalcohol [G3]-(OH)$_s$ with 24 hydroxyl end groups was prepared (Table 3).

Table 3. Selected data for 2,3-dihydroxy-1-propyl ether-based dendrimers [a]

n	Yield (%) [Gn]-(OH)$_s$ → [G(n+1)]-(OH)$_s$	Calcd MW of [Gn]-(OH)$_s$
1	–	356
2	75 [b]	801
3	–	1690

[a] Core=[G0]-(OH)$_s$=EtC($CH_2OH)_3$.
[b] Overall yield from [G0]-(OH)$_s$.

2.4
Dendritic Oligoethers Based on a 2,2-Bis(hydroxymethyl)-1-ethoxy Repeating Unit

A two-step, iterative synthetic route for the preparation of dendritic analogs of poly(ethylene glycol)s was recently disclosed by Fréchet and coworkers (Scheme 4) [8]. First, a base promoted *O*-alkylation of an oligoalcohol dendron [Gn]-OH **13** with methallyl dichloride produced the dendritic olefin [G(n+1)]-ene **14**. Subsequently, the olefin was subjected to a hydroboration-oxidation reaction to give [G(n+1)]-OH **15**. Applying the chemistry with two different surface groups, two series of oligoether dendrons up to [G4] were prepared in multigram quantities having polydispersity index (PDI) close to 1.0 (Table 4).

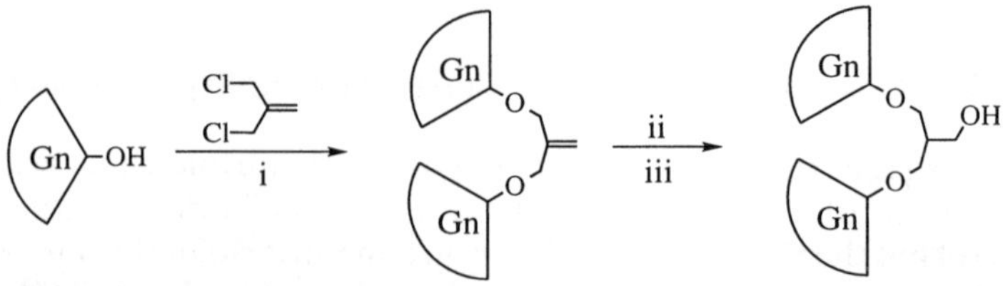

13 [Gn]-OH **14** [G(n+1)]-ene **15** [G(n+1)]-OH

Scheme 4. i) NaH, THF, KI, 18-crown-6; ii) 9-BBN or BH_3, THF; iii) H_2O_2, NaOH

Table 4. Selected data for 2,2-bis(hydroxymethyl)-1-ethyl ether-based dendrons[a]

n	Yield (%) [Gn]-OH→[G(n+1)]-OH		Calcd MW of [Gn]-OH		Calcd MW of [Gn]-ene		PDI of [Gn]-OH		PDI of [Gn]-ene	
	Series a	Series b	Series a	Series b	Series a	Series b	Series a	Series b	Series a	Series b
0	86	78	–	–	–		–	–	–	–
1	82	77	615	559	597	541	1.01	1.00	1.01	1.01
2	79	69[b]	1300	1188	1281	1170	1.01	1.00	1.01	1.00
3	72	–	2669	2446	2651	2428	1.01	1.01	1.01	1.01
4	–	–	5409	–	5391		1.01	–	1.01	–

[a] Series a: [G0]-OH= 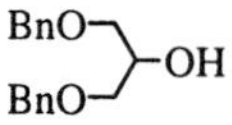 ; series b: [G0]-OH= 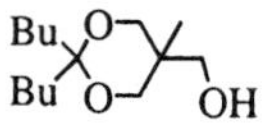 .

[b] Yield from [G2]-OH to [G3]-ene.

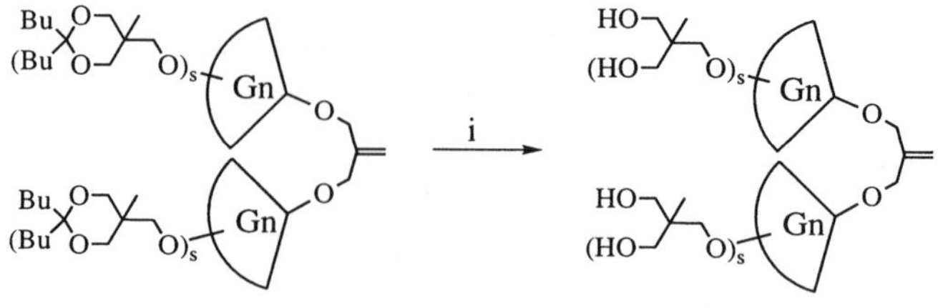

16 [G(n+1)]-ene **17**

Scheme 5. i) H^+ resin, MeOH

Upon treatment with an aqueous acid, the oligoether dendrons [G(n + 1)]-ene **16** having acetal surface groups were converted into the corresponding hydroxy-terminated dendrons **17** in nearly quantitative yield (Scheme 5). Among them, the [G4]-dendron possesses the desired water solubility that is useful in a number of biological and medicinal applications.

2.5 Dendritic Oligoethers Based on a 2,3-Dihydroxybenzyloxy Repeating Unit

In a recent publication, Weintraub and Parquette described the preparation of a series of oligoether dendrons based on a 2,3-dihydroxybenzyl ether as the repeating unit [9]. Due to the relatively congested 2,3-branching pattern, such dendrimers may exhibit restricted flexibility and will possess a more defined internal architecture that could be useful in molecular recognition and catalysis. There are three synthetic operations in the convergent iterative cycle (Scheme 6). First, bis-*O*-alkylation of the brancher 2,3-dihydroxybenzaldehyde with [Gn]-Br **18** produces the higher generation dendron [G(n + 1)]-CHO **19**. The focal aldehyde group is then converted to the corresponding bromide **20** [G(n + 1)]-Br by reduction with $NaBH_4$ or BH_3, followed by treatment with PBr_3 or PPh_3/CBr_4. Starting from 4-carbomethoxybenzyl bromide, oligoether dendrons up to [G4] were prepared (Table 5). However, this family of dendrons is acid-labile due to the presence of electron donating alkoxy group at the 2-position, which greatly facilitates the cleavage of the benzyl ether linkage. A com-

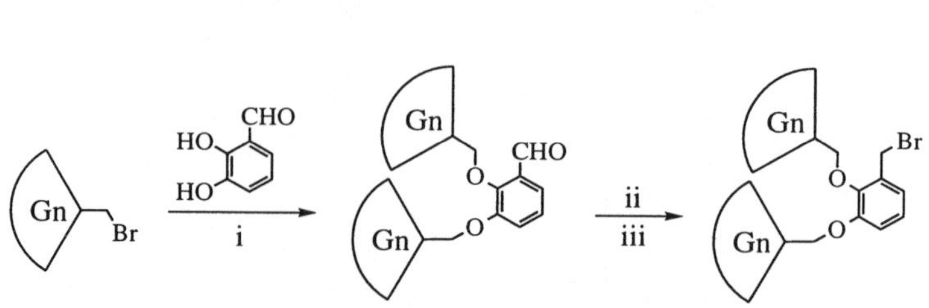

18 [Gn]-Br **19** [G(n+1)]-CHO **20** [G(n+1)]-Br

Scheme 6. i) K_2CO_3, 18-crown-6, DMF, THF, 70 °C; ii) $NaBH_4$, MeOH or BH_3, CH_2Cl_2; iii) PBr_3, CH_2Cl_2 or PPh_3, CBr_4, THF

Table 5. Selected data for 2,3-dihydroxybenzyl ether-based dendrimers[a]

n	Yield (%) [Gn]-Br→[G(n+1)]-Br	Calcd MW of [Gn]-OH	PDI of [Gn]-OH
0	70	–	–
1	61	436	1.09
2	50	977	1.03
3	98[b]	2058	1.04

[a] [G0]-Br = 4-carbomethoxybenzyl bromide.
[b] Yield from [G3]-Br to [G4]-CHO.

parative SEC analysis of this series of dendrons and the analogous 3,5-branched series (Sect. 2.6) confirmed the more compact nature of the 2,3-branched series.

2.6 Dendritic Oligoethers Based on a 3,5-Dihydroxybenzyloxy Repeating Unit

Oligoether dendrimers based on a 3,5-dihydroxybenzyl ether repeating unit reported by Hawker and Fréchet are the most widely used dendritic fragments in the literature [2, 10]. This is due to the better reaction yields, the higher product purities, and the higher generation of dendritic products that can be realized from the efficient synthetic cycle. The synthesis is based on a two-step convergent strategy starting from a selective bis-*O*-alkylation of the phenol groups of 3,5-dihydroxybenzyl alcohol **21** with an alkyl bromide **22** [Gn]-Br (Scheme 7). The resulting dendritic oligoether fragment having a focal point hydroxyl group **23** [G(n+1)]-OH is then activated to give the corresponding bromide **24** [G(n+1)]-Br. Repetition of this reaction cycle allowed the preparation of dendritic oligoether fragments up to [G6] (Table 6).

An accelerated convergent synthesis of the Fréchet's dendrons was recently developed by L'abbé and coworkers using hyperbranched AB_4 **25** or AB_8 **26** as the monomer units [12]. Thus, treatment of benzyl alcohol **27**, prepared from methyl 3,5-dihydrobenzoate via sequential silylation and reduction, with methyl 3,5-dihydroxybenzoate under Mitsunobu conditions afforded ester **28** (Scheme 8). The ester was then reduced to the hyperbranched AB_4 **25** monomer.

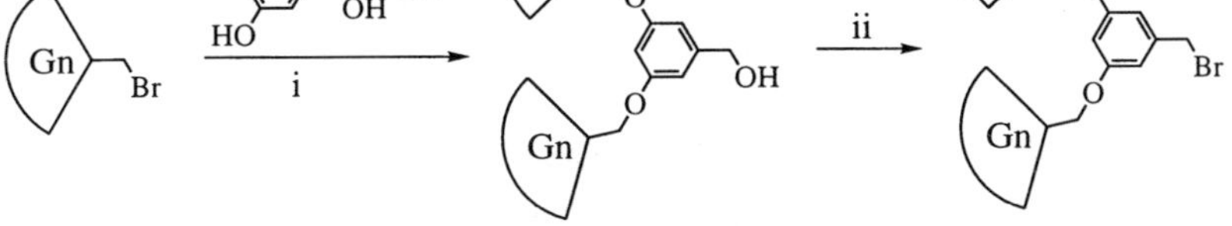

22 [Gn]-Br **23** [G(n+1)]-OH **24** [G(n+1)]-Br

Scheme 7. i) K_2CO_3, 18-crown-6, acetone, 56 °C; ii) PPh_3, CBr_4, THF

Table 6. Selected data for 3,5-dihydroxybenzyl ether-based dendrons[a]

n	Yield (%) [Gn]-Br →[G(n+1)]-Br	Calcd MW of [Gn]-OH	R_h[b] of [Gn]-OH (Å)	PDI[c] of [Gn]-OH
1	84	–	–	–
2	79	745	8	–
3	87	1594	10	–
4	71	3292	14	–
5	56	6689	18	1.02
6	–	13480	22	1.02

[a] [G0]-Br = benzyl bromide.
[b] Hydrodynamic radius determined by SEC, see [11].
[c] Polydispersity index determined by SEC, see [2].

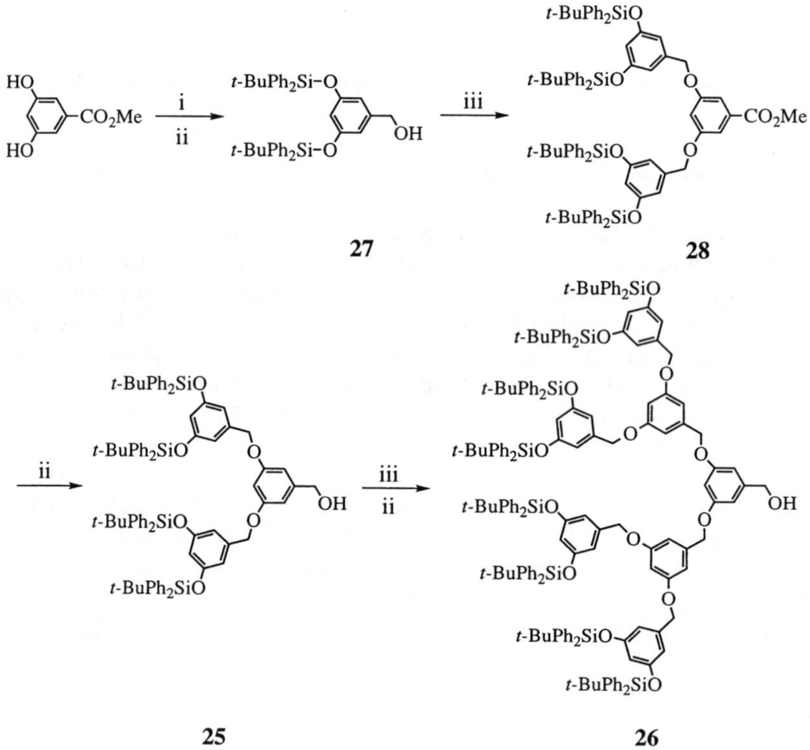

Scheme 8. i) *t*-$BuPh_2SiCl$, imidazole, DMF; ii) $LiAlH_4$, THF; iii) methyl 3,5-dihydroxybenzoate, DEAD, PPh_3, THF

Repetition of the Mitsunobu-reduction sequence on the AB_4 **25** monomer then furnished the AB_8 **26**.

The two monomers were used to prepare Fréchet's dendrons by a one-pot reaction (Scheme 9). Hence, treatment of the silylated AB_4 **25** monomer with potassium fluoride in the presence of an oligoether dendron **22** [Gn]-Br afforded

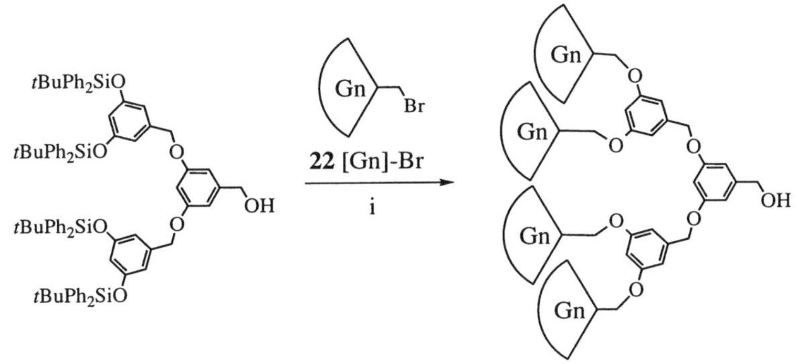

25 **29** [G(n+2)]-OH

26 → (**22** [Gn]-Br, i) **30** [G(n+3)]-OH

Scheme 9. i) KF, 18-crown-6, acetone, 56 °C

[G(n + 2)]-OH **29**, thus increasing the generation number by two in one step. Likewise, dendrimer growth could be increased by three generation by using the AB_8 **26** monomer (Table 7).

A recent report by Dehaen suggested the use of a mesylate [Gn]-OMs **31** instead of the bromide [Gn]-Br **22** in the iterative synthetic cycle (Scheme 10) [13]. By this modification, the Williamson ether products could be obtained in comparable yield in a shorter reaction time (Table 8). Furthermore, HBr-induced cleavage of the oligoether dendron during the conversion of [Gn]-OH to [Gn]-Br by PBr_3 could be avoided by adapting the mesylate route. Alternatively, dendron growth could be effected by reacting methyl 3,5-dihydroxybenzoate with a dendritic alcohol [Gn]-OH **32** under Mitsunobu conditions to furnish the higher generation dendron [G(n + 1)]-CO_2Me **33**. Subsequent reduction of the ester with $LiAlH_4$ then provided the oligoalcohol of the next generation [G(n + 1)]-OH **23**. Although the yields of this approach were inferior to those of Fréchet's original procedure, the formation of *C*-alkylation products could be avoided under the new reaction conditions.

Table 7. Accelerated synthesis of 3,5-dihydroxybenzyl ether-based dendrons [a]

n	Yield (%) [Gn]-Br → [G(n + 2)]-OH using AB_4 **25**	Yield (%) [Gn]-Br → [G(n + 3)]-OH using AB_8 **26**
1	92	82
2	90	83
3	82	–

[a] [G0]-Br = benzyl bromide.

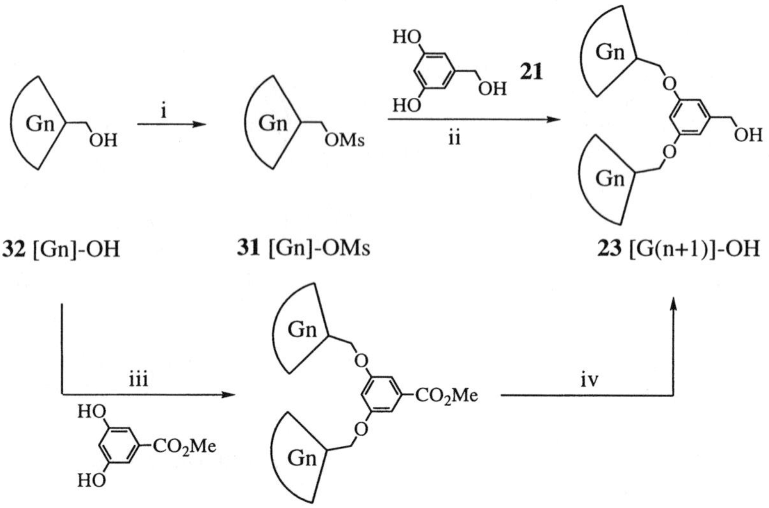

33 [G(n+1)]-CO_2Me

Scheme 10. i) MsCl, NEt_3, CH_2Cl_2, −10 °C; ii) K_2CO_3, 18-crown-6, acetone, 56 °C; iii) DEAD, PPh_3; iv) $LiAlH_4$, THF

Table 8. Alternative syntheses of 3,5-dihydroxybenzyl ether-based dendrons[a]

n	Yield (%) [Gn]-OH→[G(n+1)]-OH using the mesylate approach	Yield (%) [Gn]-OH→[G(n+1)]-OH using the Mitsunobu approach
1	70	79
2	73	82
3	71	78
4	–	59

[a] [G0]-OH = benzyl alcohol.

3,5-Dihydroxybenzyl ether-based oligoether dendrons with unsymmetrical surface character could be prepared by the convergent synthetic strategy [14]. For example, sequential mono-*O*-alkylation of 3,5-dihydroxybenzaldehyde with benzyl bromide and then with 4-cyanobenzyl bromide gave the unsymmetrical [G1]-CHO **34** (Scheme 11). The aldehyde was then converted to the corresponding bromide [G1]-Br **35** in two steps. Meanwhile, the symmetrical [G1]-Br **36** could be converted into the phenolic alcohol **37** by reacting with one equivalent of the branching agent **21**. Coupling of compound **37** with the [G1]-Br **35** then produced a [G2]-OH **38** with one cyano group at the periphery. Repetition of the process led to the formation of oligoether dendrons up to [G4] having one cyano group at the periphery.

3,5-Dihydroxybenzyl oligoether dendrons containing reactive surface functional groups could also be served as nanoscopic building blocks to form dendritic species with different surface properties. For examples, the ester-termi-

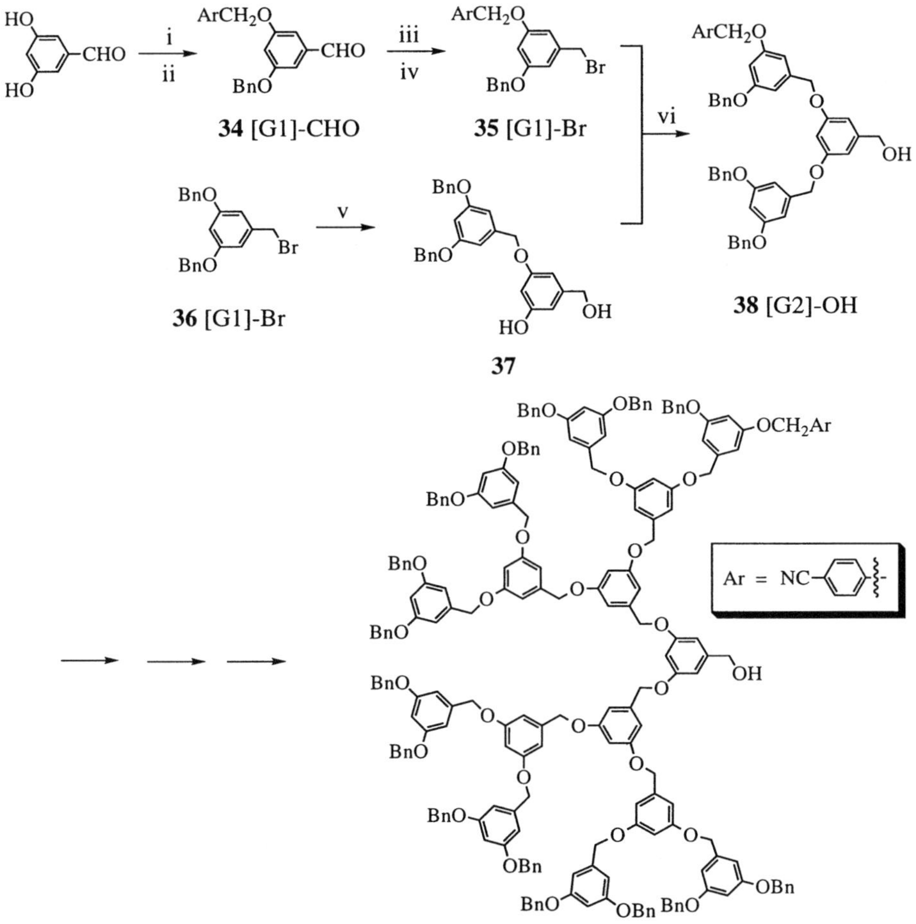

Scheme 11. i) BnBr, K_2CO_3, 18-crown-6, dioxane, 100 °C; ii) $ArCH_2Br$, K_2CO_3, 18-crown-6, dioxane, 100 °C; iii) Bu_4NBH_4, CH_2Cl_2; iv) CBr_4, PPh_3; v) 3,5-dihydroxybenzyl alcohol **21**, K_2CO_3, 18-crown-6, acetone, 56 °C; vi) K_2CO_3, 18-crown-6, acetone, 56 °C

nated oligoether dendrimer **39** could be subjected to a variety of surface modification reactions to give the corresponding carboxylic acid **40** or amide **41** dendrimers in good yields (Scheme 12) [15]. Oligoether dendrons **42** having aryl bromide surface groups could also be functionalized into dendrons having different surface properties by the Suzuki and Stille coupling reactions (Scheme 13) [16].

Instead of making use of the surface functional groups, the core functionality could also be used to prepare dendritic macromolecules of high complexity on a 3,5-dihydroxybenzyl ether-based dendritic skeleton. Hence, oligoether dendrimers **43** with an alkyne core underwent cyclotrimerization to produce a series of benzene-cored oligoether dendrimers **44** upon treatment with $Co_2(CO)_8$ (Scheme 14) [17]. As expected, the product yield decreased with increasing generation number due to increasing steric congestion of the interior core.

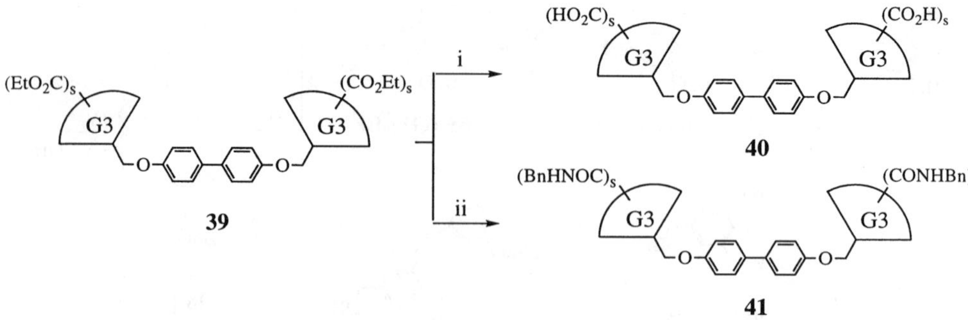

Scheme 12. i) KOH/H_2O/THF/MeOH; ii) $BnNH_2$, 140 °C

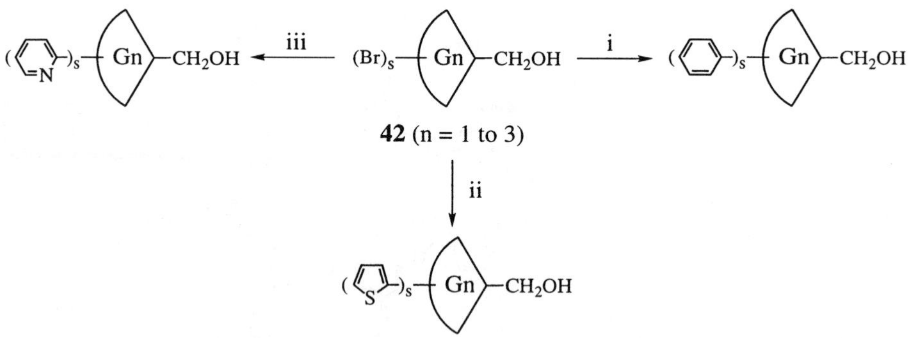

Scheme 13. i) $PhB(OH)_2$, $Pd(PPh_3)_4$, Na_2CO_3, toluene, 110 °C; ii) 2-trimethylstannylthiophene, $Pd(PPh_3)_4$, Na_2CO_3, toluene, 110 °C; iii) 2-trimethylstannylpyridine, $Pd(PPh_3)_4$, Na_2CO_3, toluene, 110 °C

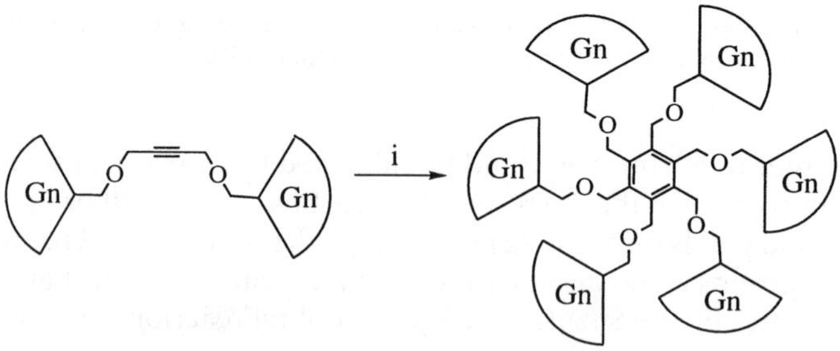

43 (n = 1 to 3) **44** (n = 1 to 3)

Scheme 14. i) $Co_2(CO)_8$, toluene, 110 °C

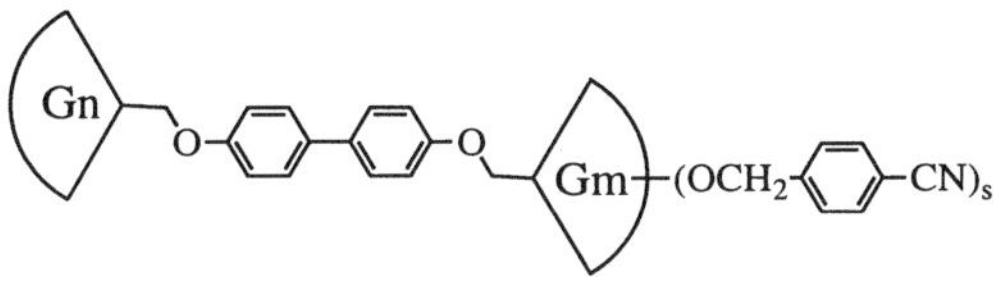

45 (n = 1 to 5, m = n–1)

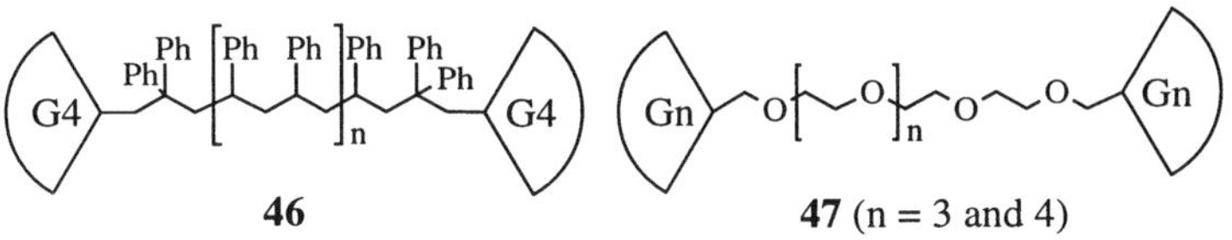

46 **47** (n = 3 and 4)

Fig. 4. Dendritic particles prepared from 3,5-dihydrobenzyl ether-based dendrons

Oligoether dendrons based on the 3,5-dihydroxybenzyl repeating units are the most commonly used dendritic fragments towards the preparation of nanoscopic materials with novel architecture. For examples, macromolecular dipoles **45** having electron donating $PhCH_2O$ groups and electron withdrawing CN groups at the opposed ends of the dendrimers were synthesized (Fig. 4) [18]. The dipole moments of the dendrimers were found to increase in a non-linear relationship with increasing molecular weight, suggesting a conformation transition from an extended to a globular shape. Novel linear-dendritic block copolymers consisting of a linear polystyrene **46** [19] or poly(ethylene glycol) chain **47** [20] end-capped with oligoether dendrons were also prepared. While the former copolymer **46** exhibited a shape transition from an extended globular structure to a statistical coil as the molecular weight of polystyrene increased, the latter **47** showed conformation switching in a solvent dependent fashion.

Oligoether dendritic fragments **48**–**50** with a polymerizable functionality at the focal point can be used as macromonomers to prepare linear polymers having dendritic side chains. Hence, dendritic oligoethers **48** having a focal point styrene functionality underwent free radical copolymerization with styrene to afford polystyrene **51** with oligoether dendritic appendages (Scheme 15) [21]. Similarly, the lower generation dendritic oligoethers **49** containing the same focal point styrene functionality could also be polymerized to produce polystyrene derivatives **52** with a highly congested environment [22]. Alternatively, an acrylate focal point functionality could also be used to produce polyacrylate derivatives **53** from oligoether macromonomeric dendrons **50** [23].

Recently Wendland and Zimmerman reported the synthesis of a novel type of dendrimer that had a covalently linked crusty shell structure (Scheme 16) [24]. An oligoether-based dendrimer with multiple olefin surface groups **54** underwent ring-closing metathesis reaction to give the intramolecular product **55**. The triester-based core was subsequently removed to produce a conceptually hollow dendritic architecture **56** with the carboxylic acid or alcohol functional groups left inside the interior. This novel synthetic strategy may be used to prepare endo-receptors with a higher degree of selectivity.

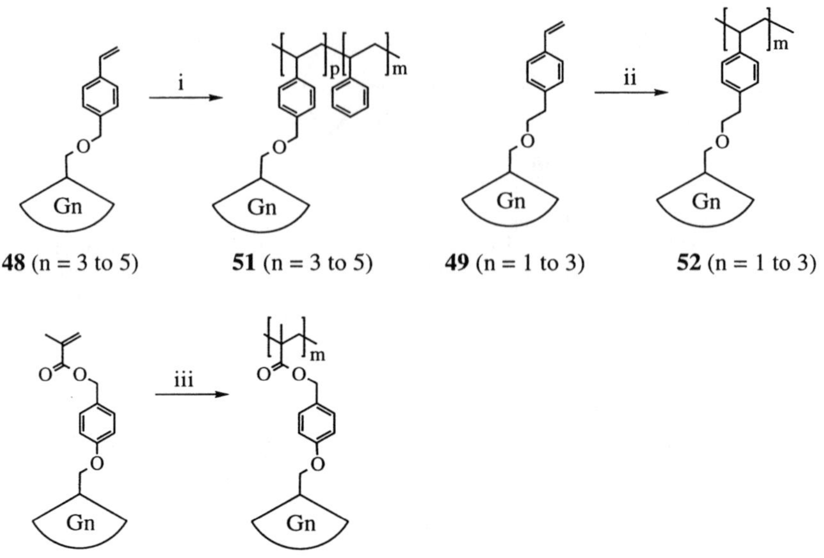

48 (n = 3 to 5) **51** (n = 3 to 5) **49** (n = 1 to 3) **52** (n = 1 to 3)

50 (n = 1, 2) **53** (n = 1, 2)

Scheme 15. i) styrene, AIBN, toluene, 70 °C; ii) *t*-butyl perbenzoate or dibenzoyl peroxide, 70 °C; iii) dibenzoyl peroxide, toluene, 60 °C

The introduction of chirality into dendritic structures represents another topic of interest with the aim to prepare dendrimers with potential applications in chiral recognition and asymmetric catalysis [25]. Kremers and Meijer reported the synthesis and characterization of a chiral dendrimer **57** having four different oligoether dendrons attached to a pentaerythritol core [26] (Fig. 5). Unfortunately, compound **57** could not be resolved which hampered any detailed chiroptical study. Subsequently another chiral dendrimer **58** based on a glycerol core was prepared in optically pure form [27]. However, the compound did not show any detectable optical activity in both CD and optical polarimetry measurements, possibly due to a lack of conformational rigidity in a dendrimer made up solely of lower generation dendrons. To force the dendrimer into a sterically more congested environment, chiral dendrimer **59** with a 2,6-branching pattern was synthesized. This compound did indeed possess a small, yet measurable specific rotation [28].

Chiral dendrimers **60**–**62** possessing Fréchet's achiral oligoether dendrons emanating from a chiral core were also prepared by Seebach and coworkers [29]. An optical dilution effect, i.e., a gradual decrease of specific rotation, was noted on going from [G0]- to [G2]-dendrimers, although their molar rotations remained constant. However, recent reports by Parquette on the chiroptical study of structurally related chiral dendrimers revealed alternative findings. Three generations of oligoether dendrimers having a (1*R*,2*S*)-2-amino-1-phenyl-1,3-propanediol chiral core **63** [30] or 2,5-anhydro-D-mannitol core **64** [31] were prepared. In both cases, a gradual decrease of absolute magnitude of the molar rotation was observed with increasing generation. Hence it appeared that the

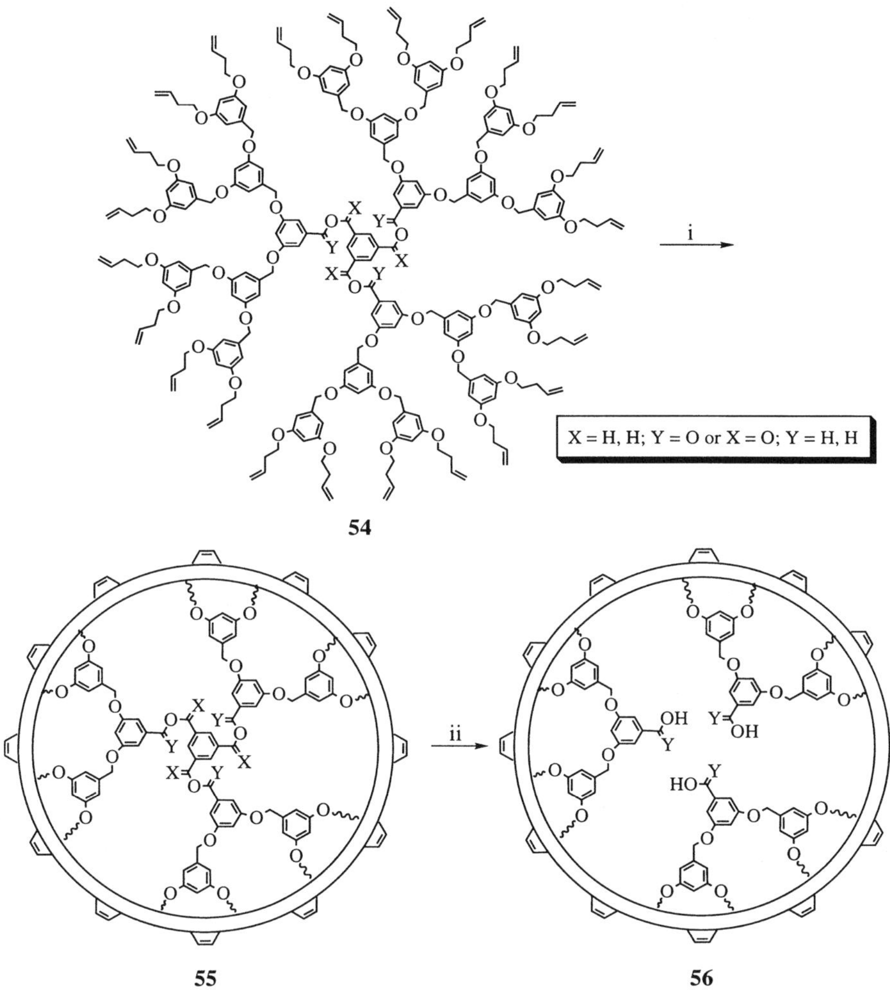

Scheme 16. i) $Ru(PCy_3)_2Cl_2(=CHPh)$, benzene; ii) KOH, EtOH, THF, H_2O

higher generation dendrimers were less chiral than the lower generation ones. Such findings were rationalized by a twisting of the conformation of the chiral core unit induced by the increasing steric crowding with increasing dendrimer generation. The dependence of molar rotation on the conformation of the chiral core had also been observed by Meijer in his study of oligoether-based dendrimers **65** containing an axially chiral binaphthol core [32].

Oligoether dendrons had also been attached to fullerenes such as C_{60} to produce soluble C_{60} derivatives **66, 67** with better processibility [33] (Fig. 6). Furthermore, the reduction potentials of the C_{60} unit in compound **67** were shifted to lower values, reflecting the insulating influence of the dendritic envelope [33b]. Alternatively, by using nucleophilic cyclopropanation, up to six Fréchet's

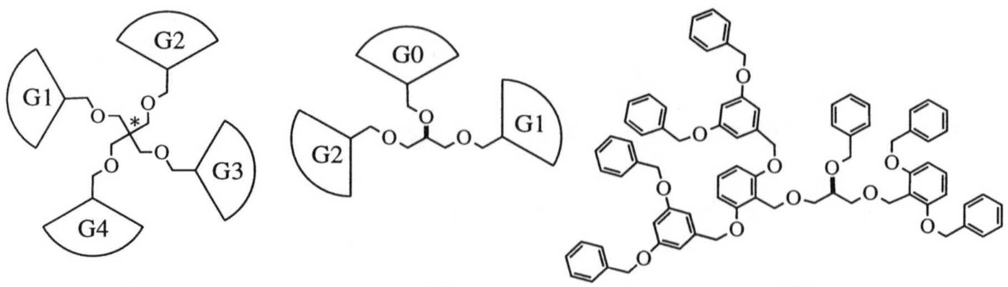

57 **58** **59**

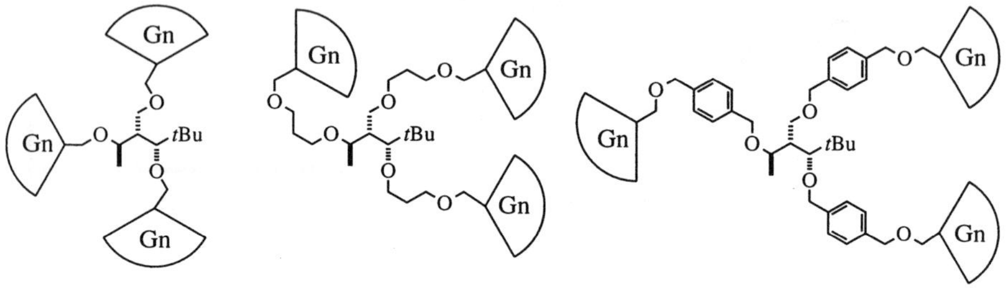

60 (n = 0 to 2) **61** (n = 0 to 2) **62** (n = 0 to 2)

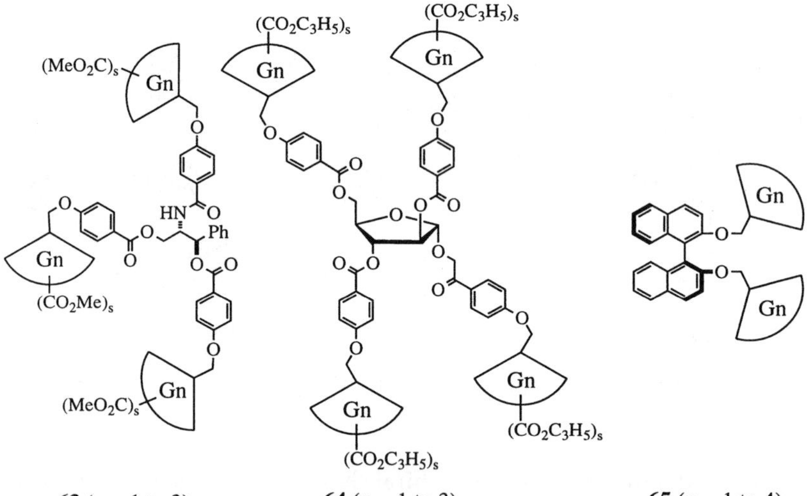

63 (n = 1 to 3) **64** (n = 1 to 3) **65** (n = 1 to 4)

Fig. 5. Chiral dendrimers derived from 3,5-dihydroxybenzyl ether-based dendrons

dendrons could be added to a C_{60} core to form achiral and chiral fullerene dendrimers with different substitution patterns and functionalities [34].

Fréchet's dendrons had been employed as stoppers for the preparation of [n]rotaxanes. For example, reaction of bipyridine with Fréchet's [G3]-Br wedge in the presence of bis-*p*-phenylene-34-crown-10 under high pressure afforded a dendritic [2]rotaxane **68** (Fig. 7) [35]. [3]- and [4]-Rotaxanes **69** and **70** were also

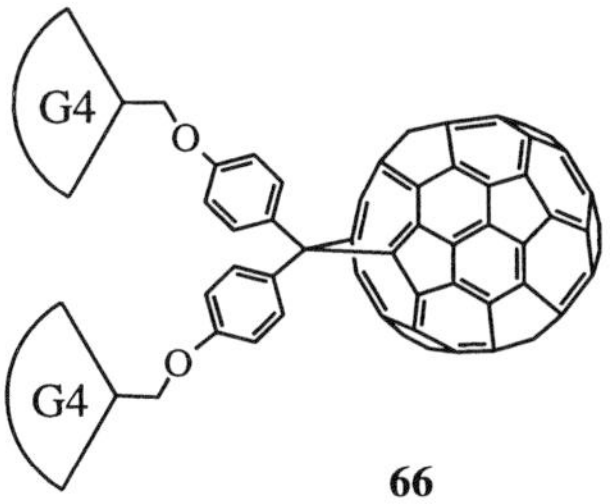

66

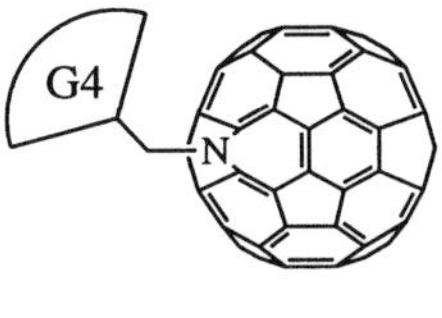

67

Fig. 6. Fullerene dendrimers containing 3,5-dihydroxybenzyl ether-based dendrons

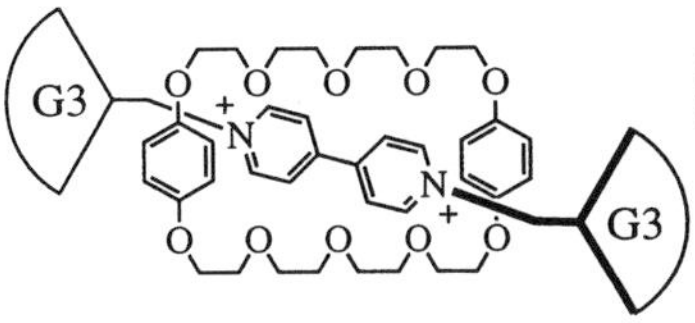

68

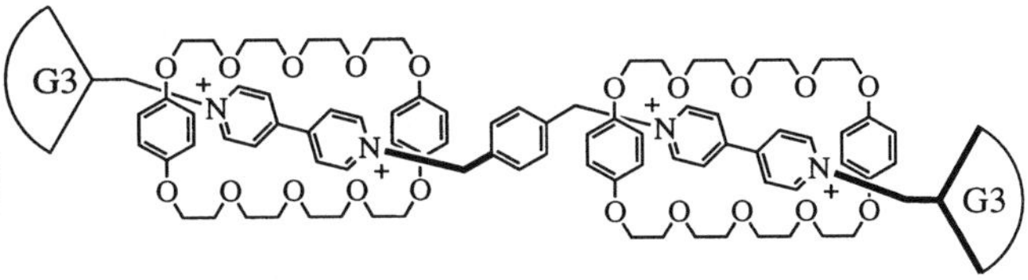

69

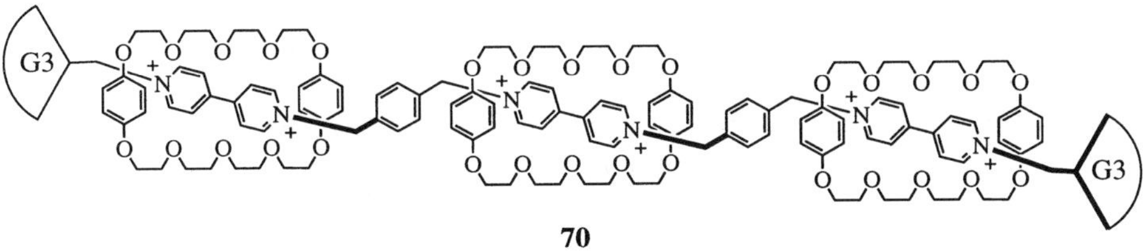

70

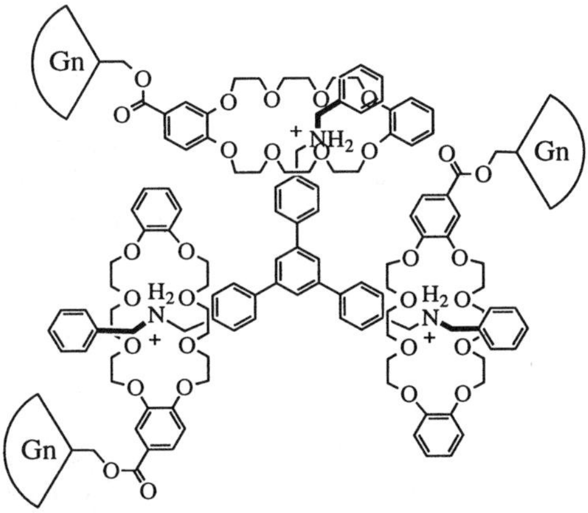

75 (n = 1 to 3)

Fig. 7. Dendritic rotaxanes and pseudorotaxanes bearing 3,5-dihydroxybenzyl ether-based dendrons

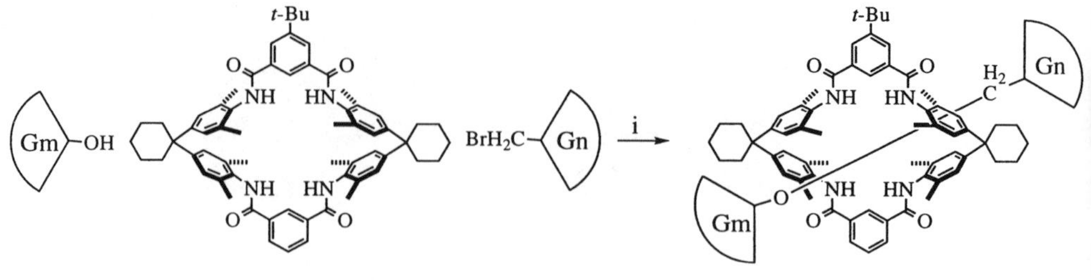

71 (m = 2, 3) **73** **72** (n = 1 to 3) **74** (n = 1 to 3, m = 2, 3)

Scheme 17. i) K_2CO_3, 18-crown-6, acetone, 25 °C

produced from bipyridine derivatives containing internal bipyridinium moieties. Interestingly, the presence of the dendritic stoppers resulted in a change of the reduction potential of the bipyridinium unit. Similarly, reaction of modified Fréchet's dendrons **71** containing a focal point phenol functionality with dendritic bromides **72** in the presence of a macrocyclic isophthalamide **73** produced [2]-rotaxanes **74** having dendritic stoppers of different sizes (Scheme 17) [36]. The spatial sizes of the various dendritic fragments were also determined by means of a deslipping experiment. Pseudorotaxanes **75** formed between dendritic dibenzo-24-crown-8 derivatives and a triply charged ammonium host was also reported [37].

Due to the flexibility and therefore poor crystallinity of Fréchet's dendritic fragments, they can be used to modify the solubility properties of structurally rigid polymers. The oligoether dendrons may be anchored as side chains along the polymer backbone [38], or as chain end stoppers to create dumbbell-shaped molecules.

Schlüter and coworkers reported the preparation of poly(*p*-phenylene) **76** having Fréchet's [G1]-oligoether dendrons as side chains [39]. It was envisaged that the dendritic fragments would wrap around the conjugated backbone to form a cylindrical structure. The polymer was prepared from a dendritic diaryl dibromide **77** and a diboronic acid **78** under the Suzuki coupling conditions (Scheme 18). Poly(*p*-phenylene)s having larger oligoether side chains but with a

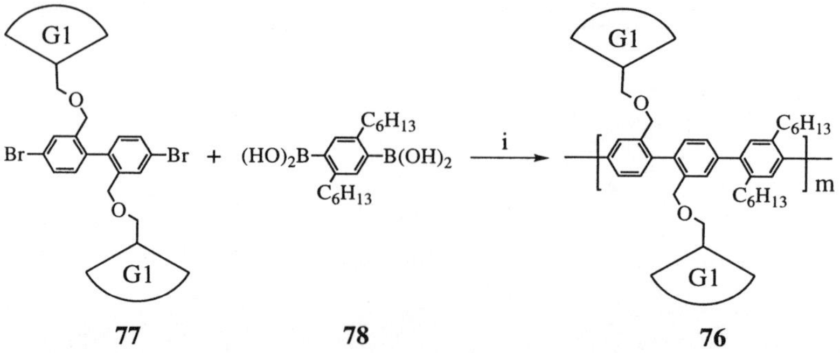

77 **78** **76**

Scheme 18. i) $Pd(PPh_3)_4$, Na_2CO_3, H_2O, toluene, 80 °C

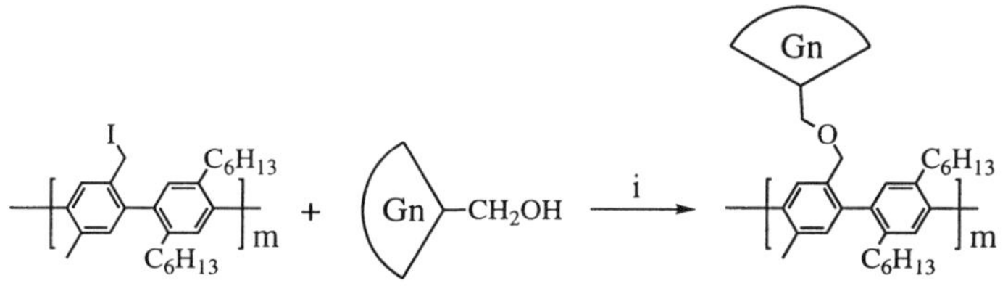

80 **79** (n = 1 to 3)

Scheme 19. i) NaH, THF, 20 °C

slightly different backbone structure were also synthesized [40]. Scanning force microscopy study on a [G3]-dendronized poly(*p*-phenylene) **79** adsorbed on graphite indicated the formation of multilayer films made up of cylindrical rods packed parallel to each other [40b]. Such dendronized poly(*p*-phenylene) polymers **79** were prepared from a direct alkylation reaction between a poly(*p*-phenylene) backbone **80** having alkyl iodide side chains with Fréchet's dendritic alcohols [40a] (Scheme 19). Poly(*p*-phenyleneethynylene)s **81** [41], polyurethanes **82** [42], and oligo(triacetylene)s **83** [43] with oligoether dendritic side chains had also been reported (Fig. 8). The conjugated polymers **81** possessing Fréchet's dendrons are blue-luminescent compounds with a quantum yield depending on the generation of the dendritic side chains. Upon excitation of the oligoether dendritic wedges, the fluorescence produced by the polymer **81** with [G4]-dendrons has a quantum yield of 100%, but the value drops significantly for those with the lower generation dendrons. Hence the larger [G4]-dendrimer framework functions as a light-harvesting envelope that encapsulates the conjugated backbone and also prevents the photoexcited state from collisional quenching.

Highly insoluble conjugated oligomers could be made soluble in common organic solvents by attaching dendritic fragments onto the chain ends. For example, the dumbell-shaped oligothiophene undecamer **84** having two oligoether [G3]-dendrons was highly soluble in CH_2Cl_2 and THF [44]. Soluble oligothienylenevinylenes **85** [45] and oligoimides **86** [46] end-capped with oligoether dendrons of different generation both exhibited generation independent redox processes having high reversibilities.

Amphiphilic dendrimer represents another new class of nanoscopic particle possessing novel conformational and physical properties. Amphiphilic dendrimer **87** containing a hydrophobic oligoether northern hemisphere and a hydrophilic polycarboxylate southern hemisphere can be used to stabilize a dichloromethane/water interface (Fig. 9) [47]. Linear-dendritic AB block copolymers **88** consisting of a hydrophobic oligoether dendritic wedge linked to a long hydrophilic poly(ethylene oxide) (PEO) chain form unimolecular micelles having the dendritic core tightly surrounded by the PEO chain in aqueous methanol solutions [48]. Multimolecular micelles, on the other hand, are formed at high concentration. Similar behavior is also noted for the linear-dendritic ABA block copolymers **89** having a hydrophilic poly(ethylene glycol) chain (PEG) end-capped with two hydrophobic oligoether dendritic fragments. More

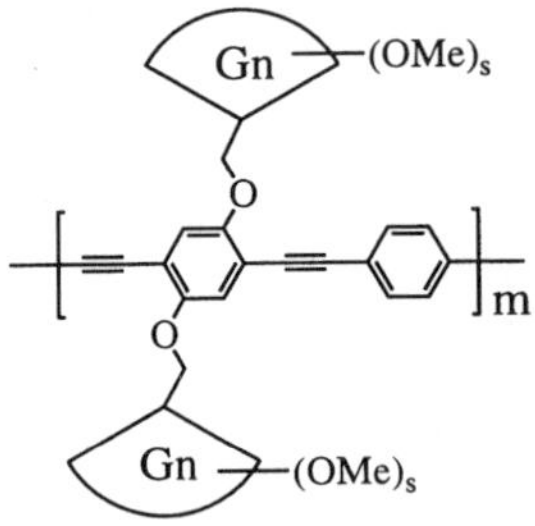

81 (n = 2 to 4)

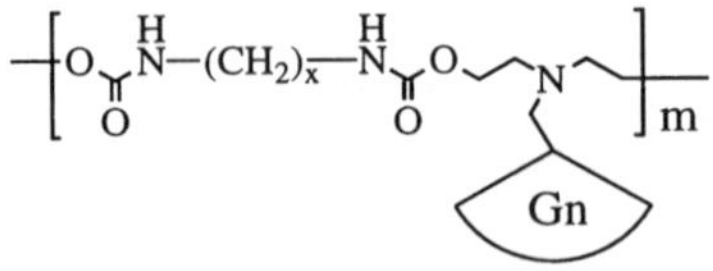

82 (n = 2 to 4, x = 6 or 12)

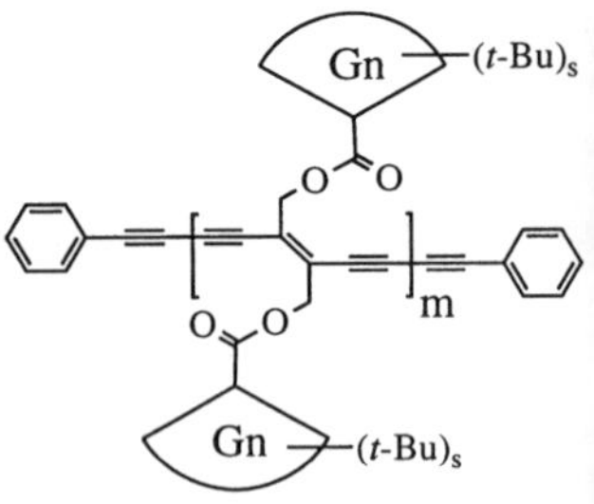

83 (n = 2, 3, m = 1 to 5)

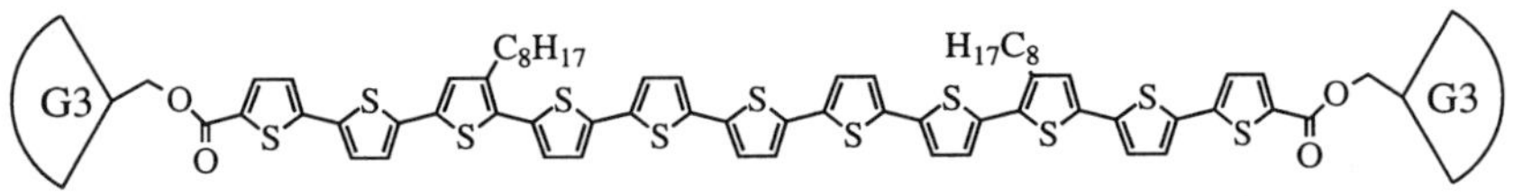

84

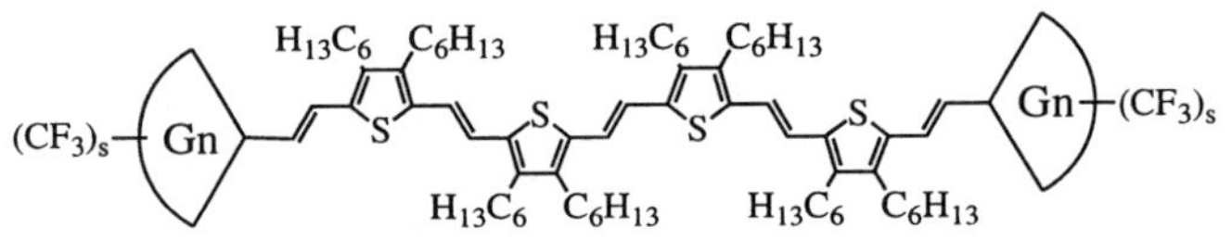

85 (n = 1 to 3)

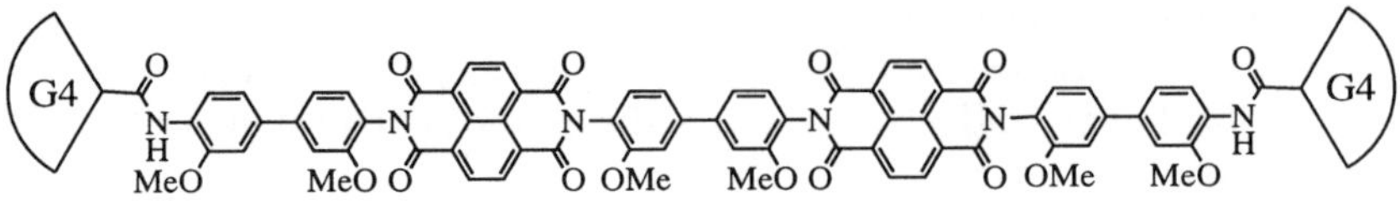

86

Fig. 8. Polymeric systems decorated with 3,5-dihydrobenzyl ether-based dendrons

interestingly, amphiphilic star copolymers **90** having four hydrophobic oligoether dendritic groups at the periphery and four internal hydrophilic PEG chains show conformational switching from one solvent system to another [49]. Surfactant-like amphiphilic oligoether-based dendrimers **91** bearing long alkyl chains on the surface and a polar focal point functional group can form stable monolayer at water surface [50]. The structurally "inverted" dendrimers **92**, having multi-hydrophilic head groups on the surface and a non-polar focal point functional group, also behave in a similar manner [51].

It is also possible to prepare polymers whose repeating unit is equipped with both hydrophobic and hydrophilic regions. In this context, Schlüter and coworkers reported the preparation of amphiphilic cylinders **93**, **94** decorated with oligoethyleneoxy-terminated Fréchet's dendrons and hydrophobic side chains along the polymer main chain [52]. As expected, these dendrimers form a stable monolayer in water, having the water soluble oligoethyleneoxy moieties segre-

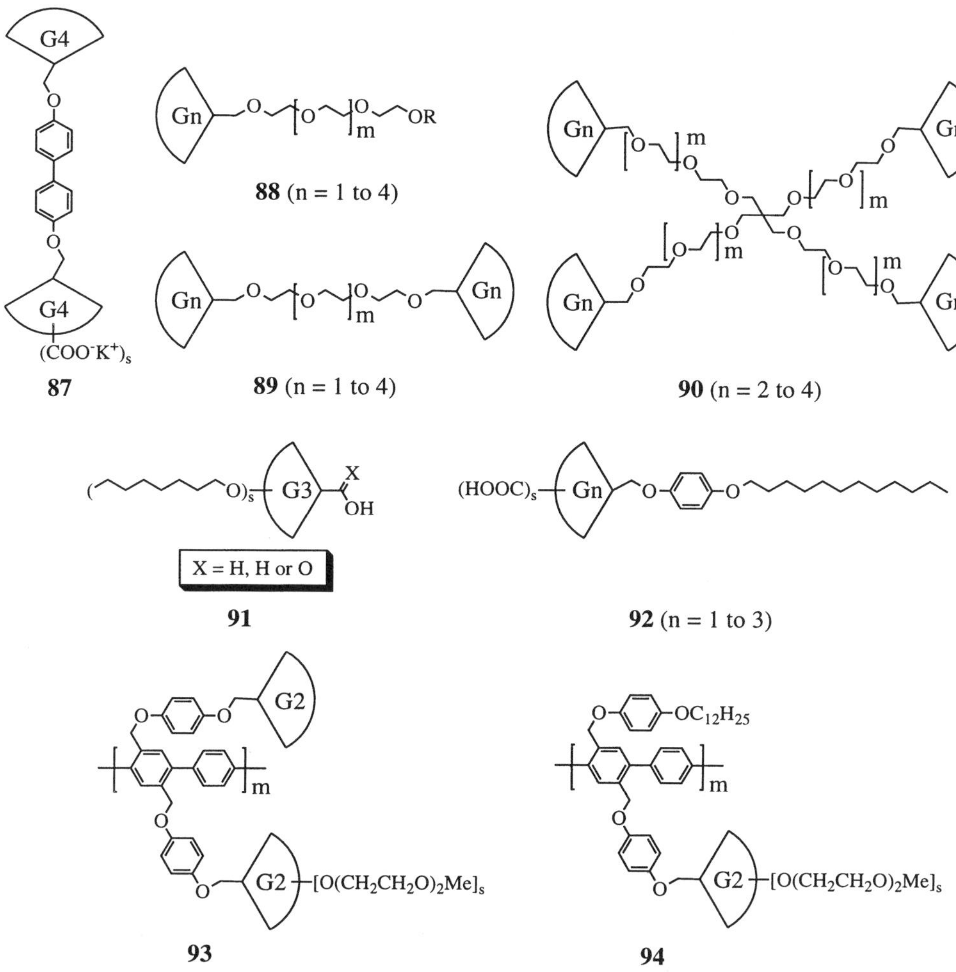

Fig. 9. Amphiphilic dendrimers based on 3,5-dihydrobenzyl-ether dendrons

gated from the hydrophobic functionalities to form nanoscopic cylindrical polymers with two distinct functional domains.

Dendrimers can also be used as macromolecular host systems to mimic the binding property of proteins and enzymes. Aida reported the synthesis of a series of dendritic porphyrins **95, 96** wherein the porphyrin nucleus was covalently encapsulated into a Fréchet's oligoether-based dendritic cage (Fig. 10). The Zn-containing porphyrin hosts **95** showed different binding affinity that was determined by the dendron generation and the size of the guest molecule [53]. Hence, the sterically less demanding [G1]-Zn-porphyrin can interact with both large and small guest molecules, whereas the sterically congested [G4]-Zn-porphyrin can only interact with smaller guest molecules. In the presence of excess 1-methylimidazole and O_2, the [G1]-Fe-porphyrin **96** is immediately and

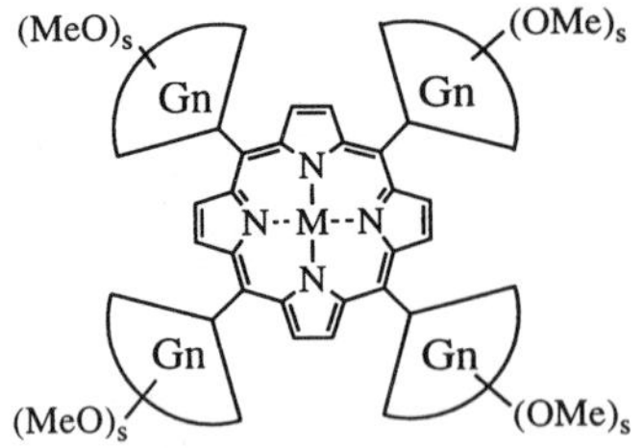

95 (M = Zn, n = 1 to 5)
96 (M = FeCl, n = 1 to 5)

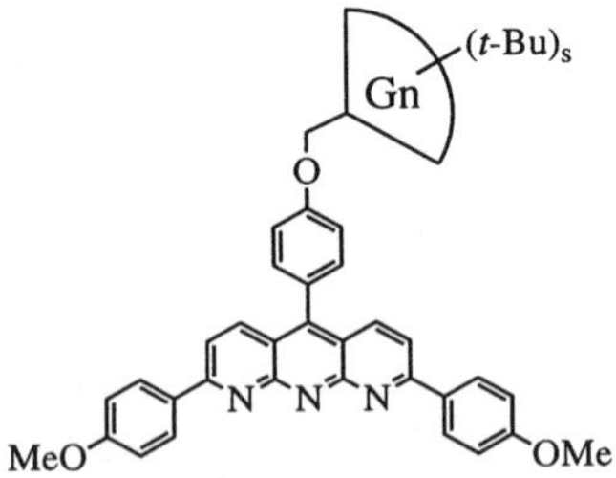

99 (n = 1 to 4)

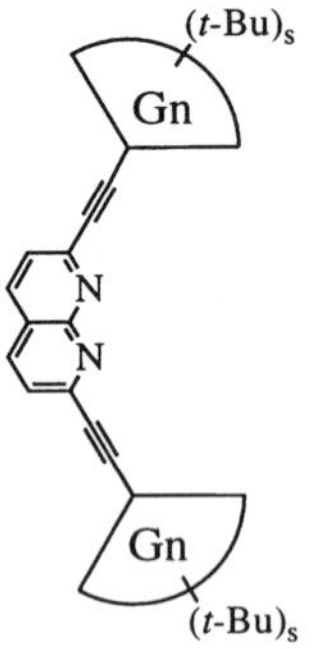

100 (n = 1 to 4)

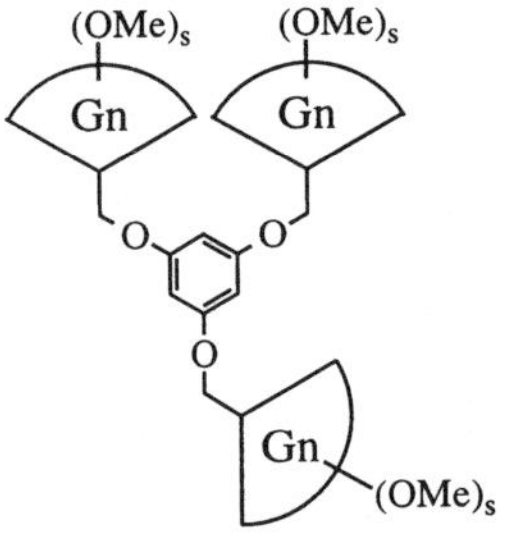

101 (n = 2 to 4)

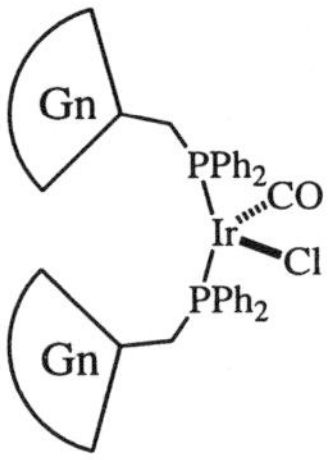

102 (n = 1, 2)

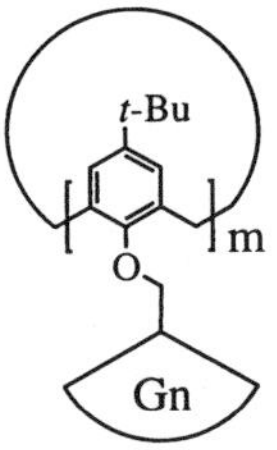

103 (n = 1 to 3, m = 4, 5)

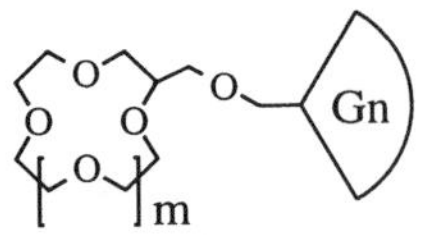

104 (n = 2 to 4, m = 1 to 3)

Fig. 10. Dendritic hosts constructed from 3,5-dihydroxybenzyl ether-based dendrons

irreversibly oxidized. On the other hand, the corresponding [G3]- to [G5]-Fe-porphyrins do not show any sign of μ-oxo dimer formation, and exhibit reversible O_2 binding profile [54]. The effect of dendrimerization on the reactivity of a series of dendritic Cu(I)-1,4,7-triazacyclononanes **97** towards the formation of unstable bis(μ-oxo)dicopper species **98** in the presence of O_2 had also been demonstrated (Scheme 20) [55]. The higher generation dendrons were shown to impede the formation of the bis(μ-oxo)dicopper species. At the same time, they also prolonged the lifetime of the unstable bis(μ-oxo)dicopper species against oxidative self-decomposition.

In addition to the examples described above, oligoether-based dendritic hosts with an anthyridine **99** [56] or naphthyridine **100** group [57] capable of binding to benzamidinium ions were reported. Oligoether-based dendrimers **101, 102** acting as hosts for C_{60} had also been described [58]. A number of calixarene-based **103** [59] or crown-ether-based [60] dendrimers **104** containing Fréchet's dendrons with potentially interesting binding affinity towards metal ions was also disclosed.

The use of dendritic fragments as basic building blocks towards the construction of well defined supramolecular architectures by self-assembly also attracts a lot of attention [61]. Zimmerman and coworkers reported the preparation and self-assembling properties of a series of bis(isophthalic acid)-based

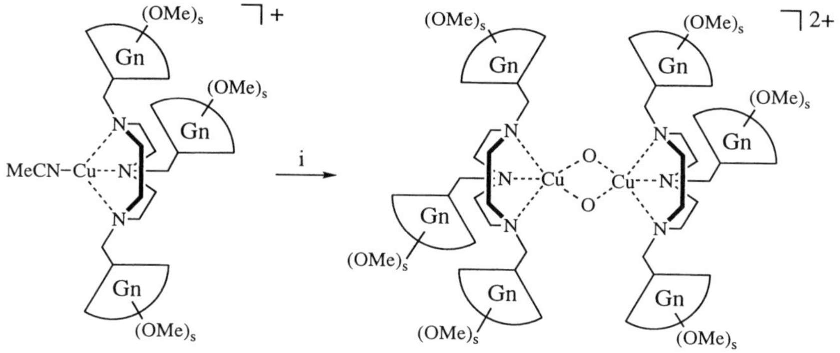

97 (n = 2 to 4) **98** (n = 2 to 4)

Scheme 20. i) O_2, CH_2Cl_2, –78 °C

dendrimers **105** substituted with Fréchet's dendrons. In weakly donating solvents such as chloroform, the sterically less hindered [G1]-dendrimer exists as an equilibrium mixture of a cyclic hexamer **106** and linear aggregates **107** (Scheme 21) [62]. On the other hand, only the cyclic hexamer **106** is formed from the [G2]- to [G4]-dendrimers, due to unfavorable steric repulsion of the larger oligoether dendrons in the linear aggregates.

Dendrimers can also form non-specific, heterogeneous aggregates or liquid crystalline (LC) materials via hydrophobic or π,π-stacking interactions. Phthalocyanines (PC) **108** substituted with one Fréchet's dendritic wedge self-organized into a hexagonal columnar mesophase (Fig. 11) [63]. It is interesting to note that the presence of the sterically bulky G3-dendron does not prohibit columnar mesophase formation. However, only the lowest generation PC **109** (n = 1) substituted with four [G1]-Fréchet's wedges displays similar optical textures. Polyether dendritic fragments **110** containing 3,4-bis-(n-dodecanyloxy)benzyloxy surface groups self-assemble into different supramolecular structures depending on the dendrimer generation [64]. The [G1]- and [G2]-tapered and the [G3]-disc-liked dendrons form supramolecular cylinders that in turn self-organize into a hexagonal columnar LC lattice. On the other hand, the much larger conical shaped [G4]-dendron self-organizes into a supramolecular spherical dendrimer and subsequently forms a cubic LC lattice.

Oligoether dendrimers **111** containing a dipeptide core undergo gelation in non-polar aprotic solvents [65]. Under scanning electron microscopy, the self-organized gelled sample was showed to have a fibrous structure with a diameter of 1–2 μm and each fiber consisted of a bundle of much thinner fibrils with a diameter of about 20 nm.

The conical shaped amphiphilic dendrons **112** containing a disaccharide core also self-assemble into large particles [66]. Interestingly, the average particle size decreases and the size distribution becomes narrower with increasing dendrimer generation. Such soft, fluid-liked particles could be immobilized into a rigid dendrimer by cross-linking of the sugar units with 1,3-phenylene diisocyanate [67].

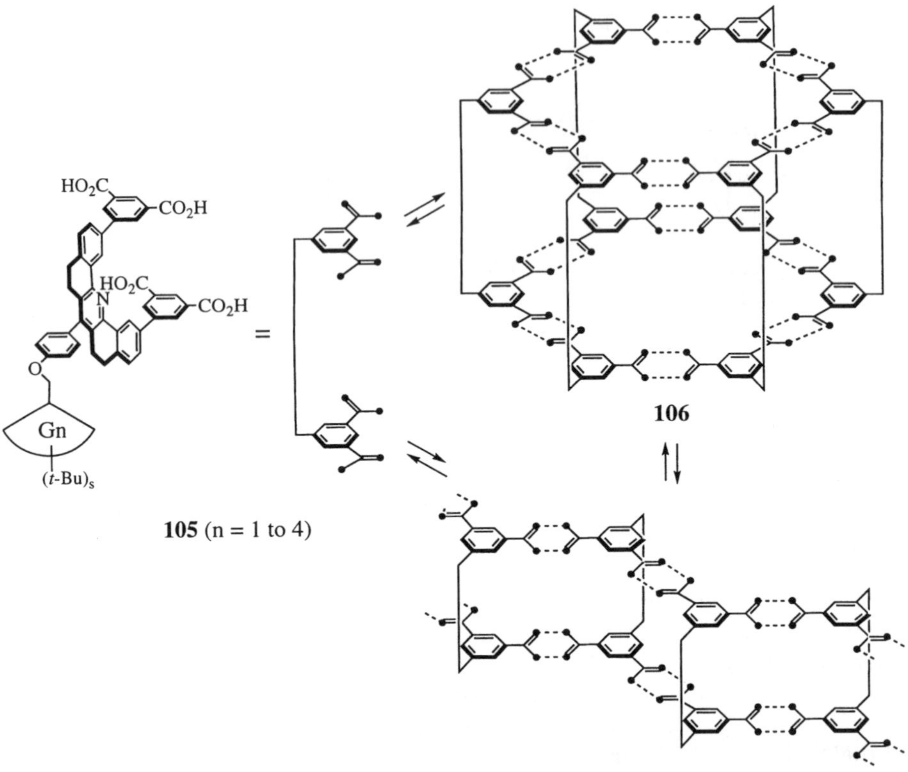

105 (n = 1 to 4)

Scheme 21

The preparation of catalytically active dendrimers with superior catalytic reactivity and better selectivity is one of the important contributions of dendrimer chemistry. Fréchet's [G4]-dendron **113** having an alkoxide focal point group has been used as macromolecular initiators for the anionic ring opening polymerization of ε-caprolactone (Fig. 12) [68]. The resulting hybrid copolymer has a narrow dispersity and a high degree of polymerization. In contrast, the corresponding [G1]-alkoxide species is a less effective initiator. The aluminum alkoxide derivatives **114** generated from the same series of polyether dendrons have also been used in the ring opening polymerization of lactones and lactides [69]. Dendritic polyether-based macro-initiators containing a tetramethylpiperidinyl-1-oxy subunit such as **115** and **116** have also been employed in living radical polymerization of vinyl monomers [70].

Dendrimeric ligands **117** having multiple chiral tetraaryl-1,3-dioxolane-4,5-dimethanol (TADDOL) units on the surface of a Fréchet's dendrimer were prepared [71]. Upon activation with the metal ions, the resulting dendritic catalysts can be used in the nucleophilic additions to carbonyl compounds and cycloaddition reactions, with enantioselectivities comparable to those promoted by non-dendritic analogs. The oligoether-based dendrimeric catalysts **118** having multiple phenylseleno surface groups can catalyze the oxidation of bromine by

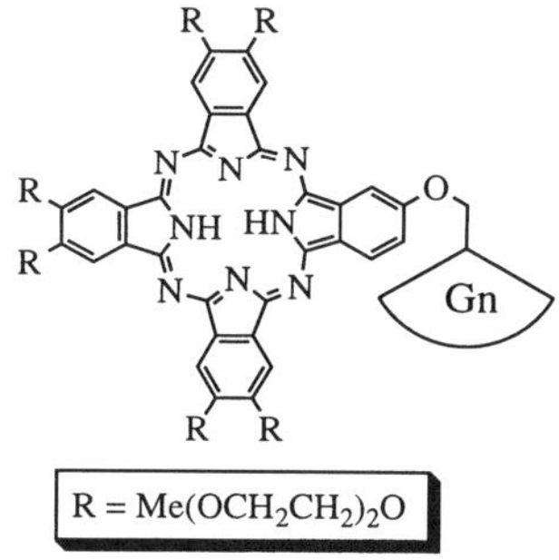

108 (n = 1 to 3)

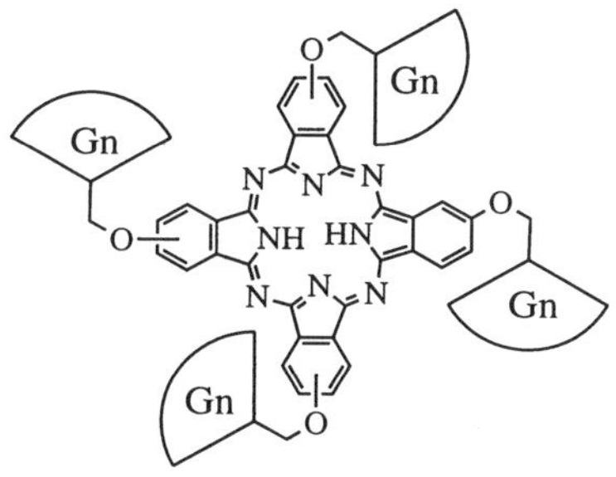

109 (n = 1 to 3)

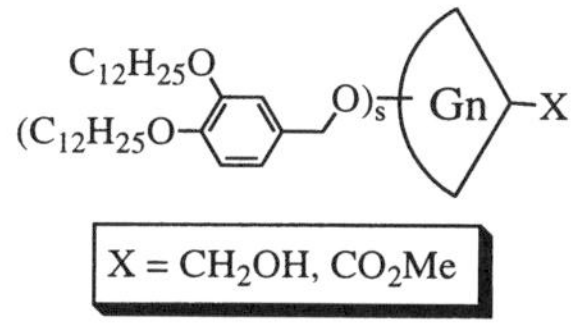

110 (n = 1 to 4)

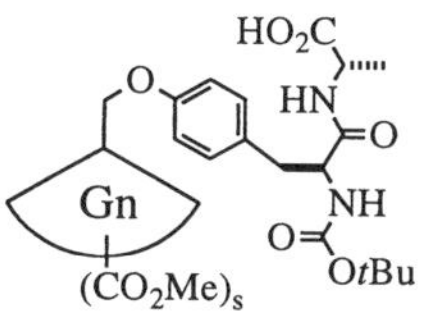

111 (n = 1 to 3)

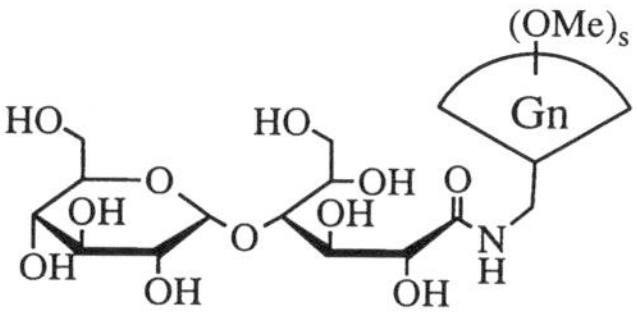

112 (n = 1 to 3)

Fig. 11. 3,5-Dihydroxybenzyl ether-based dendrons or dendrimers capable of forming self-assembled aggregates

hydrogen peroxide [72]. Most interestingly, positive cooperativity of the catalytic centers was noted for the higher generation dendritic catalysts.

Catalytically active dendrimers having the catalytic site(s) located at the focal point or near the internal core of Fréchet's dendrons were also known. For example, the previously mentioned TADDOL unit had been incorporated into the central core of oligoether dendrimers to produce ligands **119** of various generation [73]. Their corresponding dendritic titanium complexes were used to catalyze the addition of Et_2Zn to benzaldehyde. While there was little difference in terms of reactivity and enantioselectivity for the [G0]- to [G3]-dendrimers, a significant drop of reactivity was noted for the [G4]-species. When the surface groups of the these ligands were replaced by polymerizable styrene groups, the resulting dendritic TADDOL **120** ligands could be further polymerized into polymer beads which upon activation with titanium tetrapropoxide could again catalyze the enantioselective addition of Et_2Zn to aldehydes [74]. Such insoluble, dendrimer bound catalysts possess higher durability and also higher catalytic activities than those of conventional polymer bound catalysts. Similarly, Fréchet's dendrons **121** with a chiral pyridinol focal functionality can also be used to promote the enantioselective addition of Et_2Zn to benzaldehyde [75]. Other dendritic ligands having the catalytic groups located at the focal point or near the central core included chiral binaphthol derived oligoether dendrons **122** [76] and tertiary amine-based oligoether dendrimers **123** [77], which were employed to effect the enantioselective addition of allyl stanane to benzaldehyde in the presence of titanium tetrapropoxide and the nitroaldol reaction, respectively.

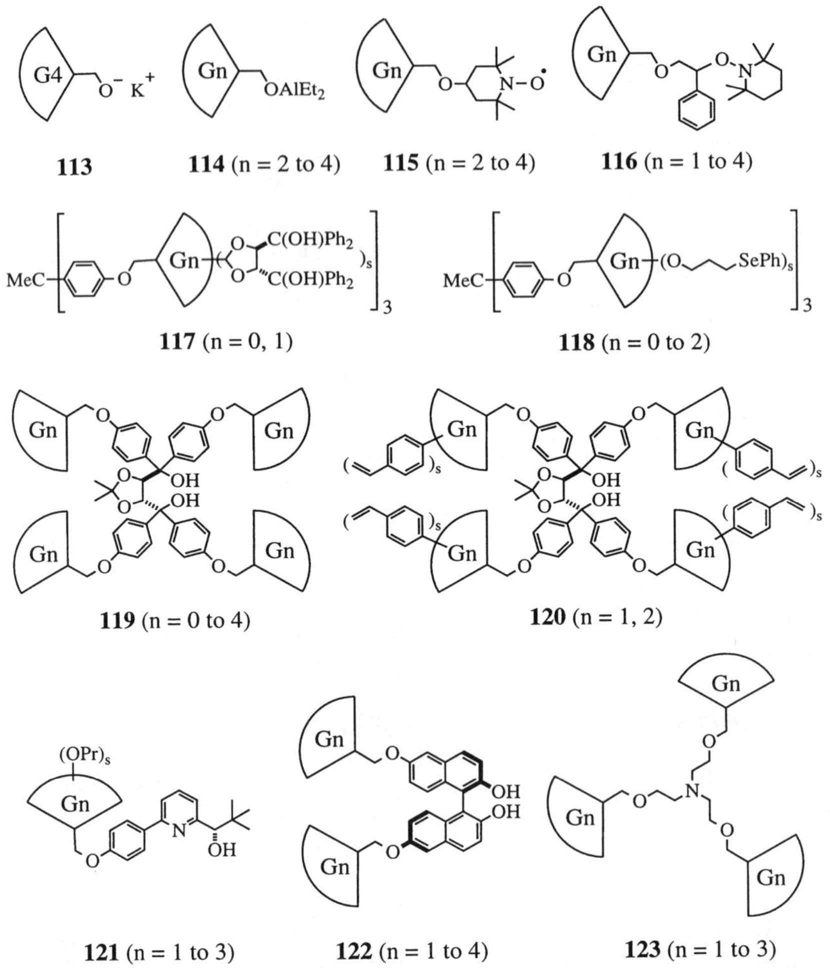

Fig. 12. Catalytically active dendrimers based on 3,5-dihydroxybenzyl ether-based dendrons

Redox active dendrimers can be prepared by attaching organometallic or electrochemically active units onto Fréchet's dendritic fragments. Ru(II)-containing dendrimers **124** were synthesized by Moss in which the dendrimer surface was covered with $CpRu(CO)_2$ residues (Fig. 13) [78]. Metallodendrimers containing an electroactive core or focal point unit surrounded by oligoether dendrons such as compounds **125** [79] and **126** [80] have also been reported. In general, the redox potentials of the redox active units were not perturbed by the oligoether dendrons; however the observed processes were not fully reversible in the presence of higher generation dendritic fragments. Similar observation was also noted for the Zn-porphyrin dendrimers **127** [81].

The photophysical properties of dendrimers can also be moderated by the presence of oligoether dendrons. In general, the dendrons can exert both steric

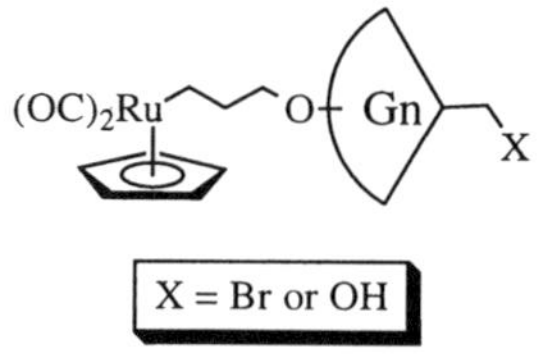

124 (n = 1 to 4)

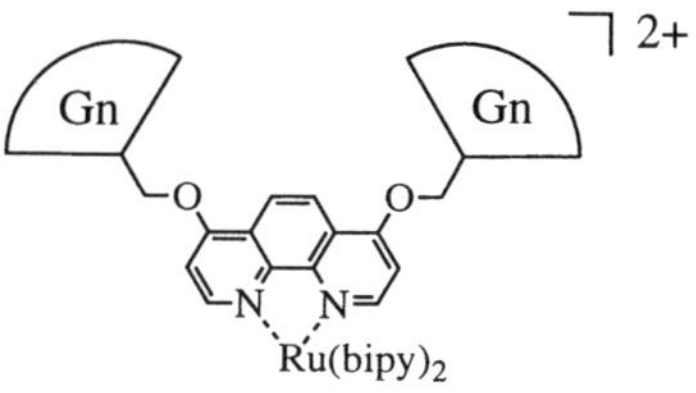

125 (n = 0, 1)

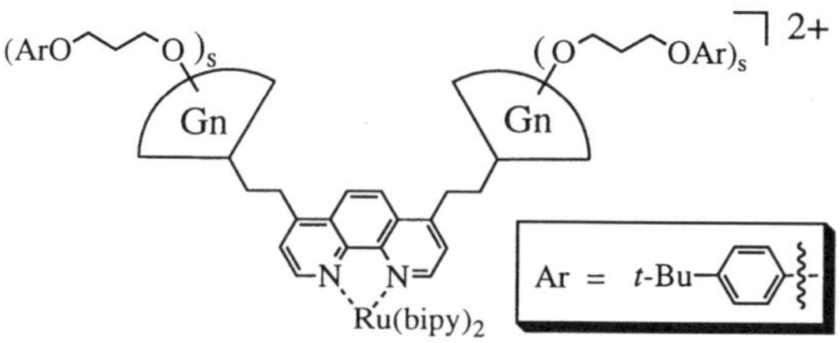

126 (n = 1, 2)

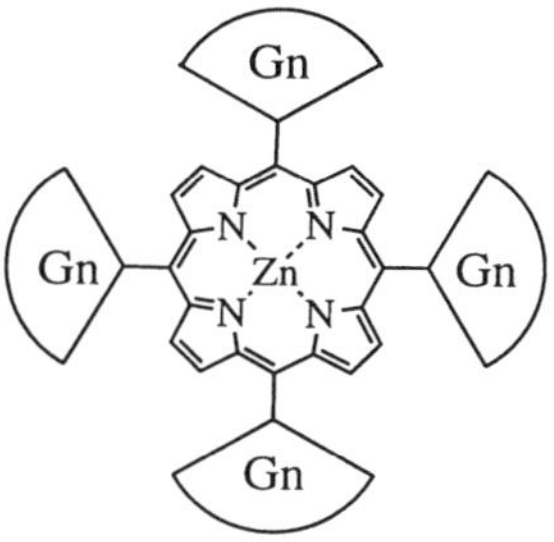

127 (n = 1 to 4)

Fig. 13. Electrochemically active dendrimers constructed from 3,5-dihydroxybenzyl ether-based dendrons

and electronic influences that may affect the lifetime of the photochemically excited state of the dendritic molecule. Hence, the lifetime of the luminescent excited state of Ru(II)-tris(bipyridine) dendrimers **128** increased gradually from the lower to the higher generation, due to a better protection of the core by the larger dendritic branches from O_2 quenching (Fig. 14) [80]. A photophysical study on the Zn-porphyrin dendrimers **129** having negatively charged carboxylate surface groups also showed that fluorescence quenching of the excited state by methyl viologen (MV^{2+}) was better protected by the [G4]-dendron and, in contrast to the lower generation dendrimers, followed a saturated profile [82]. The results suggested that the positively charged MV^{2+} molecules assembled via electrostatic force on the negatively charged dendrimer surface and the fluorescence quenching was a long-range photoinduced electron transfer process through the dendrimer framework. Surprisingly, quenching of the fluorescence of the analogous electrically neutral Zn-porphyrins **127** by benzyl viologen was slightly more effective in the presence of the larger size [G4]-dendron [81].

The layered dendritic framework also allows the controlled placement of the appropriate chromophores and quenchers inside a dendrimer and to establish a gradient for sequential photoinduced electron transfers that is essential for molecular light harvesting systems. The light harvesting antennas **130** bearing pyrene electron acceptors at the periphery and an electron donating 3-(dimethylamino)phenoxy focal point group constructed on a polyether dendritic network were reported [83]. It was found that the fluorescence of the pyrene surface groups was effectively quenched by the aminophenoxy group via an intramolecular mechanism, although quenching was less efficient for the [G2]-

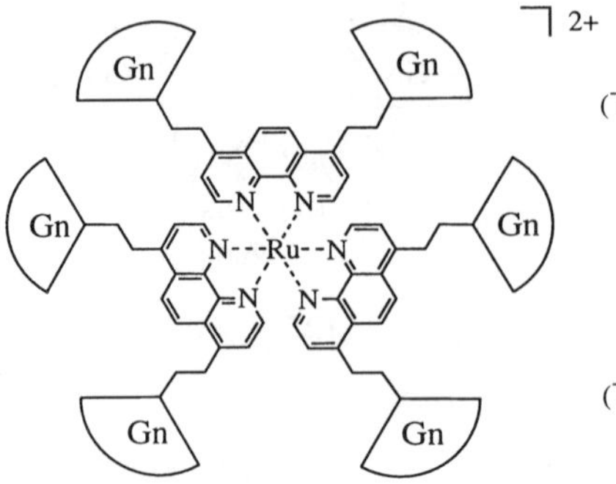

128 (n = 1 to 3)

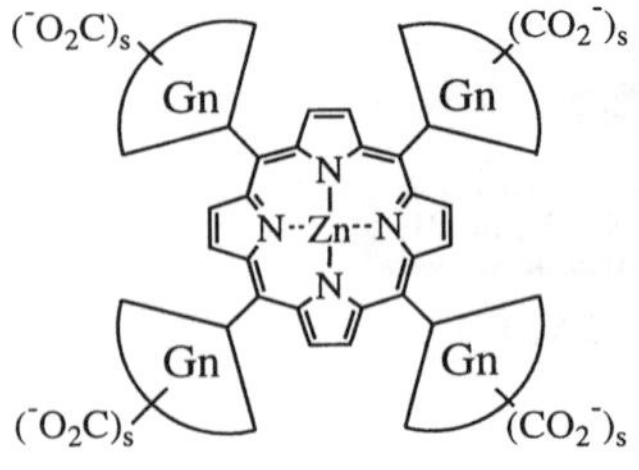

129 (n = 1 to 3)

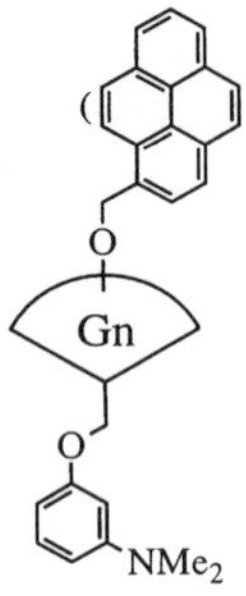

130 (n = 1, 2)

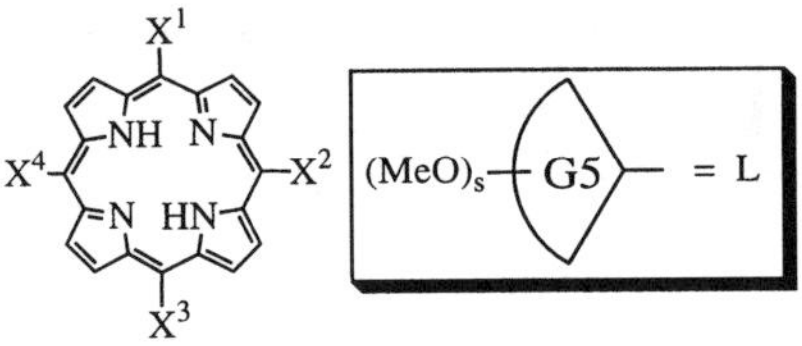

131 ($X^1 = X^2 = X^3$ = tolyl, X^4 = L)
132 ($X^1 = X^2$ = tolyl, $X^3 = X^4$ = L)
133 ($X^1 = X^3$ = tolyl, $X^2 = X^4$ = L)
134 (X^1 = tolyl, $X^2 = X^3 = X^4$ = L)
135 ($X^1 = X^2 = X^3 = X^4$ = L)

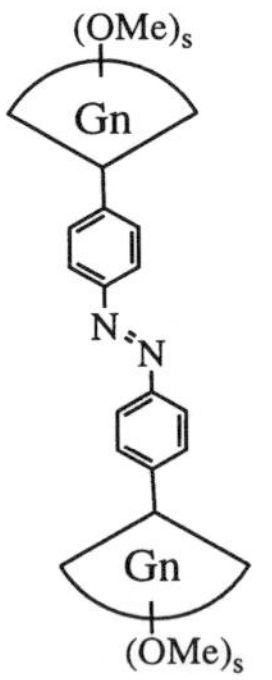

136 (n = 1 to 5)

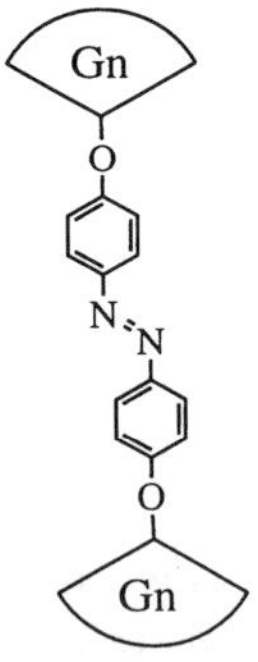

137 (n = 1, 2)

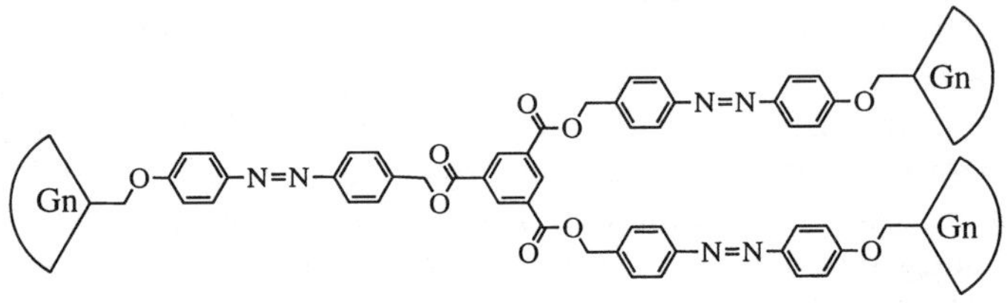

138 (n = 2, 4)

Fig. 14. Photoresponsive dendrimers constructed from 3,5-dihydroxybenzyl ether-based dendrons

wedge than for the [G1]-homologue. Quenching studies on dendritic porphyrins **131**–**135** having different number of oligoether dendrons showed a morphological dependence of the energy transduction process [84]. Upon excitation of the dendron subunits, all dendritic porphyrins showed an intramolecular energy transfer from the dendron subunits to the porphyrin core. Notably, the partially substituted porphyrins **131**–**134** had a low energy transfer quantum yield, whereas the fully substituted porphyrin **135** with a spherical architecture had a much higher value. The higher efficiency of the energy transduction process in compound **135** was attributed to the rigid shell structure that enabled the excitation energy to migrate over the dendritic network to the energy trap. In contrast, part of the excitation energy was lost by radiation due to interdendron collision of the relatively loose framework in the partially substituted porphyrins. Similarly, the Fréchet's [G5]-dendron was found to be the most effective infrared energy harvesting dendron to promote the *cis-trans* isomerization of azobenzene-cored dendrimers **136** [85]. Azobenzenes having lower generation dendritic sectors, however, are less effective in channeling the energy.

Junge and McGrath also reported the preparation of oligoether-based dendrimers **137** having an azobenzene central core and photoresponsive dendrimers **138** containing multiple azobenzene moieties [86]. The azobenzene units in such systems undergo facile UV-induced *cis-trans* isomerization independent of the steric size of the oligoether dendrons. Photolabile dendrimers **139** having three Fréchet's dendrons connected to a central core via *o*-nitrobenzyl ether linkages were also reported [87]. Upon irradiation of the sample by UV light, the oligoether dendrons were detached from the central core due to cleavage of the *o*-nitrobenzyl ether linkages (Scheme 22).

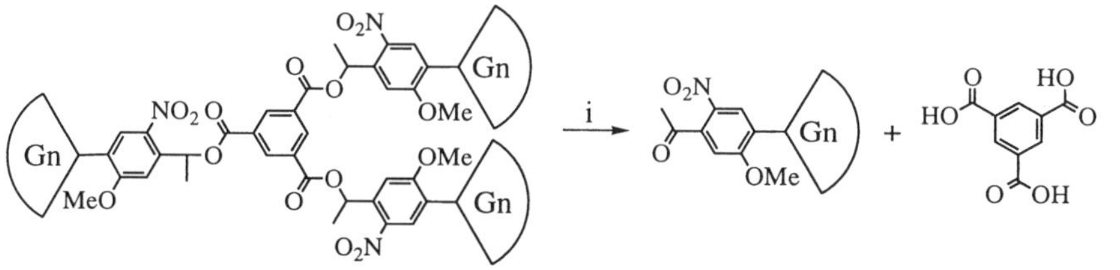

139 (n = 1 to 3)

Scheme 22. i) UV (350 nm) light

2.7 Dendritic Oligoethers Based on a 3-(3,5-Dihydroxybenzyloxy)-1-propoxy Repeating Unit

This family of oligoether dendrons may be considered as an 'elongated' Fréchet's series with an additional propoxy spacer. The convergent, iterative synthetic cycle consists of four steps (Scheme 23) [88]. Functional group conversion of a dendritic alcohol [Gn]-OH **140** produces the corresponding bromide [Gn]-Br which can then react with Fréchet's brancher – 3,5-dihydroxybenzyl alcohol **21** – to afford the dendritic alcohol of the next generation [G'(n + 1)]-OH **141**. Base-promoted allylation of the resulting alcohol then gives the [G(n + 1)]-allyl ether **142**. Finally, hydroboration of the olefin with 9-BBN followed by oxidative workup furnishes the 'elongated' alcohol [G(n + 1)]-OH **143**. Only the [G1]- and [G2]-series of compounds were synthesized (Table 9). Using this family of oligoether dendrons, multi-functional dendritic C_{60} derivatives could be prepared in better yields due to a relief of steric hindrance around the fullerene core [88].

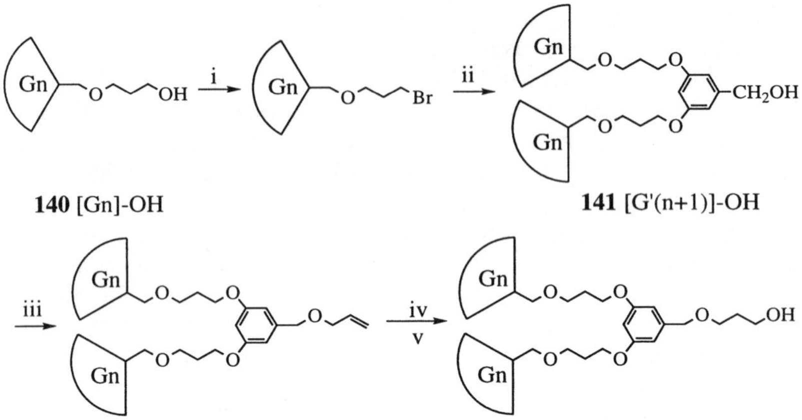

140 [Gn]-OH 141 [G'(n+1)]-OH

142 [G(n+1)]-allyl ether 143 [G(n+1)]-OH

Scheme 23. i) CBr_4, PPh_3, THF; ii) 3,5-dihydroxybenzyl alcohol, K_2CO_3, 18-crown-6, acetone; iii) NaH, allyl bromide, THF; iv) 9-BBN, THF; v) NaOH, EtOH, H_2O_2

Table 9. Selected data for 3-(3,5-dihydroxybenzyloxy)-1-propyl ether-based dendrons[a]

n	Yield (%) [Gn]-OH→[G(n+1)]-OH	Calcd MW of [Gn]-OH
1	77	378
2	–	919

[a] [G1]-OH = 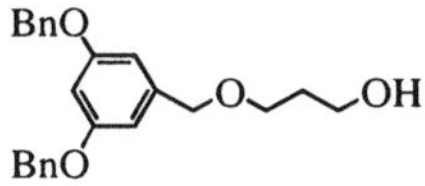 .

2.8 Dendritic Oligoethers Based on a 3,5-Bis-(4-hydroxyphenoxy)-benzyloxy Repeating Unit

This series of oligoether dendrons can also be considered as an extension of Fréchet's 3,5-dihydroxybenzyl ether series with the phenol functionalities replaced by an elongated 4-hydroxyphenoxy branch [89]. The iterative synthetic cycle is similar to that reported by Fréchet's except the branching agent is now 3,5-bis-(4-hydroxyphenoxy)-benzyl alcohol **144** (Scheme 24). Only the [G1]-series of dendrons was prepared.

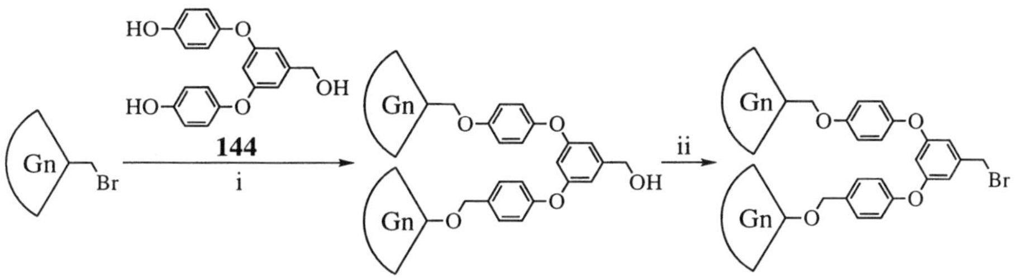

Scheme 24. i) K_2CO_3, 18-crown-6, acetone, 56 °C; ii) PPh_3, CBr_4, THF

2.9 Dendritic Oligoethers Based on a 3,5-Bis(hydroxymethyl)-phenoxy Repeating Unit

Two different routes had been developed for the preparation of this 'reversed' Fréchet series of dendrons. The first method, reported by Höger, involved the use of a protected 3,5-bis(hydroxymethyl)phenol brancher **145** that was prepared from dimethyl 3-hydroxy-isophthalate [90]. The convergent, iterative cycle for the growth of dendrons is relatively straightforward (Scheme 25). The first step is the Mitsunobu coupling of a phenol [Gn]-OH **146** with the brancher **145** to produce the silyl ether [G(n+1)]-$OSiR_3$ **147**. In the second step, the silyl group is dismantled by treatment with tetrabutylammonium fluoride to generate the phenol [G(n+1)]-OH **148** of the next generation. Using dimethyl 3-hydroxyisophthalate as the surface group, dendrons up to [G2] were synthesized (Table 10).

The second synthetic method was reported by Fréchet and coworkers [91], and employed a protected 3,5-di(bromomethyl)phenol **149** as the brancher that was prepared from dimethyl 3-hydroxyisophthalate (Scheme 26). The growth of the dendrons involves two synthetic steps. The first reaction is the base promoted alkylation of a phenol [Gn]-OH **150** with the brancher to afford the sulfonate [G(n+1)]-OSO_2R **151**. The second step is the cleavage of the sulfonate protecting group under alkaline conditions to give the phenol of the next generation **152**. To facilitate the synthesis of higher generation dendrons, a hyperbranched monomer AB_4 **153** was also prepared which allowed the growth of two generation in one cycle. Using coumarin 2 as a non-ether-based surface group, dendrons up to [G4] were prepared (Table 11).

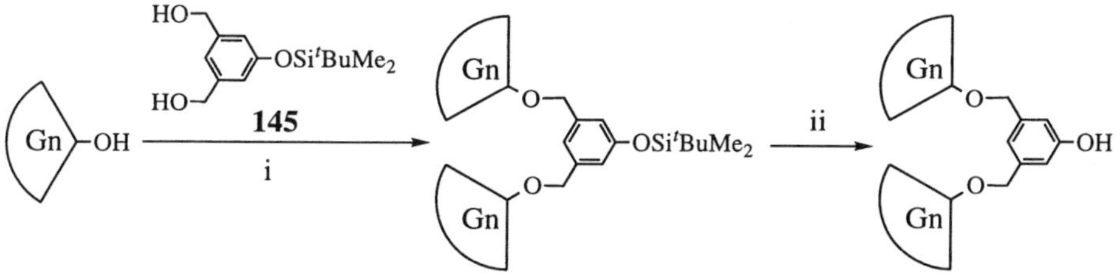

146 [Gn]-OH **147** [G(n+1)]-$OSiR_3$ **148** [G(n+1)]-OH

Scheme 25. i) PPh_3, DEAD, THF; ii) TBAF, THF

Table 10. Selected data for 3,5-bis(hydroxymethyl)phenyl ether-based dendrons synthesized by Höger's method[a]

n	Yield (%) [Gn]-OH→[G(n+1)]-OH	Calcd MW of [Gn]-OH
0	68	–
1	42	539
2	–	1195

[a] [G0]-OH= 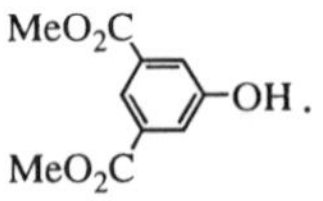 .

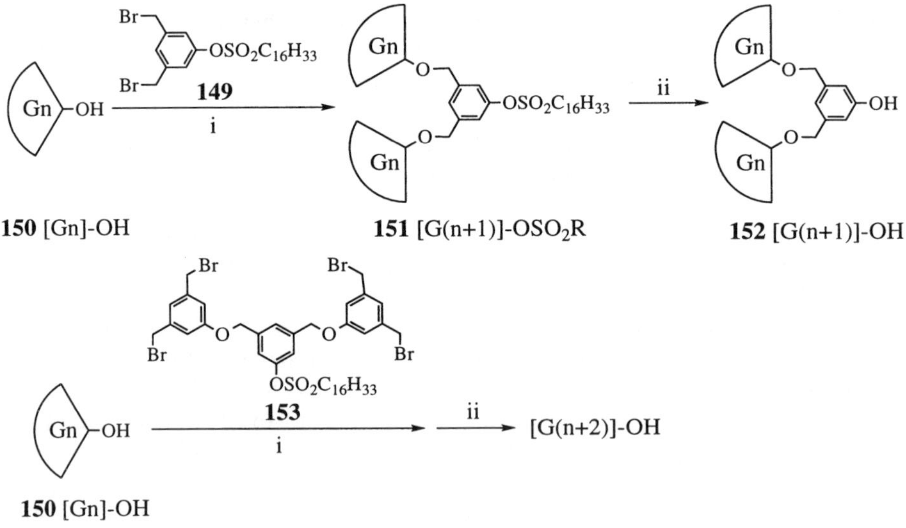

Scheme 26. i) K_2CO_3, 18-crown-6, acetone, 56 °C; ii) NaOH, EtOH, 78 °C

Table 11. Selected data for 3,5-bis(hydroxymethyl)phenyl ether-based dendrons synthesized by Fréchet's methods[a]

n	Yield (%) [Gn]-OH→ [G(n+1)]-OH using AB_2 **149**	Yield (%) [Gn]-OH→ [G(n+2)]-OH using AB_4 **153**	Calcd MW of [Gn]-OH
0	84[b]	69[c]	–
1	78	61	553
2	–	23	1224
3	–	–	2565
4	–	–	5248

[a] [G0]-NHR= 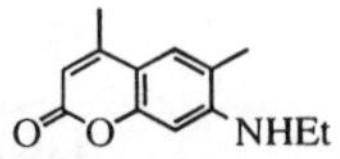 .

[b] From [G0]-NHR to [G1]-OH.

[c] From [G0]-NHR to [G2]-OH.

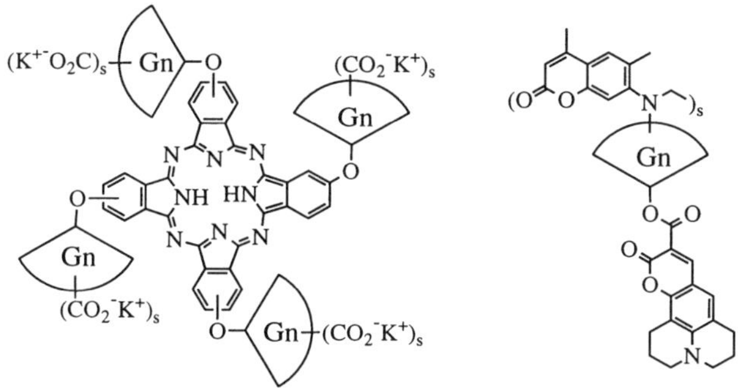

154 (n = 0 to 2) **155** (n = 1 to 4)

Fig. 15. Functional dendrimers constructed from 3,5-bis(hydroxymethyl)phenyl ether-based dendrons

Water soluble Zn-phthalocyanine (PC) derivatives **154** containing anionic surface groups built on such 'reversed' Fréchet dendrons were reported (Fig. 15) [92]. Due to larger steric inhibition, Zn-PC dendrimer decorated with [G2]-dendrons exists mainly as monomeric species in aqueous solutions. This compound, due to a lack of intermolecular aggregation and hence an absence of an efficient nonradiative energy relaxation pathway, is potentially useful for photodynamic therapy. The aggregation behavior of such species was also shown to be dependent on the presence of charged surfactant molecules [92b]. Light harvesting dendrimers **155** containing coumarin 2 surface groups and a focal point coumarin 343 functionality constructed on a 'reversed' Fréchet skeleton were also reported [93]. Energy transfer is extremely efficient (close to 100%) for the [G1]- to [G3]-dendrimers.

2.10 Dendritic Oligoethers Based on a 3,5-Dihydroxy-4-carbo-methoxybenzyloxy Repeating Unit

The synthetic route for this family of oligoether dendrons is essentially the same as that for the 3,5-dihydroxybenzyl ether-based dendrons reported by Fréchet and coworkers, except that a new brancher **156** having an extra ester group is used (Scheme 27) [94]. However, the experimental yields were not disclosed in

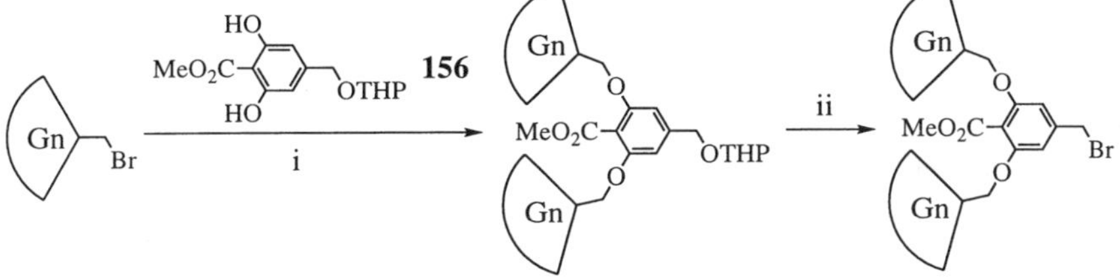

Scheme 27. i) K_2CO_3, 18-crown-6, acetone, 56 °C; ii) PPh_3, CBr_4, THF

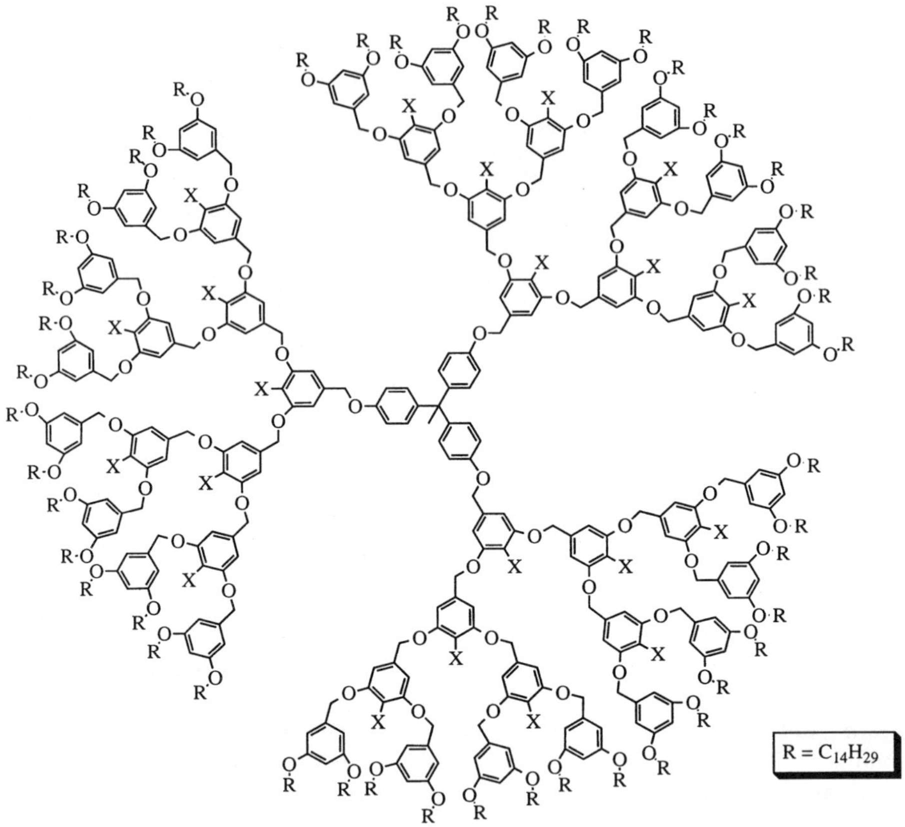

157 (X = CO_2Me)
158 (X = CH_2OH)

Fig. 16. Catalytically active dendrimers based on 3,5-dihydroxy-4-carbomethoxybenzyl ether-based dendrons

the original paper. The oligoether dendrons were used to produce two different series of oligoether dendrimers **157** and **158** having a polar internal environment encapsulated within a hydrophobic dendritic envelope (Fig. 16) [94]. Both dendrimers were used to catalyze the E1 elimination of tertiary iodides in the presence of sodium bicarbonate with catalytic turnover numbers approaching 10^4.

2.11 Dendritic Oligoethers Based on a 3,4,5-Trihydroxybenzyloxy Repeating Unit

This series of oligoether dendrons was reported by Percec and coworkers [95]. There are three synthetic steps in the convergent synthetic cycle (Scheme 28). An oligoether-based ester [Gn]-CO_2Me **159** is first reduced to the corresponding

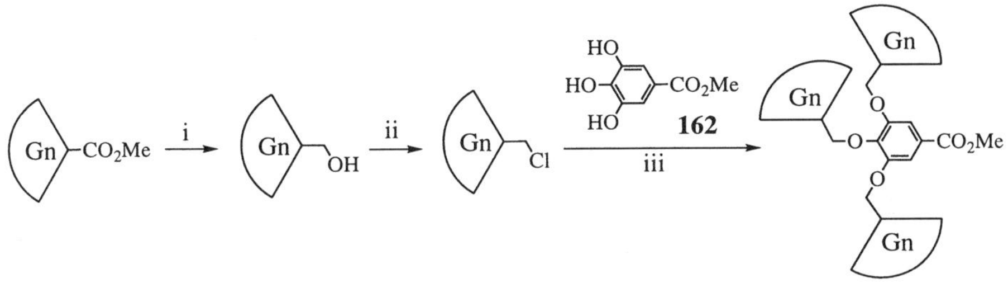

159 [Gn]-CO_2Me **160** [Gn]-OH **161** [Gn]-Cl **163** [G(n+1)]-CO_2Me

Scheme 28. i) $LiAlH_4$, Et_2O; ii) $SOCl_2$, 2,6-di-*t*-butylpyridine, DMF, CH_2Cl_2; iii) K_2O_3, NMP, 80 °C

benzyl alcohol [Gn]-OH **160** by $LiAlH_4$. The alcohol is then converted into the benzyl chloride [Gn]-Cl **161** in the presence of $SOCl_2$. The chloride is further reacted with methyl gallate **162** to provide the methyl ester [G(n + 1)]-CO_2Me **163** under Williamson synthesis conditions. Using this iterative procedure, oligoether dendrons up to [G4] were prepared (Table 12).

Using a long dodecyl hydrocarbon chain as the surface group, Percec's dendrons **164** can form supramolecular assemblies whose structures are dependent on the dendrimer generation (Fig. 17) [95, 96]. Hence, the lower generation dendrons having a flat tapered architecture self-assemble into a cylindrical structure that subsequently form a columnar hexagonal liquid crystalline assembly. On the other hand, the higher generation conical shaped dendrons self organize into spherical structures, which finally form a cubic liquid crystalline supramolecular assembly. More recently, it was shown that the sterically more congested, spherical oligoether [G4]-dendron **165** also self organized into a cubic lattice [97]. Similarly, polystyrenes **166** and polymethacrylate derivatives **167** having Percec's dendrons as side chains also self-assemble into different supramolecular architectures depending on the degree of polymerization [98]. PPV polymers **168** containing oligoether dendritic side chains were found to exhibit thermotropic nematic liquid crystalline behavior [99]. Finally, a dendrimer with an octahedral cluster $[Re_6Se_8]^{2+}$ core surrounded by six Percec's oligoether dendrons was also reported [100].

Table 12. Selected data for 3,4,5-trihydroxybenzyl ether-based dendrons[a]

n	Yield (%) [Gn]-CO_2Me → [G(n + 1)]-CO_2Me	Calcd MW of [Gn]-CO_2Me
1	74	689
2	63	2113
3	39	6386
4	–	19205

[a] [G1]-CO_2Me = methyl 3,4,5-tris(*n*-dodecyloxy)benzoate.

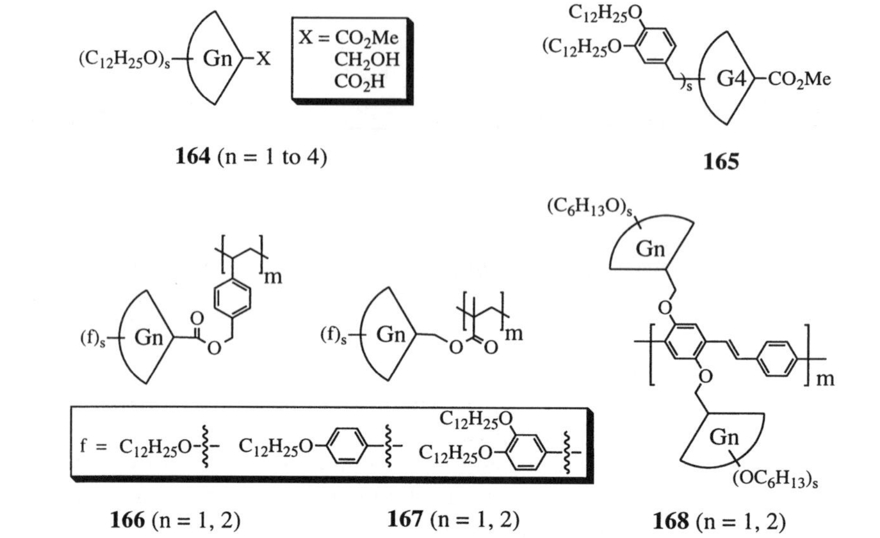

Fig. 17. Functional dendrimers derived from 3,4,5-trihydroxybenzyl ether-based dendrons

2.12 Dendritic Oligoethers Based on a 3-(3,5-Dihydroxyphenoxy)-1-propoxy Repeating Unit

The oligoether dendritic fragments based on a 3-(3,5-dihydroxyphenoxy)propyl ether repeating unit were reported by Chow and coworkers [101]. The synthesis was based on a two-step iterative cycle involving first a functional group conversion of a dendritic alcohol [Gn]-OH **169** to the corresponding bromide **170.** In the next step, the bromide was treated with a branching juncture 3-(3,5-dihydroxyphenoxy)-propan-1-ol **171** to give the dendritic alcohol of the next generation [G(n+1)]-OH **172** (Scheme 29). Using 4-*t*-butylphenol or 3,5-dimethylphenol as the surface groups, oligoether dendrons up to [G5] were prepared (Table 13).

This family of oligoether dendrons had been used to construct metallodendrimers **173** having an Fe(II)-bis(terpy) unit encapsulated inside the dendritic

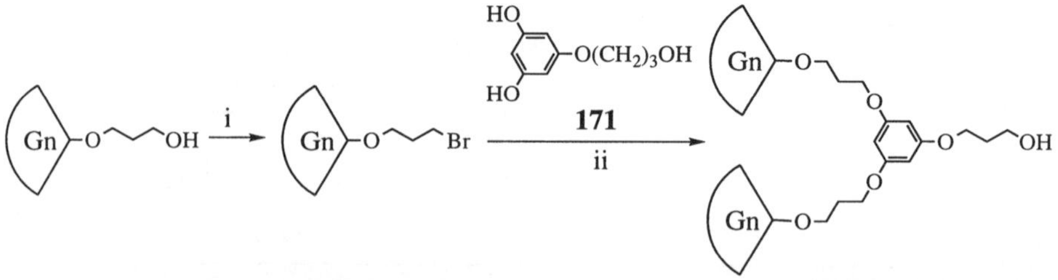

169 [Gn]-OH **170** [Gn]-Br **172** [G(n+1)]-OH

Scheme 29. i) PPh_3, CBr_4, THF; ii) K_2CO_3, 18-crown-6, THF, 66 °C

Table 13. Selected data for 3-(3,5-dihydroxyphenoxy)-1-propyl ether-based dendrons[a]

n	Yield (%) [Gn]-OH→[G(n+1)]-OH		Calcd MW of [Gn]-OH	
	Series a	Series b	Series a	Series b
0	91[b]	70[b]	–	–
1	83	48	565	509
2	88	43	1278	1166
3	65	–	2704	2479
4	58	–	5555	–
5	–	–	11259	–

[a] Series a: [G0]-OH = 4-*t*-butylphenol; series b: [G0]-OH = 3,5-dimethylphenol.
[b] From [G0]-Br to [G1]-OH.

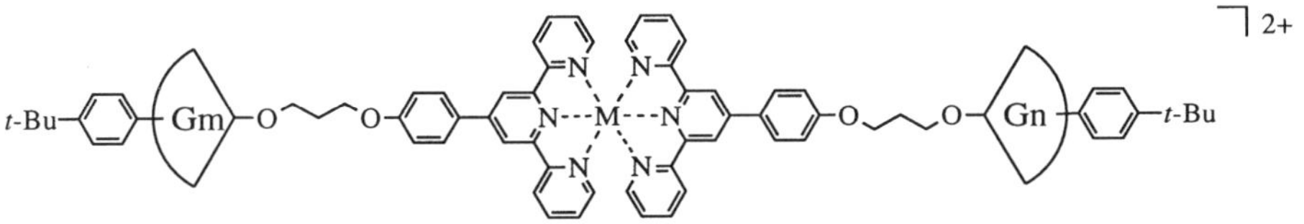

173 (M = Fe, m = n = 0 to 3)
174 (M = Ru, m = 0 to 2, n = 0 to 2)

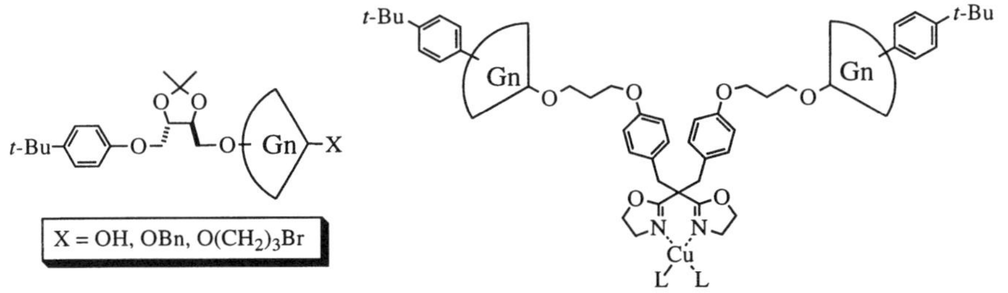

175 (n = 1 to 3) **176** (n = 0 to 3)

Fig. 18. Functional dendrimers constructed from 3-(3,5-dihydrophenoxy)-1-propylether-based dendrons

envelope (Fig. 18) [102]. The redox processes of this redox active unit became increasingly irreversible with increasing dendrimer generation, indicating that the oligoether dendrons retarded the electron transfer processes between the redox center and the electrode. Similar electrochemical behavior was also noted for the heteroleptic Ru(II)-bis(terpy) dendrimers **174** [103]. Chiral oligoether-based dendrimers **175** having multiple tartrate-derived chiral surface groups were also reported [104]. The chiroptical property of these sterically non-congested dendrons was shown to have a direct relationship with the number of chiral tartrate units. Catalytically active dendrimers **176** with a focal point Cu(II)-bis(oxazoline) active site were also prepared [105]. Slightly better substrate selectivity was noted for the [G3]-dendritic species.

2.13 Dendritic Oligoethers Based on a 4,4-Bis-(4-hydroxyphenyl)-4-methyl-1-butoxy Repeating Unit

This series of oligoether dendrons was originally reported by Fréchet and coworkers [106]. There are two steps in the iterative synthetic cycle. First, the dendritic alcohol [Gn]-OH **177** is converted into the corresponding bromide [Gn]-Br **178** by treatment with CBr_4 and PPh_3 (Scheme 30). The bromide is then reacted with a branching agent **179** to provide the higher generation dendritic alcohol [G(n+1)]-OH **180.** Using benzyl as the surface groups, dendrons up to [G3] were prepared (Table 14). Chen and Gorman later reported a modified procedure in which a dendritic mesylate [Gn]-OMs **181** was used in place of the bromide **178** [107]. After such a modification, the reactions can be conducted on a larger scale (10 g scale) and the products can be isolated in better yields.

This family of dendrons could be converted into another series of oligoether dendrons **182** containing a focal point organothiol functionality [107], which in turn was used to construct electrochemically active dendrimers **183** having an iron sulfur $[Fe_4S_4]^{2+}$ cluster core encapsulated within the dendritic envelope (Fig. 19) [108]. Hybrid organic-inorganic dendrimers containing a Mo_6Cl_8 core surrounded by oligoether dendrons were also reported [109].

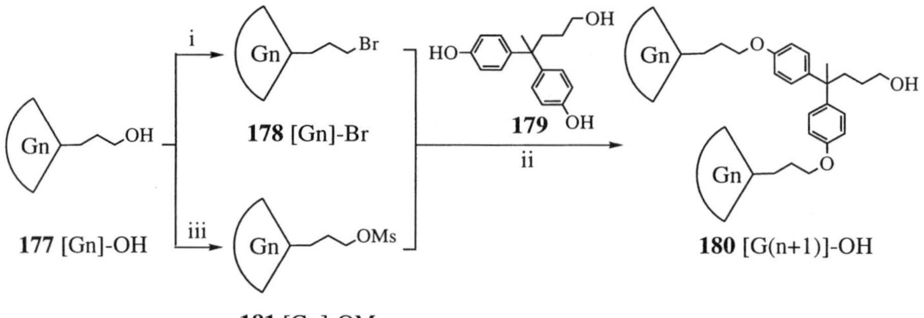

Scheme 30. i) PPh_3, CBr_4, THF; ii) K_2CO_3, 18-crown-6, THF, 66 °C; iii) MsCl, DMAP, NEt_3, CH_2Cl_2

Table 14. Selected data for 4,4-Bis-(4-hydroxyphenyl)-4-methyl-1-butyl ether-based dendrons[a]

n	Yield (%) [Gn]-OH→ [G(n+1)]-OH via [Gn]-Br	Yield (%) [Gn]-OH→ [G(n+1)]-OH via [Gn]-OMs	Calcd MW of [Gn]-OH
0	95[b]	–	–
1	82	86	453
2	79	85	1142
3	–	89	2519
4	–	–	5275

[a] [G0]-OH = benzyl alcohol.
[b] From [G0]-Br to [G1]-OH.

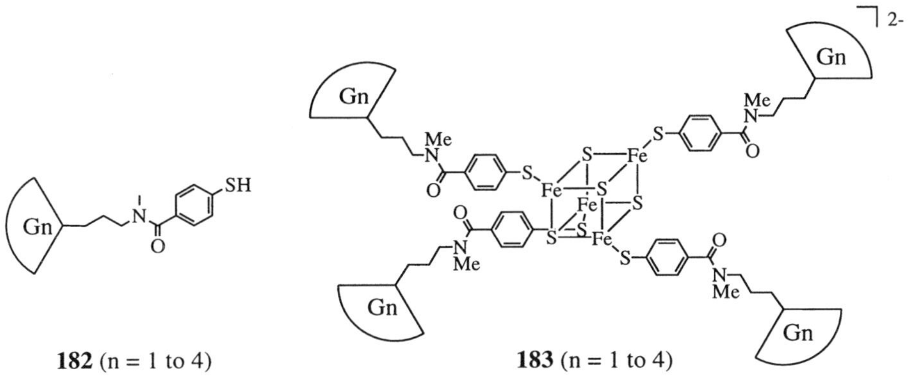

182 (n = 1 to 4)

Fig. 19. Functional dendrimers constructed from 4,4-bis-(4-hydroxyphenyl)-4-methyl-1-butyl ether-based dendrons

2.14 Dendritic Oligoethers Based on a 4-[1,1,1-Tris-(3-hydroxy-propyl)methyl]phenoxy Repeating Unit

Recently Astruc reported the synthesis of oligoether dendrons having a branching multiplicity of four using either the divergent or convergent strategy (Scheme 31) [110]. The divergent route began with a functional group conversion of a polyene **184** [Gn]-(ene)$_s$ to the corresponding polymesylate **185** [Gn]-(OMs)$_s$ via sequential hydroboration-oxidation-mesylation reactions, followed by excessive *O*-alkylation with a branching phenol **186** to give a polyene **187** [G(n+1)]-(ene)$_s$ of the next generation. Repetition of this method afforded

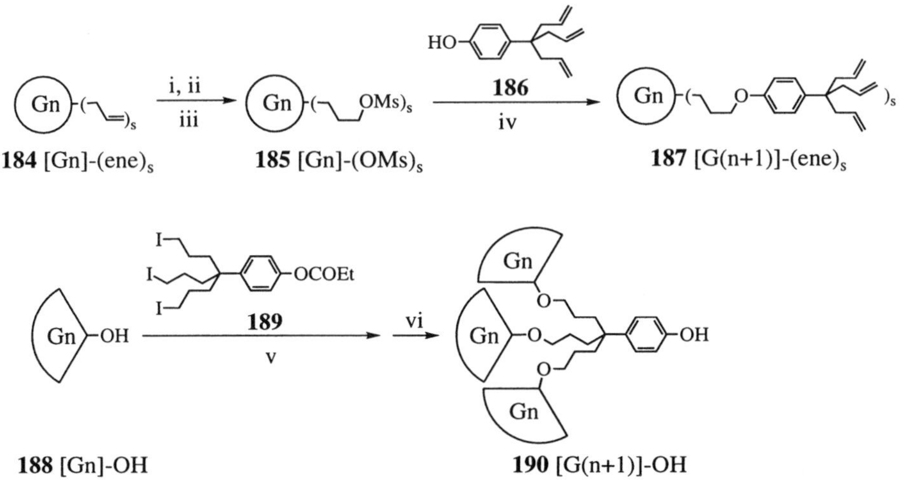

Scheme 31. i) disiamylborane, THF, 0 °C; ii) H_2O_2, NaOH, 50 °C; iii) MsCl, py, 0 °C; iv) CsF, DMF; v) K_2CO_3, DMF; vi) K_2CO_3, H_2O

Table 15. Selected data for 4-[1,1,1-tris-(3-hydroxypropyl)methyl]phenyl ether-based dendrimers[a] and dendrons[b]

n	Yield (%) [Gn]-(ene)$_s$ →[G(n+1)]-(ene)$_s$	Calcd MW of [Gn]-(ene)$_s$	Yield (%) [Gn]-OH →[G(n+1)]-OH	Calcd MW of [Gn]-OH
0	–	–	30	–
1	5 (26)[c]	2536	–	913
2	5	8701	–	–
3	–	27196	–	–

[a] Core = 1,3,5-tris-[1,1,1-tris-(3-hydroxypropyl)methyl]benzene.
[b] [G0]-OH = 4-[1,1,1-tris-(prop-2-enyl)methyl]phenol.
[c] Via the corresponding [G1]-iodide.

symmetrically substituted dendrimers having a maximum of 243 olefin groups (Table 15). Alternatively, oligoether dendrons of a similar structural skeleton could be assembled using a convergent method. Hence, an oligoether phenol [Gn]-OH **188** was reacted with a triiodo branching derivative **189** to give an intermediate [G(n+1)]-ester which was then hydrolyzed to afford the phenol dendron [G(n+1)]-OH **190.** However, further growth of the dendrons beyond [G1] was prohibited due to eliminative side reactions. Based on this dendritic skeleton, a family of electrochemically active dendrimers bearing ferrocenyl groups on the dendrimer surface was reported [111]. As expected, all the ferrocenyl groups are oxidized at the same potential.

2.15 Dendritic Oligoethers Based on a 13-(4-Hydroxyphenyl)-12-(4-hydroxy-4″-*p*-terphenylyl)-1-tridecoxy Repeating Unit

Percec and coworkers reported the preparation of liquid crystalline dendrimers based on a mesogenic 13-(4-hydroxyphenyl)-12-(4-hydroxy-4″-*p*-terphenylyl)-1-tridecanol **191** branching unit (Scheme 32) [112]. An oligoether dendritic alcohol [Gn]-OH **192** was converted into the corresponding dendritic bromide

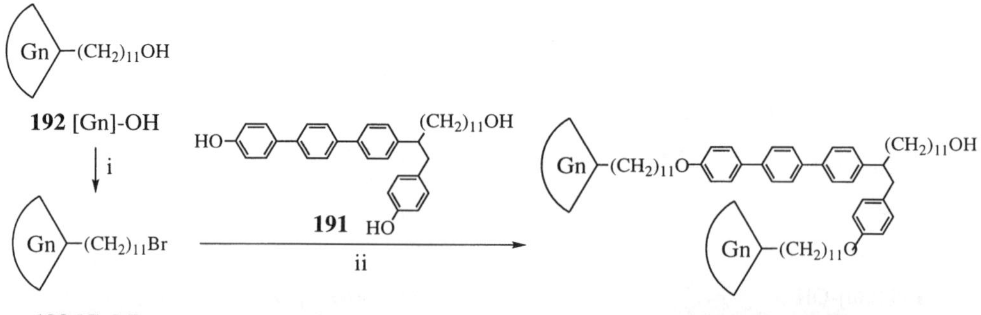

Scheme 32. i) PPh_3, CBr_4, THF; ii) NaOH, 1,2-dichlorobenzene, H_2O, Bu_4NOH, 80 °C

Table 16. Selected data for 13-(4-hydroxyphenyl)-12-(4-hydroxy-4″-*p*-terphenylyl)-1-tridecyl ether-based dendrons[a]

n	Yield (%) [Gn]-OH → [G(n+1)]-OH	Calcd MW of [Gn]-OH
0	80[b]	–
1	61	841
2	48	2183
3	42	4867
4	–	10236

[a] [G0]-Br = $H_2C{=}CH(CH_2)_9Br$.
[b] From [G0]-Br.

[Gn]-Br **193** by treatment with CBr_4 and PPh_3. The bromide was then reacted with the brancher **191** in the presence of NaOH to provide [G(n+1)]-OH **194.** Using 10-undecenol as the surface group, liquid crystalline oligoether dendrons with a focal point hydroxy functionality up to [G4] were prepared (Table 16). Finally, all the various dendrons were anchored to a trimesic acid core to give nonspherical dendrimers that displayed calamitic nematic and smectic thermotropic liquid crystalline phases [112].

2.16 Chiral Dendritic Oligoethers Based on an Optically Active (2S,3R)-4-[(3-Hydroxy-2-hydroxymethyl)butyl]benzyloxy Repeating Unit

Optically active (2*S*,3*R*)-4-[(3-hydroxy-2-hydroxymethyl)butyl]benzyl trialkylsilyl ether **195**, available in several steps from (*R*)-3-hydroxy-butanoic acid, had been used as a chiral branching unit in Seebach's synthesis of chiral oligoether dendrons [113]. There are three steps in the iterative synthetic cycle (Scheme 33). A dendritic alcohol [Gn]-OH **196** is first converted to the corresponding bromide [Gn]-Br **197.** The product is then reacted with the branching agent **195** to give a dendritic silyl ether [G(n+1)]-$OSiR_3$ in which the silyl protecting group is subsequently removed to give the dendritic alcohol [G(n+1)]-OH **198.** Chiral oligoether dendrons up to [G4] were synthesized by this method (Table 17). Using an elongated branching unit **199** or elongated chiral core **200,** layer-block oligoether dendrons and dendrimers up to [G5] could be obtained (Fig. 20) [113b].

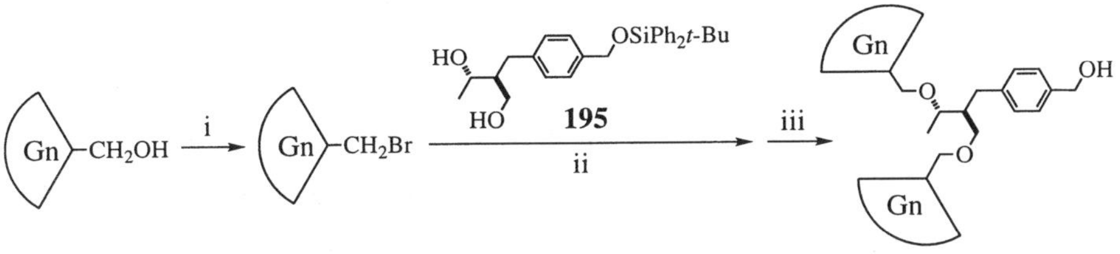

196 [Gn]-OH **197** [Gn]-Br **198** [G(n+1)]-OH

Scheme 33. i) PPh_3, CBr_4, THF; ii) NaH, THF 70 °C; iii) Bu_4NF, THF

Table 17. Selected data for (2*S*,3*R*)-4-[(3-hydroxy-2-hydroxymethyl)butyl]benzyl ether-based dendrons [a]

n	Yield (%) [Gn]-OH →[G(n+1)]-OH	Calcd MW of [Gn]-OH
0	100 [b]	–
1	88	238
2	76	651
3	–	1476

[a] [G0]-I = MeI
[b] From [G0]-I

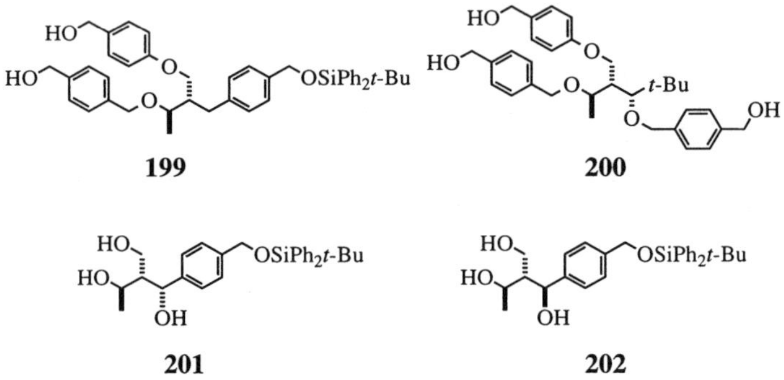

Fig. 20. Chiral branching units and chiral core reported by Seebach

2.17
Chiral Dendritic Oligoethers Based on an Optically Active (1*R*,2*S*,3*R*)- or (1*S*,2*S*,3*R*)-4-[(1,3-Dihydroxy-2-hydroxymethyl)butyl]benzyloxy Repeating Unit

Chiral oligoether dendrons based on an AB_3 branching pattern had also been reported by Seebach and coworkers [113a, 114]. The iterative sequence for their synthesis was essentially the same as that described in Scheme 33, except that the branching agent used was either (1*R*,2*S*,3*R*)-4-[(1,3-dihydroxy-2-hydroxymethyl)butyl]benzyl trialkylsilyl ether **201** or its C-1 epimer **202**. Using these branching units, several diastereomeric series of layer-block oligoether dendrons were prepared. Due to the relatively congested structure of the branching pattern, only dendrons up to [G2] could be synthesized. Rather unexpectedly, these diastereomeric dendrons exhibited significantly different reactivity towards the alkylation of a chiral core [114].

2.18
Chiral Dendritic Oligoethers Based on an Optically Active (2*R*,3*R*)- or (2*S*,3*S*)-2,3-Dihydroxy-2,3-*O*-isopropylidene-4-(3,5-dihydroxyphenoxy)-1-butoxy Repeating Unit

These two series of chiral oligoether dendrons have a chiral branch derived from optically active tartrate acid. A convergent iterative synthetic approach was used for the construction of these dendrons (Scheme 34) [115]. An oligoether dendritic

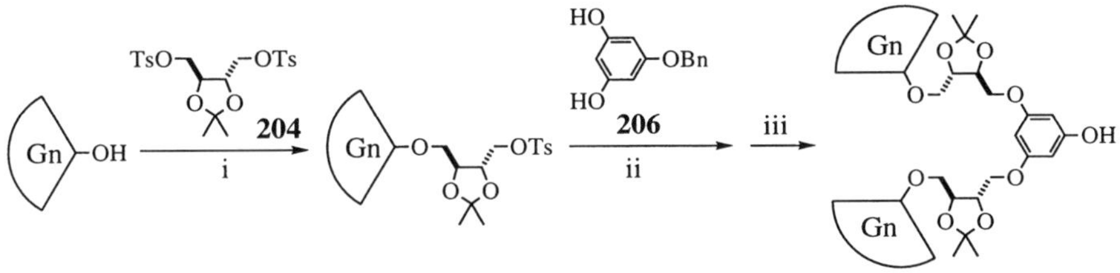

203 [Gn]-OH **205** [Gn]-OTs **207** [G(n+1)]-OH

Scheme 34. i) K_2CO_3, DMF, 110 °C; ii) K_2CO_3, DMF, 115 °C; iii) H_2, 10% Pd-C, EtOH, EtOAc

phenol [Gn]-OH **203** was subject to mono-*O*-alkylation with optically active (2*S*,3*S*)-1,4-di-*O*-tosyl-2,3-*O*-isoproylidene-L-threitol **204** to give the tosylate [Gn]-OTs **205**. Subsequently the tosylate was reacted with 5-benzyloxyresorcinol **206** to give a [G(n+1)]-OBn which was subsequently hydrogenolyzed to produce [G(n+1)]-OH **207**. The dendrons were than finally anchored to a phloroglucinol central core to provide homochiral [G0]- and [G1]-dendrimers of C_3-symmetry. Layer-block dendrimers containing both D- and L-threitol derived chiral branches were also prepared [116]. The chiroptical property of these dendrimers was shown to be the sum of the contribution originated from each chiral component.

2.19
Chiral Dendritic Oligoethers Based on an Optically Active (1*R*,2*R*)-4-[(1,2-Dihydroxy-2-phenyl)ethyl]benzyloxy, (*R*)-4-(1,2-Dihydroxyethyl)benzyloxy or (1*R*,2*R*)-3-[(1,2-Dihydroxy-2-cyclohexyl)ethyl]benzyloxy Repeating Unit

The chiral repeating branching units of these oligoether dendrons were all prepared from an asymmetric dihydroxylation of prochiral styrenes. A divergent approach was adopted for their synthesis. Hence, alkylation of a dendritic polyol [Gn]-$(OH)_s$ **208** with a benzyl bromide branching agent **209** gives a dendritic acetonide [G(n + 1)]-$(acetonide)_s$ **210** which can then be converted into the corresponding dendritic polyol of the next generation [G(n + 1)]-$(OH)_s$ **211** upon acid treatment (Scheme 35) [117]. Alternatively, alkylation of the dendritic polyol [Gn]-$(OH)_s$ **208** with a hyperbranched benzyl bromide monomer **212** afforded a dendritic acetonide [G(n + 2)]-$(acetonide)_s$ **213**. Chiral oligoether dendrons up to [G4] were prepared conveniently using this method.

2.20
Chiral Dendritic Oligoethers Based on an Optically Active (2*R*,3*R*)-[2,3-Dihydroxy-2,3-*O*-isopropylidene-3-(3,n-dihydroxy-phenyl)]-1-propoxy Repeating Unit (n = 4 or 5)

The branching units **214** for these chiral oligoether dendrons were also prepared by the asymmetric dihydroxylation of prochiral styrene derivatives followed by protection of the resulting 1,2-diols as the acetonide derivatives [118]. The iterative synthetic cycle consists of two steps (Scheme 36). First, functional group conversion of a dendritic alcohol [G(n + 1)]-OH **215** gives either the corre-

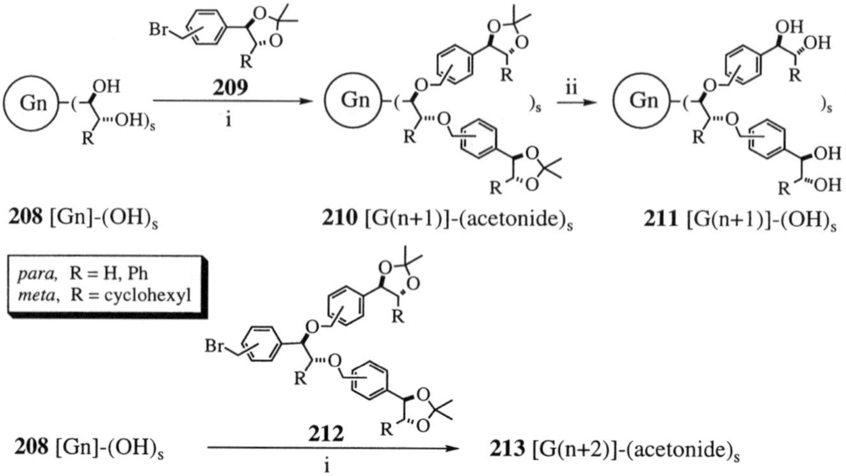

Scheme 35. i) KOH, toluene, 110 °C; ii) HCl, CH_3CN

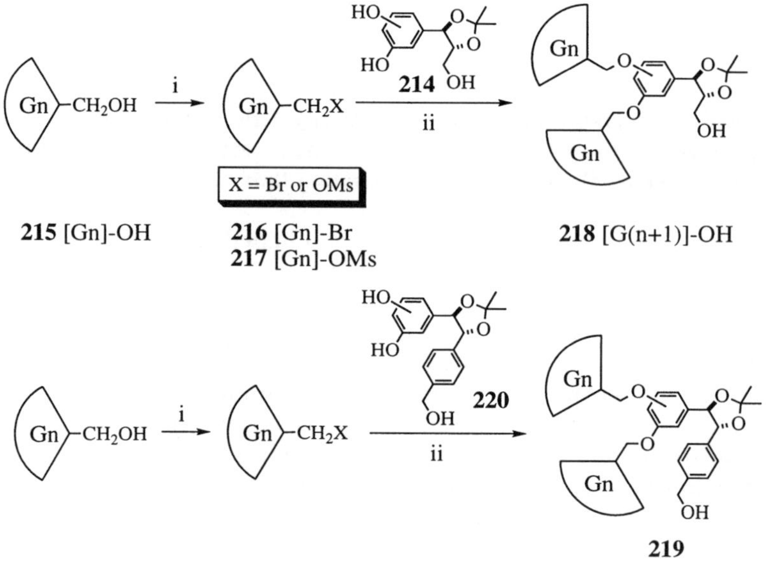

Scheme 36. i) CBr_4, PPh_3 or MsCl, NEt_3, CH_2Cl_2; ii) K_2CO_3, 18-crown-6, acetone, 56 °C

sponding bromide **216** [G(n + 1)]-Br or mesylate **217** [G(n + 1)]-OMs. Second, alkylation of **216** or **217** with the branching agent **214** provides dendritic alcohol of the next generation **218**. Using benzyl alcohol as the surface group, optically active oligoether dendrons up to [G3] were prepared. These chiral dendrons were then attached to a trimesic acid core to produce homochiral, symmetrical dendrimers [119]. The acetonide groups in these dendrimers could then be taken off by acid treatment to produce chiral polyhydroxylated dendrimers [120].

2.21 Chiral Dendritic Oligoethers Based on an Optically Active (1*R*,2*R*)-4-{[1,2-Dihydroxy-1,2-*O*-isopropylidene-2-(3,n-dihydroxyphenyl)]ethyl}benzyloxy Repeating Unit (n = 4 or 5)

The preparation of these two series of chiral oligoether dendrons **219** is similar to the one described in Sect. 2.20, except that the chiral branching units **220** were prepared from the asymmetric dihydroxylation of the prochiral stilbenes followed by protection of the resulting 1,2-diols as their acetonide derivatives (Scheme 36) [118, 121].

3 Perspectives

The field of dendrimer chemistry has undergone a rapid change in the past decade and will continue to grow at even greater pace in the future. New and exciting dendritic molecules with unusual properties will emerge and will find potential applications in material and biomedical chemistry.

Acknowledgements. We thank the Research Grants Council, HKSAR for financial support (CUHK307/96P).

References

1. For some reviews of dendrimer chemistry, see (a) Tomalia DA, Naylor AM, Goddard WA III (1990) Angew Chem Int Ed Engl 29:138; (b) Mekelburger H-B, Jaworek W, Vögtle F (1992) Angew Chem Int Ed Engl 31:1571; (c) Tomalia DA, Durst HD (1993) Top Curr Chem 165:193; (d) Issberner J, Moors R, Vögtle F (1994) Angew Chem Int Ed Engl 33:2413; (e) Moorefield CN, Newkome GR (ed) (1994) Advances in dendritic macromolecules, vols 1–4. JAI Press, Greenwich; (f) Newkome GR, Moorefield CN, Vögtle F (1996) Dendritic molecules. VCH, Weinheim; (g) Chow H-F, Mong TK-K, Nongrum MF, Wan C-W (1998) Tetrahedron 54:8543; (h) Archut A, Vögtle F (1998) Chem Soc Rev 27:233; (i) Smith DK, Diederich F (1998) Chem Eur J 4:1353; (j) Hearshaw MA, Moss JR (1999) Chem Commun 1; (k) Fischer M, Vögtle F (1999) Angew Chem Int Ed Engl 38:884; (l) Bosman AW, Jannsen HM, Meijer EW (1999) Chem Rev 99:1665; (m) Newkome GR, He E, Moorefield CN (1999) Chem Rev 99:1689
2. Hawker CJ, Fréchet JMJ (1990) J Am Chem Soc 112:7638
3. Padias AB, Hall HK Jr, Tomalia DA, McConnell JR (1987) J Org Chem 52:5305
4. (a) Buhleier E, Wehner W, Vögtle F (1978) Synthesis 155; (b) Tomalia DA, Baker H, Dewald JR, Hall M, Kallos G, Martin S, Roeck J, Ryder J, Smith P (1985) Polym J 17:117; (c) Newkome GR, Baker GR (1986) Org Prep Proced Int 18:117
5. Lee J-J, Ford WT, Moore JA, Li Y (1994) Macromolecules 27:4632
6. Nemoto H, Wilson JG, Nakamura H, Yamamoto Y (1992) J Org Chem 57:435
7. Haag R, Sunder A, Stumbé J-F (2000) J Am Chem Soc 122:2954
8. (a) Jayaraman M, Fréchet JMJ (1998) J Am Chem Soc 120:12,996; (b) Grayson SM, Jayaraman M, Fréchet JMJ (1999) Chem Commun 1329
9. Weintraub JG, Parquette JR (1999) J Org Chem 64:3796
10. Hawker C, Fréchet JMJ (1990) J Chem Soc Chem Commun 1010
11. Mourey TH, Turner SR, Rubinstein M, Fréchet JMJ, Hawker CJ, Wooley KL (1992) Macromolecules 25:2401

12. L'abbé G, Forier B, Dehaen W (1996) Chem Commun 2143
13. Forier B, Dehaen W (1999) Tetrahedron 55:9829
14. (a) Hawker CJ, Fréchet JMJ (1990) Macromolecules 23:4726; (b) Wooley KL, Hawker CJ, Fréchet JMJ (1991) J Chem Soc Perkin Trans 1 1059
15. Leon JW, Kawa M, Fréchet JMJ (1996) J Am Chem Soc 118:8847
16. Groenendaal L, Fréchet JMJ (1998) J Org Chem 63:5675
17. Hecht S, Fréchet JMJ (1999) J Am Chem Soc 121:4084
18. Wooley KL, Hawker CJ, Fréchet JMJ (1993) J Am Chem Soc 115:11,496
19. Gitsov I, Fréchet JMJ (1994) Macromolecules 27:7309
20. (a) Gitsov I, Wooley KL, Fréchet JMJ (1992) Angew Chem Int Ed Engl 31:1200; (b) Gitsov I, Wooley KL, Hawker CJ, Ivanova PT, Fréchet JMJ (1993) Macromolecules 26:5621
21. Hawker CJ, Fréchet JMJ (1992) Polymer 33:1507
22. Neubert I, Amoulong-Kirstein E, Schlüter A-D, Dautzenberg H (1996) Macromol Rapid Commun 17:517
23. Neubert I, Klopsch R, Claussen W, Schlüter A-D (1996) Acta Polymer 47:455
24. Wendland MS, Zimmerman SC (1999) J Am Chem Soc 121:1389
25. For reviews, see (a) Peerlings HWI, Meijer EW (1997) Chem Eur J 3:1563; (b) Seebach D, Rheiner PB, Greiveldinger G, Butz T, Sellner H (1998) Top Curr Chem 197:125
26. (a) Kremers JA, Meijer EW (1994) J Org Chem 59:4262; (b) Kremers JA, Meijer EW (1995) React Funct Polym 26:137
27. Peerlings HWI, Struijk MP, Meijer EW (1998) Chirality 10:46
28. Peerlings HWI, Trimbach DC, Meijer EW (1998) Chem Commun 497
29. (a) Seebach D, Lapierre J-M, Skobridis K, Greiveldinger G (1994) Angew Chem Int Ed Engl 33:440; (b) Seebach D, Lapierre J-M, Greiveldinger G, Skobridis K (1994) Helv Chim Acta 77:1673
30. Chaumette J-L, Laufersweiler MJ, Parquette JR (1998) J Org Chem 63:9399
31. Rohde JM, Parquette JR (1998) Tetrahedron Lett 39:9161
32. Rosini C, Superchi S, Peerlings HWI, Meijer EW (2000) Eur J Org Chem 61
33. (a) Wooley KL, Hawker CJ, Fréchet JMJ, Wudl F, Srdanov G, Shi S, Li C, Kao M (1993) J Am Chem Soc 115:9836; (b) Hawker CJ, Wooley KL, Fréchet JMJ (1994) J Chem Soc Chem Commun 925
34. (a) Camps X, Schönberger H, Hirsch A (1997) Chem Eur J 3:561; (b) Herzog A, Hirsch A, Vostrowsky O (2000) Eur J Org Chem 171; (c) Djojo F, Ravanelli E, Vostrowsky O, Hirsch A (2000) Eur J Org Chem 1051
35. Amabilino DB, Ashton PR, Balzani V, Brown CL, Credi A, Fréchet JMJ, Leon JW, Raymo FM, Spencer N, Stoddart JF, Venturi M (1996) J Am Chem Soc 118:12,012
36. Hübner GM, Nachtsheim G, Li QY, Seel C, Vögtle F (2000) Angew Chem Int Ed Engl 39:1269
37. Yamaguchi N, Hamilton LM, Gibson HW (1998) Angew Chem Int Ed Engl 37:3275
38. For a review on dendronized polymers, see Schlüter AD, Rabe JP (2000) Angew Chem Int Ed Engl 39:865
39. Claussen W, Schulte N, Schlüter A-D (1995) Macromol Rapid Commun 16:89
40. (a) Karakaya B, Claussen W, Gessler K, Saenger W, Schlüter A-D (1997) J Am Chem Soc 119:3296; (b) Stocker W, Karakaya B, Schürmann BL, Rabe JP, Schlüter AD (1998) J Am Chem Soc 120:7691; (c) Bo Z, Schlüter AD (1999) Macromol Rapid Commun 20:21
41. Sato T, Jiang D-L, Aida T (1999) J Am Chem Soc 121:10,658
42. Jahromi S, Coussens B, Meijerink N, Braam AWM (1998) J Am Chem Soc 120:9753
43. Schenning APHJ, Martin RE, Ito M, Diederich F, Boudon C, Gisselbrecht J-P, Gross M (1998) Chem Commun 1013
44. Malenfant PRL, Groenendaal L, Fréchet JMJ (1998) J Am Chem Soc 120:10,990
45. Jestin I, Levillain E, Roncali J (1998) Chem Commun 2655
46. Miller LL, Zinger B, Schlechte JS (1999) Chem Mater 11:2313
47. Hawker CJ, Wooley KL, Fréchet JMJ (1993) J Chem Soc Perkin Trans 1 1287
48. Gitsov I, Fréchet JMJ (1993) Macromolecules 26:6536
49. Gitsov I, Fréchet JMJ (1996) J Am Chem Soc 118:3785

50. Bo Z, Zhang X, Yi X, Yang M, Shen J, Rehn Y, Xi S (1997) Polym Bull 38:257
51. Cui G, Xu Y, Liu M, Fang F, Ji T, Chen Y, Li Y (1999) Macromol Rapid Commun 20:71
52. (a) Bo Z, Rabe JP, Schlüter AD (1999) Angew Chem Int Ed Engl 38:2370; (b) Bo Z, Zhang C, Severin N, Rabe JP, Schlüter AD (2000) Macromolecules 33:2688
53. (a) Jin R-H, Aida T, Inoue S (1993) J Chem Soc Chem Commun 1260; (b) Tomoyose Y, Jiang D-L, Jin R-H, Aida T, Yamashita T, Horie K, Yashima E, Okamoto Y (1996) Macromolecules 29:5236
54. Jiang D-L, Aida T (1996) Chem Commun 1523
55. Enomoto M, Aida T (1999) J Am Chem Soc 121:874
56. Wang Y, Zeng F, Zimmerman SC (1997) Tetrahedron Lett 38:5459
57. Zimmerman SC, Wang Y, Bharathi P, Moore JS (1998) J Am Chem Soc 120:2172
58. (a) Catalano VJ, Parodi N (1997) Inorg Chem 36:537; (b) Numata M, Ikeda A, Fukuhara C, Shinkai S (1999) Tetrahedron Lett 40:6945
59. (a) Ferguson G, Gallagher JF, McKervey MA, Madigan E (1996) J Chem Soc Perkin Trans 1 599; (b) Yamakawa Y, Ueda M, Nagahata R, Takeuchi K, Asai M (1998) J Chem Soc Perkin Trans 1 4135
60. Gitsov I, Ivanova PT (2000) Chem Commun 269
61. For a review, see Zeng F, Zimmerman SC (1997) Chem Rev 97:1681
62. Zimmerman SC, Zeng F, Reichert DEC, Kolotuchin SV (1996) Science 271:1095
63. Brewis M, Clarkson GJ, Holder AM, McKeown NB (1998) Chem Commun 969
64. Percec V, Cho W-D, Ungar G, Yeardley DJP (2000) Angew Chem Int Ed Engl 39:1598
65. Jang W-D, Jiang D-L, Aida T (2000) J Am Chem Soc 122:3232
66. Ikeda A, Numata M, Shinkai S (1999) Chem Lett 929
67. Numata M, Ikeda A, Shinkai S (2000) Chem Lett 370
68. Gitsov I, Ivanova PT, Fréchet JMJ (1994) Macromol Rapid Commun 15:387
69. Mecerreyes D, Dubois P, Jérôme R, Hedrick JL, Hawker CJ (1999) J Polym Sci Part A Polym Chem 37:1923
70. (a) Leduc MR, Hawker CJ, Dao J, Fréchet JMJ (1996) J Am Chem Soc 118:11,111; (b) Matyjaszewski K, Shigemoto T, Fréchet JMJ, Leduc M (1996) Macromolecules 29:4167
71. Seebach D, Marti RE, Hintermann T (1996) Helv Chim Acta 79:1710
72. Francavilla C, Bright FV, Detty MR (1999) Org Lett 1:1043
73. Rheiner PB, Seebach D (1999) Chem Eur J 5:3221
74. (a) Rheiner PB, Sellner H, Seebach D (1997) Helv Chim Acta 80:2027; (b) Sellner H, Seebach D (1999) Angew Chem Int Ed Engl 38:1918
75. Bolm C, Derrien N, Seger A (1996) Synlett 387
76. Yamago S, Furukawa M, Azuma A, Yoshida J-I (1998) Tetrahedron Lett 39:3783
77. Morao I, Cossío FP (1997) Tetrahedron Lett 38:6461
78. (a) Liao Y-H, Moss JR (1993) J Chem Soc Chem Commun 1774; (b) Liao Y-H, Moss JR (1995) Organometallics 14:2130; (c) Liao Y-H, Moss JR (1996) Organometallics 15:4307
79. Serroni S, Campagna S, Juris A, Venturi M, Balzani V, Denti G (1994) Gazz Chim Ital 124:423
80. Vögtle F, Plevoets M, Nieger M, Azzellini GC, Credi A, De Cola L, De Marchis V, Venturi M, Balzani V (1999) J Am Chem Soc 121:6290
81. (a) Pollak KW, Leon JW, Fréchet JMJ, Maskus M, Abruña HD (1998) Chem Mater 10:30; (b) Pollak KW, Sanford EM, Fréchet JMJ (1998) J Mater Chem 8:519
82. Sadamoto R, Tomioka N, Aida T (1996) J Am Chem Soc 118:3978
83. Stewart GM, Fox MA (1996) J Am Chem Soc 118:4354
84. Jiang D-L, Aida T (1998) J Am Chem Soc 120:10,895
85. Jiang D-L, Aida T (1997) Nature 388:454
86. (a) Junge DM, McGrath DV (1997) Chem Commun 857; (b) Junge DM, McGrath DV (1999) J Am Chem Soc 121:4912
87. Smet M, Liao L-X, Dehaen W, McGrath DV (2000) Org Lett 2:511
88. Camps X, Dietel E, Hirsch A, Pyo S, Echegoyen L, Hackbarth S, Röder B (1999) Chem Eur J 5:2362
89. Chen Y-M, Chen C-F, Liu W-H, Xi F (1996) Polym Bull 37:557

90. Höger S (1997) Synthesis 20
91. Gilat SL, Adronov A, Fréchet JMJ (1999) J Org Chem 64:7474
92. (a) Ng ACH, Li X-Y, Ng DKP (1999) Macromolecules 32:5292; (b)Li X-Y, He X, Ng ACH, Wu C, Ng DKP (2000) Macromolecules 33:2119
93. (a) Gilat SL, Adronov A, Fréchet JMJ (1999) Angew Chem Int Ed Engl 38:1422; (b) Adronov A, Gilat SL, Fréchet JMJ, Ohta K, Neuwahl FVR, Fleming GR (2000) J Am Chem Soc 122:1175
94. Piotti ME, Rivera F Jr, Bond R, Hawker CJ, Fréchet JMJ (1999) J Am Chem Soc 121:9471
95. Balagurusamy VSK, Ungar G, Percec V, Johansson G (1997) J Am Chem Soc 119:1539
96. Ungar G, Percec V, Holerca MN, Johansson G, Heck JA (2000) Chem Eur J 6:1258
97. Percec V, Cho W-D, Möller M, Prokhorova SA, Ungar G, Yeardley DJP (2000) J Am Chem Soc 122:4249
98. (a) Percec V, Ahn C-H, Barboiu B (1997) J Am Chem Soc 119:12,978; (b) Percec V, Ahn C-H, Ungar G, Yeardley DJP, Möller M, Sheiko S S (1998) Nature 391:161; (c) Percec V, Ahn C-H, Cho W-D, Jamieson AM, Kim J, Leman T, Schmidt M, Gerle M, Möller M, Prokhorova SA, Sheiko SS, Cheng SZD, Zhang A, Ungar G, Yeardley DJP (1998) J Am Chem Soc 120:8619
99. Bao Z, Amundson KR, Lovinger AJ (1998) Macromolecules 31:8647
100. Wang R, Zheng Z (1999) J Am Chem Soc 121:3549
101. Chow H-F, Wang Z-Y, Lau Y-F (1998) Tetrahedron 54:13,813
102. Chow H-F, Chan IY-K, Chan DTW, Kwok RWM (1996) Chem Eur J 2:1085
103. Fung PS (1999) MPhil Thesis, The Chinese University of Hong Kong
104. Chow H-F, Mak CC (1996) Tetrahedron Lett 37:5935
105. (a) Mak CC, Chow H-F (1997) Macromolecules 30:1228; (b) Chow H-F, Mak CC (1997) J Org Chem 62:5116
106. Wooley KL, Hawker CJ, Fréchet JMJ (1991) J Am Chem Soc 113:4252
107. Chen K-Y, Gorman CB (1996) J Org Chem 61:9229
108. (a) Gorman CB, Parkhurst BL, Su WY, Chen K-Y (1997) J Am Chem Soc 119:1141; (b) Gorman CB, Hager MW, Parkhurst BL, Smith JC (1998) Macromolecules 31:815
109. Gorman CB, Su WY, Jiang H, Watson CM, Boyle P (1999) Chem Commun 877
110. Sartor V, Djakovitch L, Fillaut J-L, Moulines F, Neveu F, Marvaud V, Guittard J, Blais J-C, Astruc D (1999) J Am Chem Soc 121:2929
111. Nlate S, Ruiz J, Blais J-C, Astruc D (2000) Chem Commun 417
112. Percec V, Chu P, Ungar G, Zhou J (1995) J Am Chem Soc 117:11,441
113. (a) Murer P, Seebach D (1995) Angew Chem Int Ed Engl 34:2116; (b) Murer P, Seebach D (1998) Helv Chim Acta 81:603
114. Murer PK, Lapierre J-M, Greiveldinger G, Seebach D (1997) Helv Chim Acta 80:1648
115. (a) Chow H-F, Fok LF, Mak CC (1994) Tetrahedron Lett 35:3547; (b) Chow H-F, Mak CC (1994) J Chem Soc Perkin Trans 1 2223
116. (a) Mak CC, Chow H-F (1996) Chem Commun 1185; (b) Chow H-F, Mak CC (1997) J Chem Soc Perkin Trans 1 91; (c) Chow H-F, Mak CC (1997) Pure Appl Chem 69:483
117. Chang H-T, Chen C-T, Kondo T, Siuzdak G, Sharpless KB (1996) Angew Chem Int Ed Engl 35:182
118. (a) McElhanon JR, Wu M-J, Escobar M, McGrath DV (1996) Macromolecules 29:8979; (b) McElhanon JR, Wu M-J, Escobar M, Chaudhry U, Hu C-L, McGrath DV (1997) J Org Chem 62:908
119. McElhanon JR, McGrath DV (1998) J Am Chem Soc 120:1647
120. McElhanon JR, McGrath DV (2000) J Org Chem 65:3525
121. McGrath DV, Wu M-J, Chaudhry U (1996) Tetrahedron Lett 37:6077

Dendrimers with Carbon Rich-Cores

Andreas Hirsch[1], Otto Vostrowsky[2]

Institut für Organische Chemie der Universität Erlangen-Nürnberg, Henkestrasse 42, 91054 Erlangen, Germany

[1] *E-mail: andreas.hirsch@chemie.uni-erlangen.de*

[2] *E-mail: otto.vostrowsky@organik.uni-erlangen.de*

Carbon rich compounds such as C_{60} buckminsterfullerene, conjugated oligoynes, and single-walled carbon nanotubes offer advantages as core templates for the design of dendrimers with a predefined shape because of their rigid structures. I_h-symmetrical C_{60} permits the stereochemically-controlled attachment of anchor groups for the addition of dendrons and allows the realization of the formation of perfect spherical dendrimers and of variable addition patterns. The dendronization of fullerenes improves their solubility and provides the carbon sphere with additional chemical and physical properties. Medium-chain oligoynes are used as one-dimensional core tectons, decorated with dendritically branched end-caps. Single-walled carbon nanotubes represent tubular templates for cylindrical dendrimeric nanostructures.

Keywords: Dendrimer, Carbon rich-cores, Fullerenes, Oligoynes, Carbon nanotubes

1 Introduction

A dendrimer's core can be considered as the "origin" of the dendritic structure, representing the center of attachment of a number of molecular branches. Practically all dendrimers known have cores with a few functional anchors as focal points to which the corresponding number of dendrons (dendritic wedges) are attached. An oligovalent atom or a multifunctional molecule may be considered as such a core. A tetravalent carbon atom is the central core of the 36-cascade compound micellanoic acid $C[(CH_2)_8C[(CH_2)_8C[CH_2CH_2CO_2H]_3]_3]_3$ and its derivatives [1]. Similarly, silicon has been used as atomic core for the synthesis of a number of four-armed tetraalkylsilane-based dendrimers, mixed Si/P-based dendrimers, carbosilane dendrimers, polysiloxane and poly(siloxysilane) dendrimers. Applying a trivalent nitrogen atom as central core, three-armed dendritic structures of polyamide type and PAMAM (polyamidoamine) starburst polymers are obtained. A corresponding tetrahedral ammonium moiety as center leads to four-armed dendrimers. A series of three-armed neutral pentavalent P-based dendrimers with P^V-atoms as cores is known [1].

Bifunctional α,ω-diaminoalkanes represent the central cores of poly(propylene imine) dendrimers [1]. Polyols and polyethereal compounds like 1,1,1-tris(hydroxymethyl)alkanes and 1,1,1,1-tetrakis(hydroxymethyl)methane derivatives were used as trifunctional and tetrafunctional molecular core compounds in the synthesis of a number of three-armed branched architectures of the "tree-like" polyamide arborol-type with fractal geometry and of the four-armed polyamide and polyether type dendrimers [1]. Frequently applied as oligofunctional aromatic cores in dendrimers construction are 1,3,5-substituted benzene derivatives like, e.g., 1,3,5-triarylbenzenes, 1,3,5-trialkylbenzenes, 1,3,5-trihydroxybenzene, 3,5-dihydroxybenzyl alcohol, 1,3,5-benzenetricarboxylic acid, and 1,1,1-tri(4-hydroxyphenyl)ethane [1]. Rigid core structures are represented by 1,3,5-trialkynylsubstituted benzenes and the tetravalent core 1,3,5,7-adamantane tetracarboxylic acid. Zn-porphyrins provide unique cores for the study of electron transport through dendritic superstructures [1].

At the other end of this core-scale, very large and multiatomic molecules, nanostructures, and polymers may be considered as cores as well. Using, e.g., a polymeric filament core with a repetitive number of anchor sites along its extension, macromolecules of the kind of "rod-like dendrimers" are obtained by attachment of branched side chains and may lead to nanocylinders [2,3]. Materials like this may be considered as either dendrimers with a polymeric core or alternatively as dendronized polymers.

Somewhere between a dot-like core consisting of a small molecule and the polymeric core mentioned above, a number of intermediate structures are conceivable as cores, which can bear a defined amount of focal points in a geometrically well-defined arrangement. With the fixed number of attached dendrons (equal to the number of focal points) and their branching multiplicity, the molecular shape of such cores can serve as an architectural template, forcing the attached dendrons into canonical arrangements.

However, the central core does not just act as the structure-determining tecton. A change of the core's shape also causes a change of the dendron-filled volume of the target dendrimer. As a consequence, the inner core also determines the outer surface structure. This brings about a change of chemical and physical properties with respect to surface characteristics, increasing the importance of the core with respect to function and properties of dendrimers. For example, chirality of such a core or an inherently chiral addition pattern will induce chirality into the spatial arrangement of branching units and consequently into the molecule in its entirety. In particular, rigid molecules with minute thermal mobility and flexibility can therefore offer advantages as building core templates for the design of dendrimers with a predefined shape.

2 Carbon Rich-Cores

Typical members of the class of compounds mentioned above are found in the field of carbon-rich compounds and include all-carbon compounds such as fullerenes, carbon nanotubes, and polyynes. Since all these compounds exhibit a rigid architecture, the potential of fullerenes, carbon nanostructures and, polyynes as central cores for dendrimers is obvious. The football-like C_{60} fullerene represents a perfect spherical template, a tubular single-walled carbon nanotube can be considered as a template for a hollow cylindrical dendrimeric structure, and a polyyne can serve as a rod-like tecton. All these carbon-rich core compounds have in common a highly symmetrical, aesthetically-pleasing structure (Fig. 1), and all of them lack the flexibility, associated with common aliphatic and/or aromatic core compounds.

The spherical framework of C_{60} is an ideal core tecton for dendrimers [4–12], leading to perfect globular systems even with low-generation dendrons. Since its regiochemistry is well established [4, 5, 13–19], C_{60} can easily be multiply-functionalized with anchor points in topologically-defined positions, opening synthetic routes to tailor-made designed functional dendrimers [20]. Realizing variable attachment patterns with a given degree of addition and the addition of both similar and dissimilar addends in a stereochemically-controlled way permits the combination of different dendrons and of dendrons with selected functionalities. The functionalization of C_{60} with a controlled number of dendrons dramatically improves the solubility of the fullerenes and provides a compact insulating layer around the carbon sphere. Incorporation of fullerenes into well-ordered structures can be easily achieved. Cavities and clefts within the dendritic structure can be utilized for the insertion of functional groups or to form host-guest-complexes with other molecules. At present, interest is growing in fullerene-functionalized dendrimers, or *fullerenodendrimers* [12]. Such

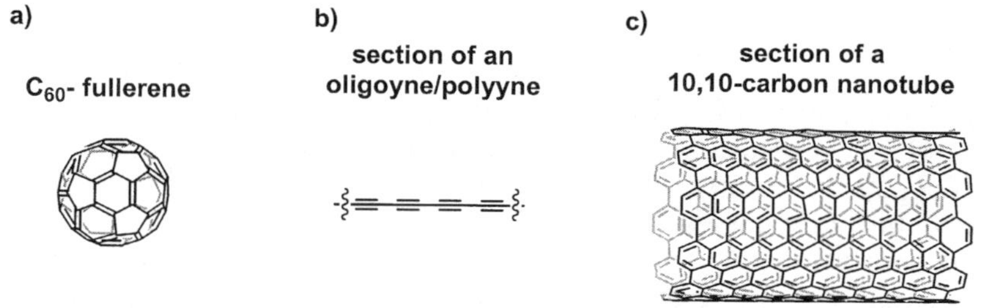

Fig. 1. **a** The I_h-symmetrical C_{60} fullerene sphere. **b** A rod-like oligoyne chain. **c** A tubular single-walled (10,10)-carbon nanotube, carbon-rich compounds to be used as central cores for dendrimer synthesis

fullerenodendrimer structures represent versatile building blocks for further functional supramolecular architectures such as artificial enzymes, catalysts, etc. and appear to be promising candidates for a variety of interesting applications in supramolecular chemistry and materials science.

Dendrimers resulting from the attachment of dendrons to the terminal anchor groups of a rod-like oligoyne core will have rather a double arborol-like propagation, in contrast to the 3D-structures with C_{60}. Large dendritic wedges prevent a close approach between the polyunsaturated carbon rods and thus additionally act as protecting groups for polyynes, preventing polymerization. The use of a single-walled carbon nanotube as core, the dendrons attached terminally or along the cylindrical wall of carbon hexagons, will give rise to the formation of hollow tubular dendrimer structures. With such functionalization the solubility of the previously insoluble nanotube can be dramatically enhanced and specifically tuned to the specific solvents by varying the nature of the dendrons.

3
Dendrimers with C_{60}-Based Cores

I_h-symmetrical C_{60} represents a kinetically-stable carbon cluster and is considered to be a versatile building block and a topologically well-defined three-dimensional tecton in organic synthesis [21]. It consists of 12 fused pentagons and 20 hexagons. C_{60} reacts as a polyolefin and is susceptible to many useful preparative addition reactions [22–24]. Many principles of fullerene reactivity are now well-established [16]. Monoadducts and stereochemically-defined multiple adducts having two to six addends have been synthesized. Among the multiple addition products, hexakisadducts with a T_h-symmetrical octahedral addition pattern are of special interest [25]. With six malonate addends as anchors for the dendritic wedges, the exceptionally high multiplicity of 12 for the initiator core C_{60} is found. Only a small number of generations of the dendrimer are necessary to reach sterically overburdened, compact structures.

3.1
Dendrimers with C_{60} Monoadduct Cores

In 1993, Wooley et al. [26] synthesized the first dendrimer containing a C_{60} core by coupling 3,5-dihydroxybenzyl bromide dendrons of the Frechet-type [27] with a bisphenol prefunctionalized C_{60}. Ether cleavage of the 6,6-methano-bridged fullerene **1** and treatment with 2.7 equivalents of fourth-generation poly(aryl ether) dendron **2** according to a Williamson synthesis afforded the highly soluble C_{60}-monoadduct dendrimer **3** with $(1 \rightarrow 2)$ *phenyl*-branching and *ether* connectivity, possessing two dendritic branches (Scheme 1) [26]. Size-exclusion chromatography (SEC) appeared to be ideally suited for monitoring the coupling and separating the target molecule from the monocoupling product and higher molecular weight impurities. Similarly, by treatment of C_{60} in refluxing dry chlorobenzene with the terminally perdeuterated D_{112}-dendron **4** possessing an azide focal point, the azafulleroid **5** (68 % after flash chromatography)

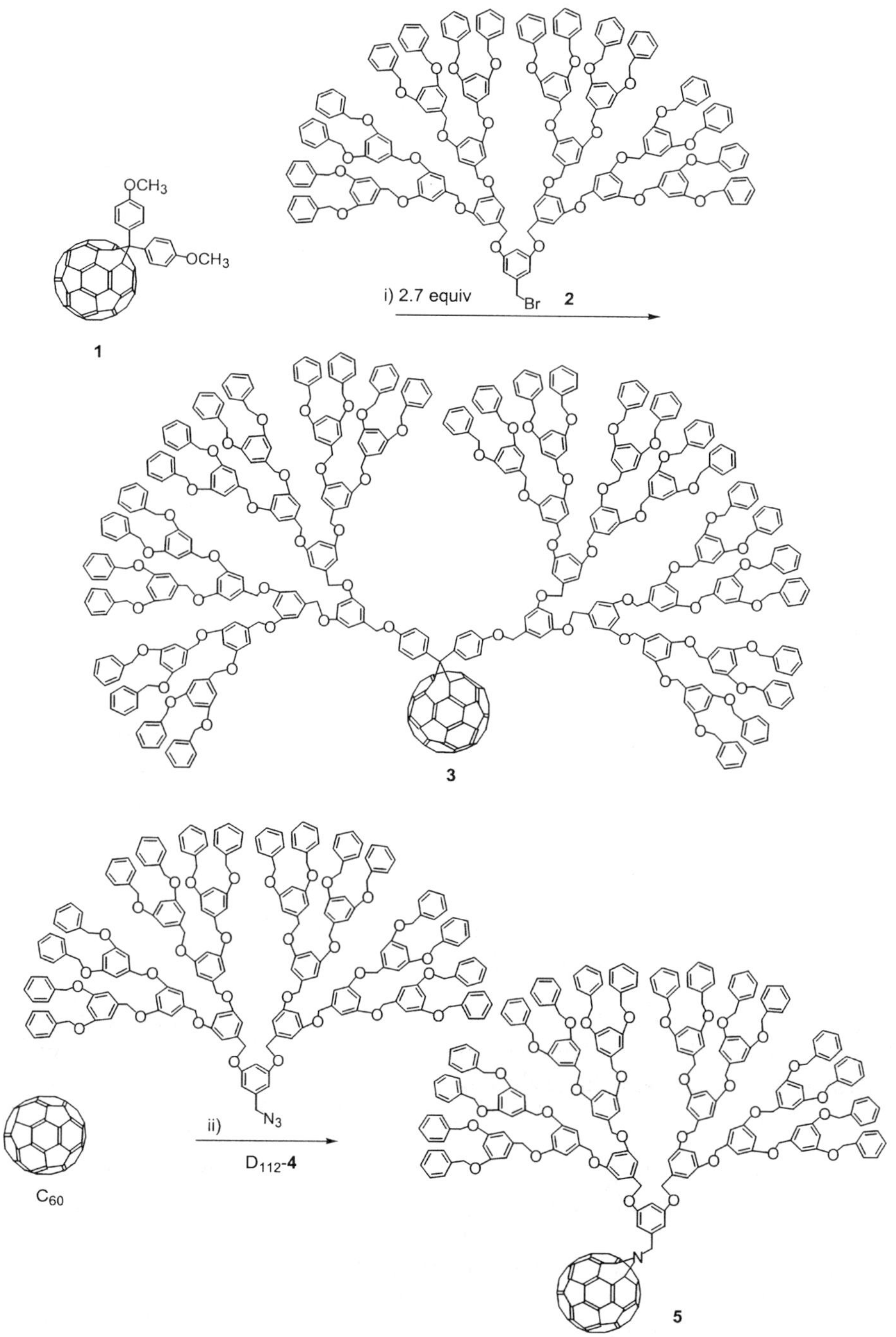

Scheme 1. Synthesis of two-armed Frechet-type poly(benzyl ether) dendrimer **3** and one-armed poly(benzyl ether) dendrimer **5** with a C_{60} fullerene as core: i) (a) BBr_3; (b) 2.7 equiv. **2**, K_2CO_3; ii) terminally deuterated D_{112}-**4**, 24 h reflux in dry chlorobenzene

with one fourth-generation dendritic arm was obtained (Scheme 1) [28]. Both dendrimers were fully characterized and the authors reported on the encapsulation and coverage of the C_{60} core by the dendrimer shell.

The fullerodendrimer **3** is a light brown-colored glass and the dendritic addend dramatically improves the solubility of the fullerene subunit [26]. Similarly, the dendritic fullerene **5** proved to be extremely soluble in a variety of organic solvents and to have a glass transition temperature of 325 K, 13 K higher than the starting dendrimer. Investigations of the redox properties of **5** revealed low reduction potentials for the first three reduction waves in the cyclic voltammograms which may reflect the insulating influence of the globular dendritic macromolecule [28].

The nucleophilic cyclopropanation of C_{60} with α-bromomalonates in the presence of a base according to Bingel [29] is one of the most efficient reactions in fullerene chemistry, providing [6,6]-addition products in fairly good yields. The reaction of malonyl dichloride with benzyl-protected Frechet-type [27] dendritic benzylic alcohols [G1]-OH (first) to [G3]-OH (third) generation and subsequent bromination [30] gave rise to the formation of dendritic bromomalonates [5]. The treatment of C_{60} with these bromomalonates in the presence of sodium hydride afforded three C_{60} monoadducts **6**, **7**, and **8** with two dendritic arms of first- (**6**), second- (**7**), and third-generation (**8**) in 52%, 20%, and 43% yield, respectively (Fig. 2) [5]. Nucleophilic cyclopropanation of C_{60} in comparable or even better yields can also be achieved by allowing dendritic malonates to react directly with C_{60} in the presence of CBr_4 and DBU, as demonstrated with the synthesis of **7** [6].

The isolation of the products from unreacted C_{60} and of undesired bisadducts was achieved by flash chromatography on silica gel. The dendrimers were completely characterized by ^{1}H- and ^{13}C-NMR, IR, UV/Vis, and FAB mass spectrometry. Due to their C_{2v}-symmetry, the dendrimers **6–8** show 15 ^{13}C-NMR signals between $\delta = 139$ and 145 and one signal at $\delta = 71$ corresponding to 15 different types of sp^2-carbon atoms and the two equivalent sp^3-carbons of the fullerene core, respectively [5]. Molecular mechanics and molecular dynamics calculations were performed to explore the geometries and energetics of these dendrimers [31].

In order to avoid steric hindrance among the dendritic branches, the classical Frechet-type dendrons mentioned above have been modified by introduction of a C_3 spacer unit between the aryl-benzyl bond [8]. Thus, the typical aryl-benzyl cadence of Frechet-dendrons [27] changes to an aryl-alkyl-benzyl motif and as a result the dendrons become more flexible and less bulky. These new dendra were prepared according to Scheme 2 in a convergent synthesis starting from benzyl-protected 3,5-dihydroxybenzyl alcohol **9**. Allylation, hydroboration to C_3-elongated benzylic ether **11**, and bromination gave protected 3-benzyloxypropyl bromide **12**, which was grafted twice onto 3,5-dihydroxybenzyl alcohol **13**. Using the same reaction sequence again afforded the second generation chain elongated dendron **16**. Transforming the alcohols **11** and **16** with NaH/THF into the corresponding alkoxides **17** and **19** and reacting them with malonyl dichloride afforded the two-armed C_3-elongated malonate dendrons **18** and **20** (Scheme 2) [8].

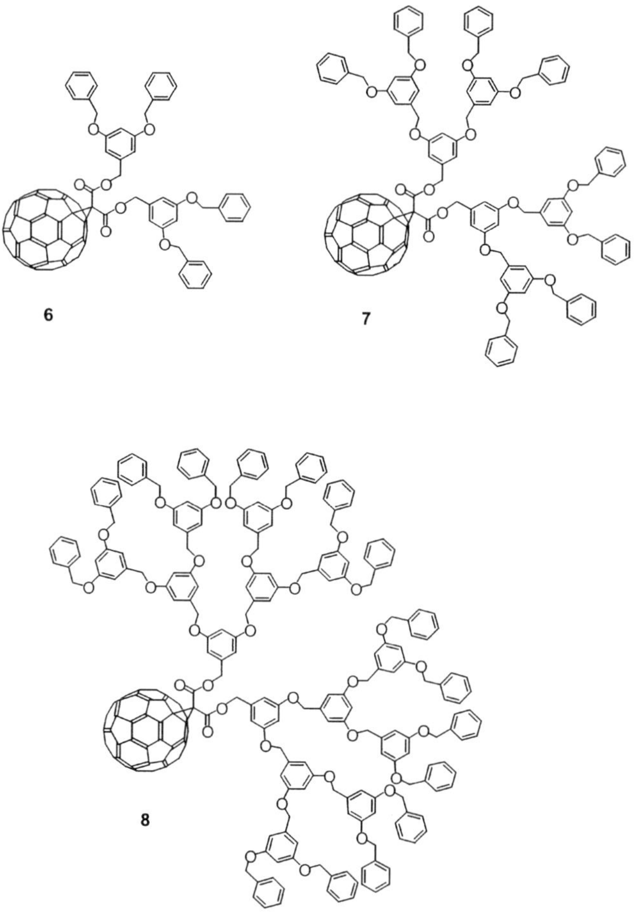

6 7

8

Fig. 2. Monoadduct fullerenodendrimers **6**, **7**, and **8** with first-, second-, and third-generation Frechet-type [27] benzylether dendrons attached to a C_{60} core

Using the classical Bingel conditions [29], we achieved the synthesis of the dendritic first-generation monoadduct **22** [8]. The product was isolated by flash chromatography in 42% yield and separated from unreacted C_{60} (23%) and a regiomeric mixture of bisadducts (15%). The second-generation (1 → 2) *aryl*-branched adduct **23** with *ether* connectivity was obtained under modified cyclopropanation conditions [6] using C_{60}, equimolar amounts of the corresponding malonate, CBr_4, and a small excess of DBU. The dendrimeric product **23** was also isolated by flash chromatography in 42% yield (Scheme 3) [8].

To use fullerene derivatives in screenings for biological activity and in pharmaceutical investigations it is necessary to make them accessible to an organism through enhanced solubility in water. This is possible via covalent attachment of hydrophilic addends, especially through accumulation of carboxylic functions. For this reason, we decided to decorate C_{60} with dendritic addends containing a

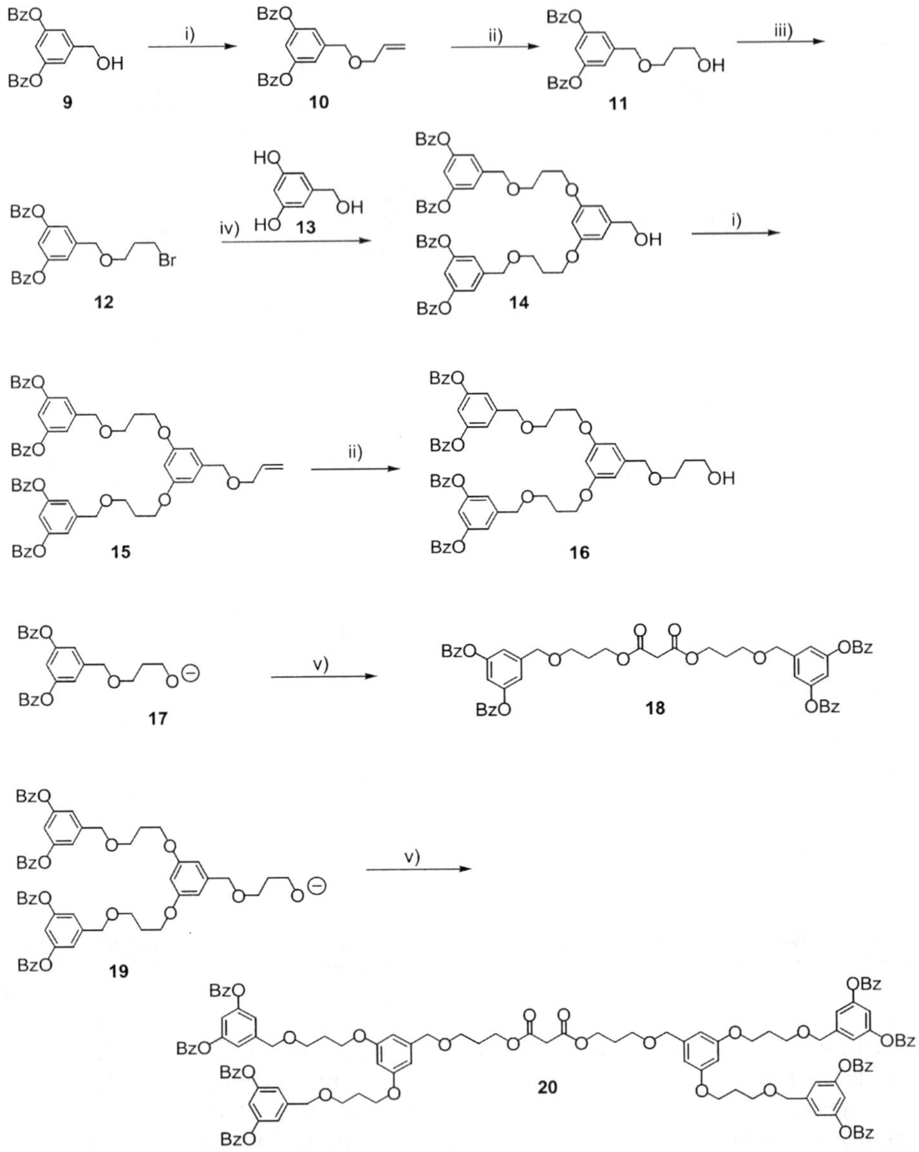

Scheme 2. Synthesis of the first- and second-generation dendritic malonates **18** and **20**: i) allyl bromide, NaH/THF; ii) (a) 9-BBN/THF; (b) EtOH, H_2O_2, NaOH; iii) CBr_4, PPh_3/THF; iv) **13**, K_2CO_3, [18]crown-6/acetone; v) allyl bromide, NaH/THF

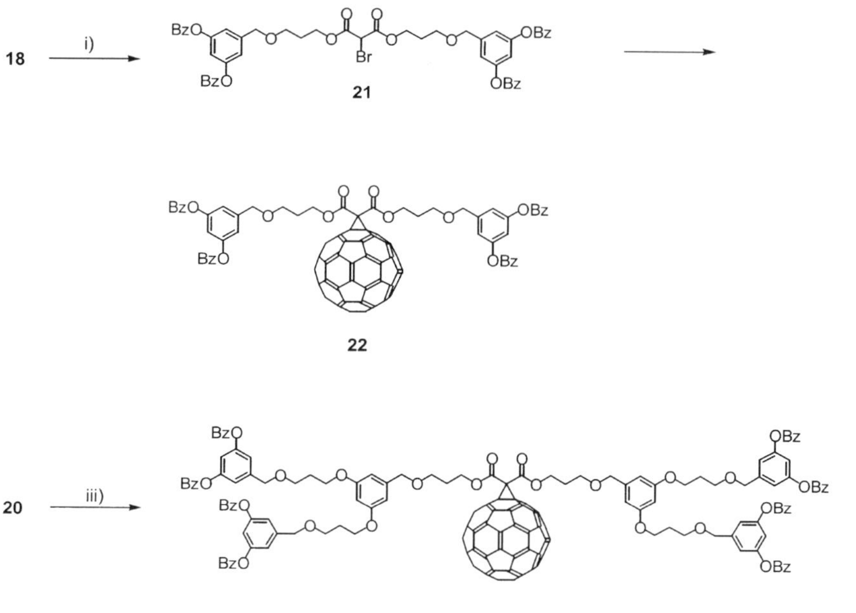

Scheme 3. Synthesis of first- and second-generation (1 $\rightarrow$ 2) *aryl*-branched monoadduct fullerenodendrimers **22** and **23** under "classical" (**22**) and "modified" (**23**) Bingel reaction conditions: i) CBr_4, DBU/THF; ii) C_{60}, NaH/toluene; iii) C_{60}, CBr_4, DBU/toluene; Bz = benzyl

sufficient number of carboxylic acids in their periphery [7, 32]. Such water-soluble dendro[60]fullerenes were obtained by the synthesis of a bis[3-(*tert*-butoxycarbonyl)propyl]malonate **25** from (*tert*-butyl) 4-hydroxybutyrate **24** and malonyl dichloride, and subsequent cyclopropanation of C_{60}. Compound **26** was deprotected and the fullerodicarboxylic acid **27** condensed with the first- (**28** "Behera's amine") [33, 34] and second-generation Newkome-type dendrons (**31**) [33, 34] to yield the polyamide dendrimers **29** and **32** (branching multiplicity of 3). The six and eighteen terminal ester functions of **29** and **32** were cleaved by hydrolysis and the water-soluble dendritic hexaacid **30** and octadecaacid **33** with (1 $\rightarrow$ 3) *C*-branching and *amide* connectivity were isolated (Scheme 4) [32].

In another pathway leading to **33**, the second-generation Newkome polyamide [G2]-NH_2 **31** [34] was subjected to coupling with the adapter di(3-carboxypropyl) malonate **34** affording the didendro malonate **35** which is suitable for direct nucleophilic cyclopropanation [6] of C_{60} (Scheme 5) [7]. The *tert*-butyl protected fullerodendrimer **32** was isolated in 29% yield after repeatedly purifying with flash chromatography as a red-brown amorphous solid, soluble in most organic solvents. The deprotection was achieved by stirring in formic acid and the red-brown powder **33** spectroscopically completely characterized [7].

Due to the presence of 18 carboxy groups, the polycarboxylate **33** is soluble in water and methanol (red solution) and insoluble in most organic solvents. In a buffer-solution at pH 7.4 at least 34 mg/ml of the acid **33** is soluble. This corre-

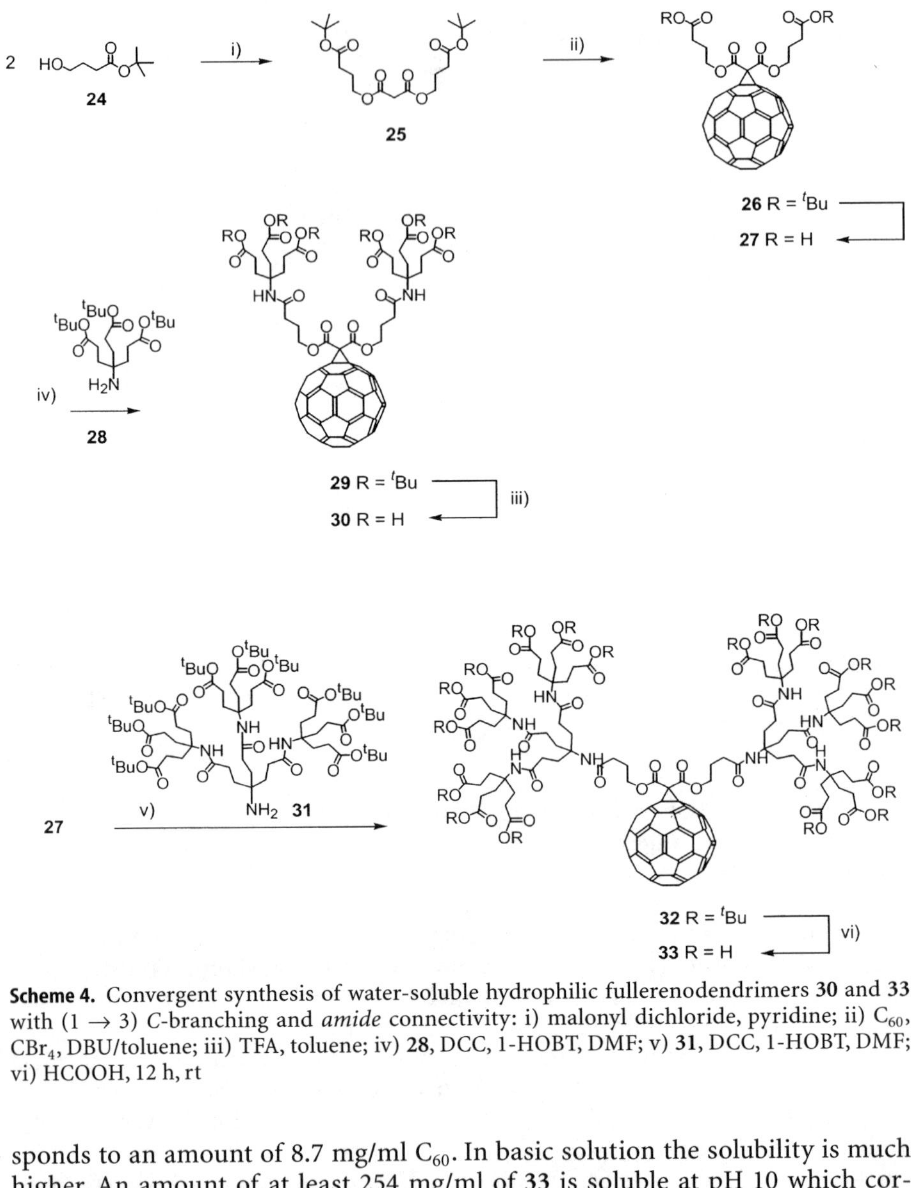

Scheme 4. Convergent synthesis of water-soluble hydrophilic fullerenodendrimers **30** and **33** with (1 → 3) *C*-branching and *amide* connectivity: i) malonyl dichloride, pyridine; ii) C_{60}, CBr_4, DBU/toluene; iii) TFA, toluene; iv) **28**, DCC, 1-HOBT, DMF; v) **31**, DCC, 1-HOBT, DMF; vi) HCOOH, 12 h, rt

sponds to an amount of 8.7 mg/ml C_{60}. In basic solution the solubility is much higher. An amount of at least 254 mg/ml of **33** is soluble at pH 10 which corresponds to an equivalent of 64.7 mg of C_{60} per milliliter [7]. From small angle neutron scattering (SANS) we could deduce that **33** up to pH 5 and in a concentration range from 10^{-3} to 10^{-5} M forms tetrameric micellary aggregates of a diameter of ~60 Å [32]. With higher pH values most of the aggregates dissociate to a monomeric solution state [32].

The highly water-soluble **33** appeared to be one of the most active antiviral fullerene derivative studied to date [32, 35]. An aqueous solution showed an EC_{50} of 0.22 μmol/l against HIV-infected human lymphocytes, and also several infec-

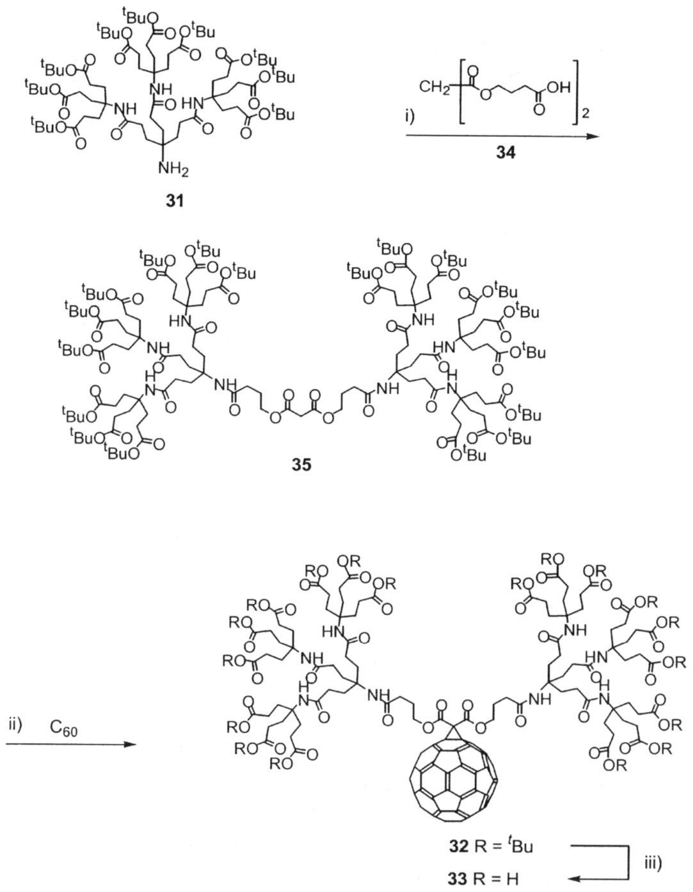

Scheme 5. Synthesis of the water-soluble (1 → 3) *C*-branched and *amide* connected dendro[60]fullerene **33**: i) **34**, DCC, 1-HOBT/DMF; ii) C_{60}, CBr_4, DBU, toluene, 29% yield; iii) HCOOH, 12 h, rt

tious clones of HIV-1 are susceptible to compound **33**. The dendrimer had no apparent toxicity in human cells [32, 35] and showed rather weak toxicity in in-vivo studies with mice [32, 36] when applied by intra-peritoneal injections.

To increase the photovoltaic effect in single layer poly(*p*-phenylenevinylene) (PPV) cells, effective charge separation is necessary and the recombination of photogenerated charge carriers has to be suppressed. This was achieved using blends of the conjugated polymer with buckminsterfullerene C_{60} [37]. For this purpose we have built blend systems with various mixtures of solutions of the methanol soluble fullerodendrimer **32** and a PPV precursor solution [38]. An aluminum top electrode was used and the samples were illuminated through the ITO-electrode by a W-Xenon lamp and monochromator. Short-circuit photocurrent spectra were measured, and a scanning electron microscope (SEM) used to obtain details of the surface structures of the samples. Although an electron transfer from PPV to the second-generation fullerenodendrimer **32** occurred in the blend systems, as indicated by photoluminescence quenching and photocurrent spectra, the efficiencies in the PPV/fullerodendrimer blends were about one order of magnitude lower than in PPV/C_{60} heterostructures. The charge carrier transport was lower than in pure PPV and C_{60} layers [36]. Investigation of the morphology of the PPV blends films by SEM revealed a structure like a cratered landscape, the crater structures being apparently responsible for the bad electrical transport properties of the blend system [38].

Methanofullerene acid **34** was used by Diederich et al. as the initiator core for the synthesis of first- and second-generation (1 $\rightarrow$ 3) *C*-branched monoadduct fullerodendrimers **36** and **38** [39]. The coupling reactions of the first- and second-generation amine dendrons **35** and **37** with **34** under peptide coupling conditions yielded the expected dendrimers **36** and **38** in 40 % and 19 % yield, respectively (Scheme 6) [39]. The synthesis of the dendrons followed the branching methodology of Newkome et al. [40, 41]. The reaction to the corresponding third-generation (1 $\rightarrow$ 3) *C*-branched dendrimer applying similar synthetic routes failed.

The steric bulk of substituents at the N-adjacent quaternary C-atom of the dendrons apparently caused the inaccessibility of this focal point in the third-generation dendron. Therefore, the authors extended the distance between the poly(ether-amide) wedge and the primary amine by introducing a short spacer [39]. By coupling the third-generation amine dendron **41** with Z-protected (*N*-benzyloxycarbonyl)-glycine as a spacer under forcing conditions, the Z-protected elongated dendron **40** was obtained in 67 % yield. Deprotection gave the new third-generation dendron **41** which was coupled to methanofullerene acid **34** to give the third-generation fullerodendrimer **44** in 42 % yield (Scheme 7) [39]. The remarkably improved yield compared to the 19 % for the second-generation dendrimer **38** obviously demonstrated the advantage of introducing short spacers into bulky and sterically-crowded dendritic branches.

Hydrophilic dendrimers with a fullerene core and peripheral *O*-acetylated glucose units have been prepared by Diederich et al. [42] and subjected to Langmuir film experiments. The fullerene amphiphiles **43** and **44** with their glycodendron headgroups were shown to form layers at the air-water interface that were monomolecular, stable, and reversibly formed as proven by the lack of hysteresis in compression/expansion cycles (Fig. 3) [42]. The dendritic wedges seemed to limit the packing and prevent the fullerenes from aggregating irreversibly.

Chuard and Deschenaux have shown that the functionalization of C_{60} with a malonate bearing two mesogenic cholesterol subunits resulted in a fullerene derivative with liquid crystalline properties [43]. Recently, the same group demonstrated that the fullerene-functionalized dendrimer **45** exhibited mesogenic properties similar to those of the corresponding dendritic addend only (Fig. 4) [44]. Similarly, as in the examples above, the C_{60} core of **45** is buried in the center of the dendritic structure. Thus, unfavorable effects of the C_{60} unit such as aggregation or steric hindrance are prevented.

Nierengarten et al. recently succeeded in the convergent synthesis of a dendrimer **46** with a C_{60} core and with C_{60} spheres at each branching unit, using

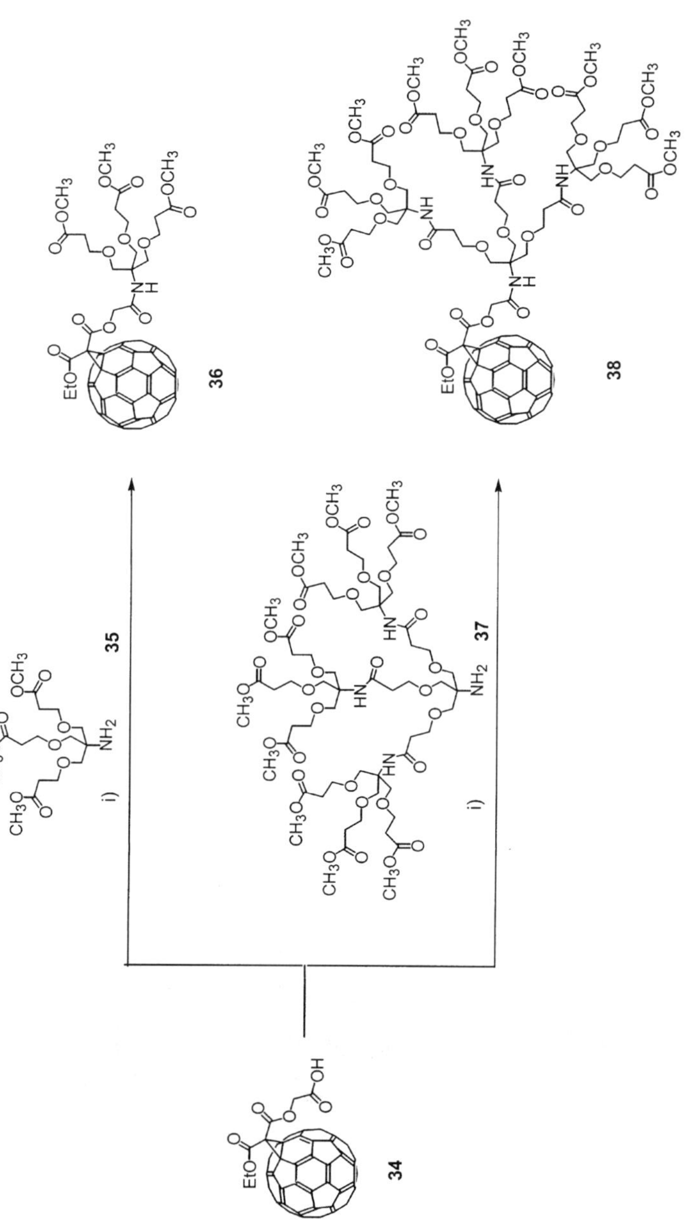

Scheme 6. Preparation of first- and second-generation (1 → 3) C-branched dendritic C_{60}-derivatives **36** and **38**: i) DCC, 1-HOBT, CH_2Cl_2

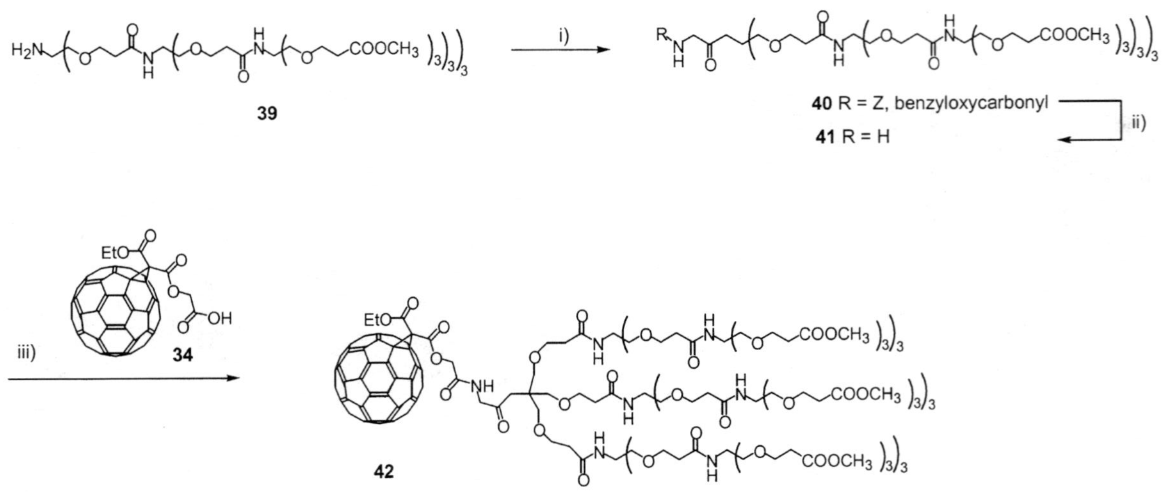

Scheme 7. Preparation of third-generation one-armed fullerenodendrimer **42** with *amide* connectivity: i) *Z*-glycine (10 equiv.), DCC, 1-HOBT/THF, 67% yield; ii) $HCOONH_2$, 10% Pd/C, EtOH; iii) **34**, DCC, 1-HOBT, CH_2Cl_2, 42% yield

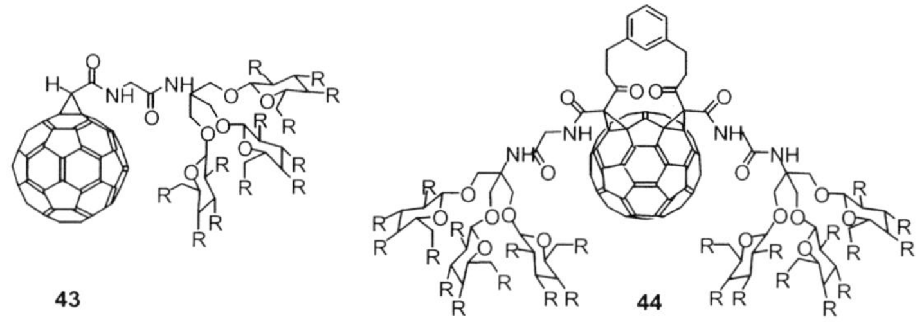

43 44

Fig. 3. Amphiphilic fullerodendrimers **43** and **44** with one and two glycodendron headgroups, respectively: R = OAc

DCC-mediated esterifications, followed by the cleavage of a *tert*-butyl ester moiety under acidic conditions (Fig. 5) [45]. These fullerodendrons/dendrimers are interesting building blocks for the preparation of monodisperse fullerene-rich macromolecules and capable of forming Langmuir-films at the air-water interface [45].

3.2 Dendrimers with C_{60} Multiple Adduct Cores

3.2.1 *Dendrimers with C_{60} Multiple Adduct Cores with One Type of Addend*

Apart from one-armed and two-armed dendritic C_{60} monoaddition products, higher adducts of C_{60} up to T_h-symmetrical hexakisadducts can also be achieved by nucleophilic cyclopropanation [25]. Examples of fullerene bisadducts with dendritic structure include the first- and second-generation fullerodendrimers **50** (Scheme 8) and **52** and the spacer-elongated amine dendron **51** (Fig. 6) [39]. Diederich et al. obtained the C_s-symmetrical *cis*-2-bisadduct **48** from the bismalonate derivative **47** in 22% yield by macrocyclization of C_{60} [39]. Subsequent selective cleavage of the *tert*-butyl ester functions (71% yield) provided the dicarboxylic acid core **49**. The coupling of the diacid core **49** to the dendritic amine branch **35** afforded the two-armed first-generation dendrimer **50** in 66% yield (Scheme 8). The corresponding two-armed second-generation dendrimer **52** was obtained in 76% yield by condensation of **49** with the "glycine-elongated" amine dendron **51** (Fig. 6) [39].

A related four-armed second-generation dendrimer **58** was synthesized by Diederich and coworkers, starting with bisadduct **57** with four carboxylic acid residues (Scheme 9) [39]. Diacid **53**, obtained from *m*-benzenedimethanol and Meldrum's acid, with DCC-mediated esterification with benzylic alcohol derivative **54**, gave the bismalonate derivative **55**. Reaction with C_{60}, DBU, and I_2 in toluene led to the macrocyclic *cis*-2-bisadduct **56**. The free tetraacid **57** was subsequently generated by hydrolysis of **56** with TFA. Finally, the four-armed second-generation fullerodendrimer **58** with (1 → 3) *C*-branching was obtained as a dark-orange glassy compound by DCC-coupling with the "glycine elongated" amine dendron **51** in 34% yield (Scheme 9) [39].

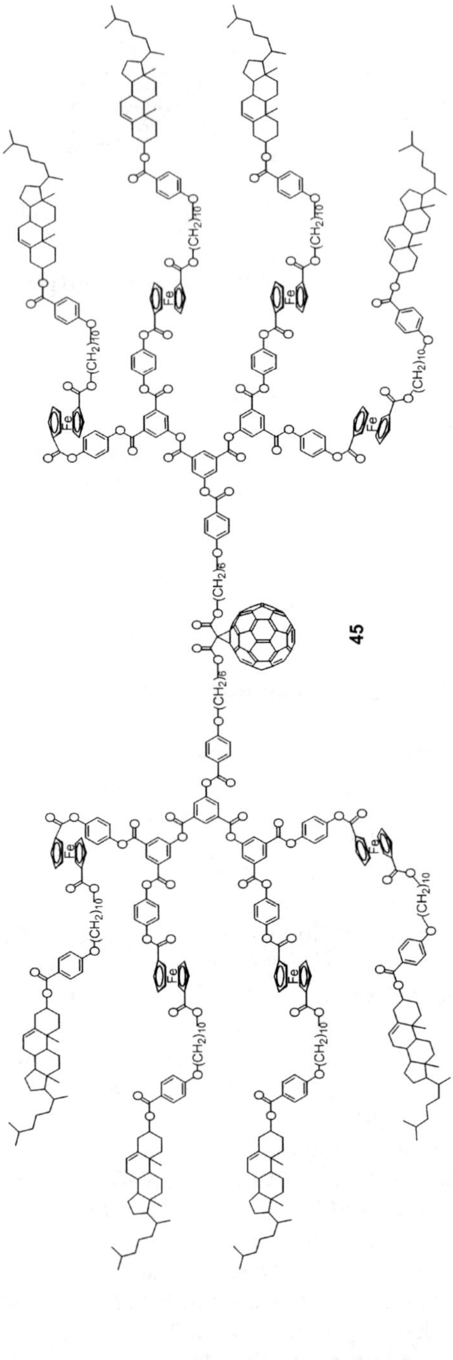

Fig. 4. Liquid crystalline fullerenodendrimer 45

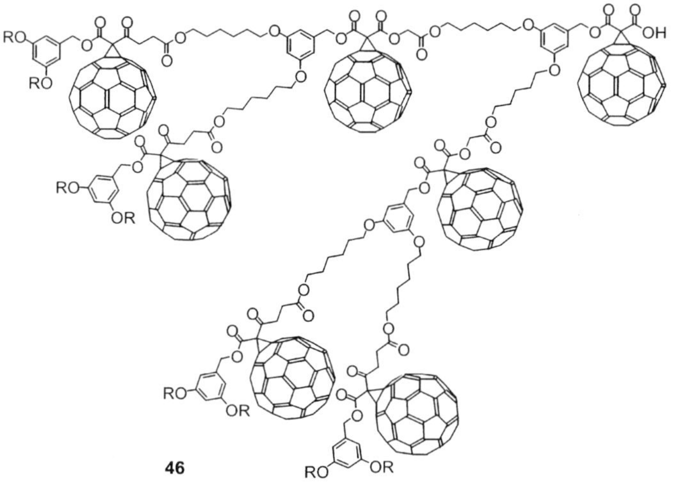

46

Fig. 5. Fullerenodendron/dendrimer **46** containing a C_{60} core as well as C_{60} spheres at each branching unit: $R = C_8H_{17}$

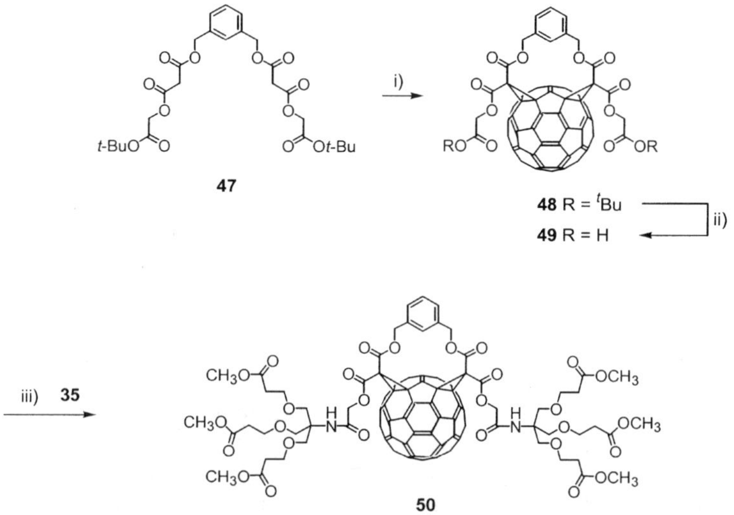

Scheme 8. Synthesis of dicarboxylic acid **49** and the two-armed first-generation bisadduct fullerenodendrimer **50** with (1 → 3) *C*-branching and *ether* connectivity: i) C_{60}, DBU, I_2, toluene, rt, 22% yield; ii) TosOH, toluene, heating, 71% yield; iii) DCC, 1-HOBT, THF, 0 °C, 66% yield

The UV/Vis spectra of the three (1 → 3) *C*-branched fullerodendrimers **50**, **52**, and **58** displayed the characteristic absorptions of *cis*-2-bisadducts [39]. Their NMR spectra revealed clearly the C_s symmetry of the compounds. The MALDI-TOF MS-spectrum displaying the sodium complex of **58** as base peak at $m/z = 7368$ ($^{13}C_4{}^{12}C_{344}H_{450}N_{20}O_{125}Na$) provides clear evidence for the monodis-

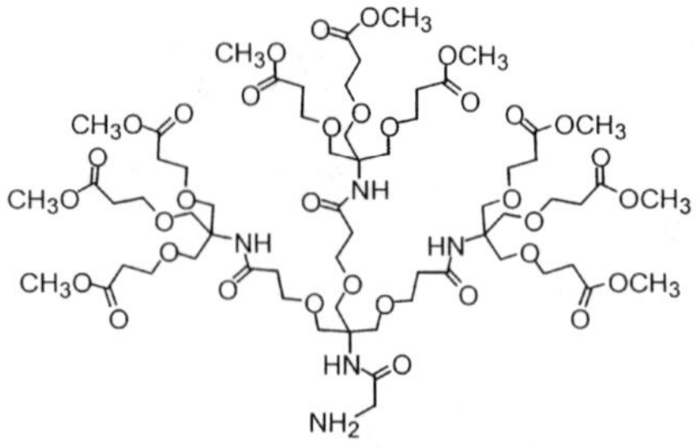

51

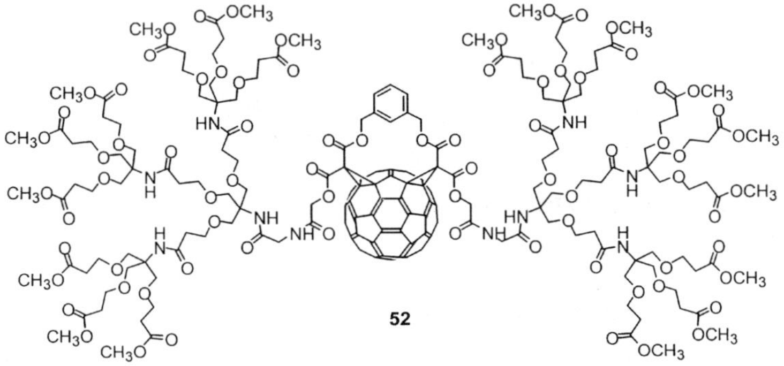

52

Fig. 6. The second-generation "glycine-elongated" amine dendron **51**, derived from **35**, dendritic wedge for the second-generation dendrimer **52** (*cis*-2, C_s)

persity of **58**. Cyclic voltammetric studies revealed that the dendritic *cis*-2-bisadducts undergo multiple reductions. In CH_2Cl_2 the redox potential of the fullerene core is not affected by size and density of the surrounding dendritic shell. Interestingly, the first reduction step is irreversible in the case of **50** and **52**, whereas it is reversible in **58** even though the fullerene core is more encapsulated [39].

Since the regiochemistry of bromomalonate addition is well established [13, 14, 16, 17], spherical dendrimers with high symmetries and a core branching multiplicity of 12 can be envisaged, even if low-generation dendra are employed. As examples for this new dendrimer prototype we synthesized the twelve-armed benzylether-based dendrimers **59** (Fig. 7), **60** and **61** [5, 8] (Fig. 8) in one step starting with the corresponding dendritic malonates and using the template-mediation technique [4, 46]. Since the dendrons within **60** and **61** give rise to less steric hindrance due to an additional C_3-spacer unit, much higher yields were obtained [8] upon the convergent malonate addition by comparison with **59** [5].

The high symmetry of the dendrimers **59**–**61** becomes obvious by the extreme simplicity of the ^{1}H- and ^{13}C-NMR spectra of these macromolecules with molecular weights of 4956, 5652, and 12,132, respectively. For example, only three signals due to the C-atoms of the fullerene core are found at about $\delta = 146$, 141, and 69, nicely reflecting the T_h-symmetry [5, 8].

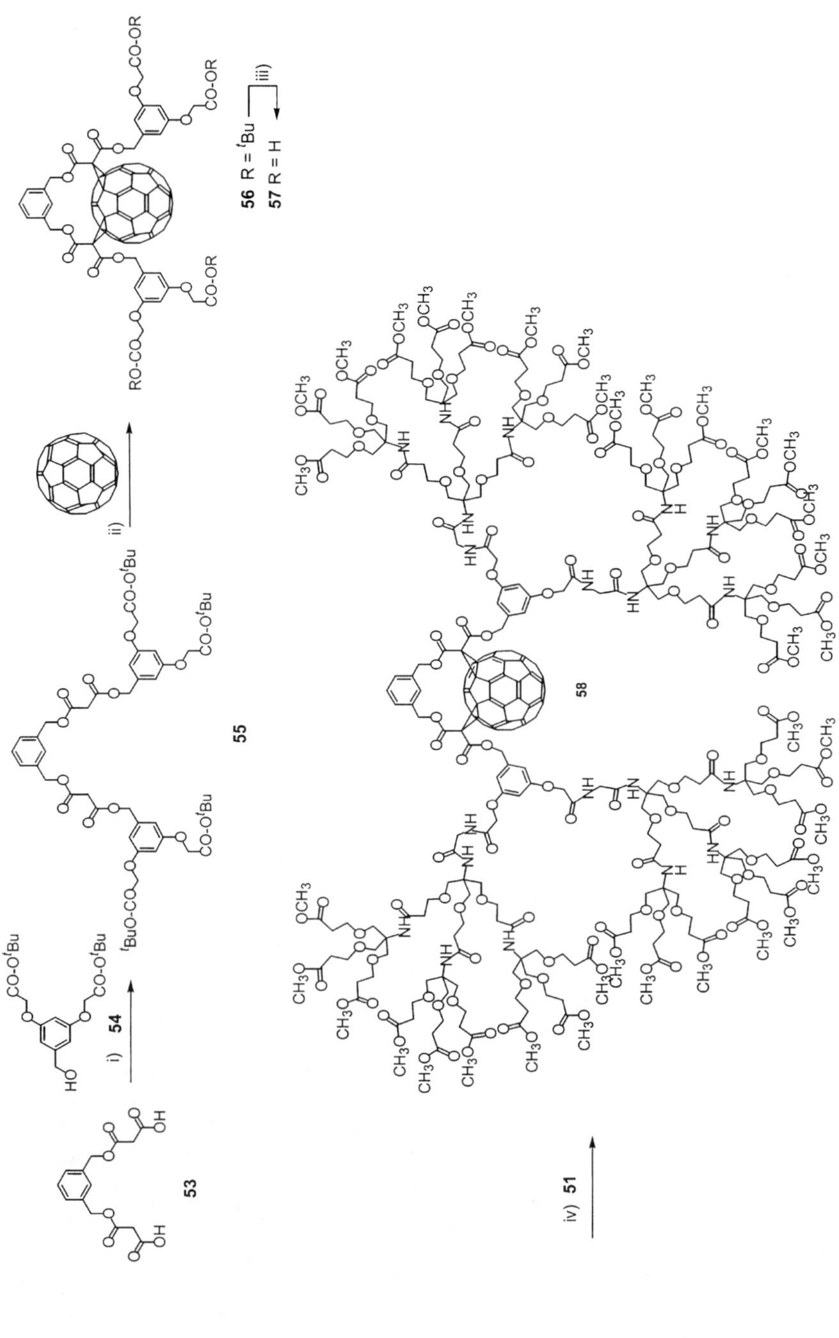

Scheme 9. Synthesis of initiator core tetraacid **57** and second-generation fullerenodendrimer **58**: i) **54**, DCC, DMAP, THF; ii) C_{60}, DBU, I_2, toluene; iii) TFA, CH_2Cl_2; iv) **51**, DCC, 1-HOBT, THF

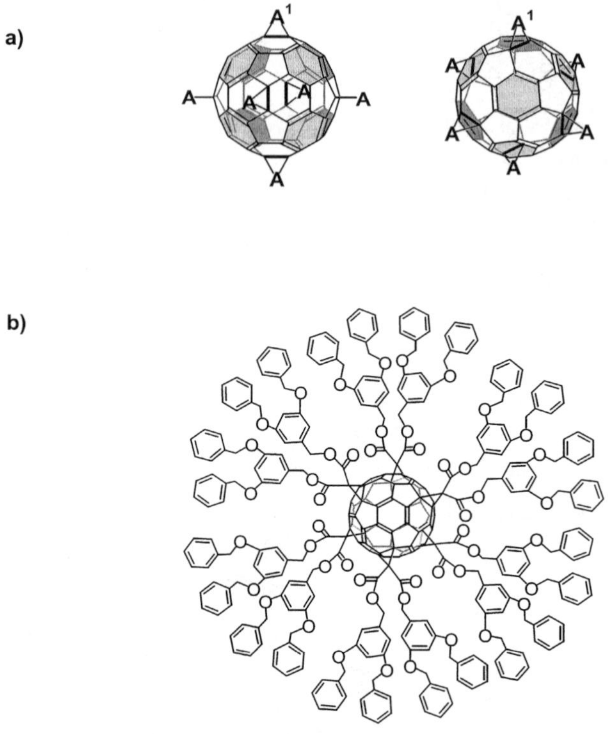

59

Fig. 7. **a** Two different views of octahedral positions relative to the first addend A^1 in a C_{2v} symmetrical hexaadduct of C_{60}. **b** T_h-symmetrical ($1 \rightarrow 2$) *aryl*-branched hexakisadduct fullerenodendrimer **59** with original Frechet-type [27] first-generation dendritic wedges

As the diameter of a dendrimer grows linearly with the number of repetition steps while the volume of the outer sphere grows exponentially, a "self-limiting-generation" may appear [47]. This self-limiting-generation is determined by the core and branching multiplicity as well as by the dimensions of the building blocks and should be characterized by a rapidly increasing energy due to close atom contacts within the molecule. In order to determine the self-limiting generation for such dendrimers, the PM3-heat of formation of C_{60} (527.88 kcal mol^{-1}) was subtracted from the energy of the entire structure and the resulting energy of the dendrimeric shell was divided by the number of its atoms. For the dendrimeric series starting with the first generation system **59** the third generation was estimated to be the self-limiting-generation [31]. Practically, however, using the short Frechet benzyl ether dendrons, with the successive cyclopropanation with bromomalonates it was not possible to prepare even the second-generation dendrimer.

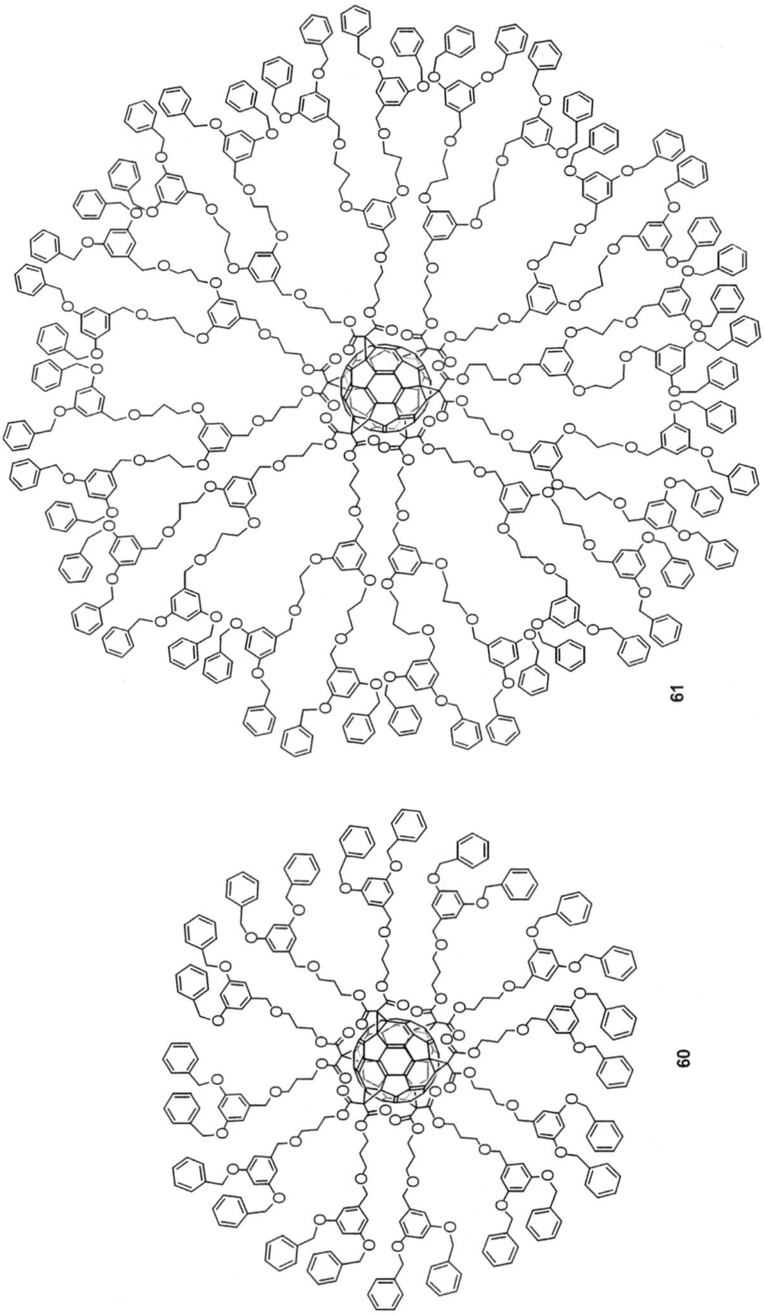

Fig. 8. First- **60** and second-generation (1 → 2) *aryl*-branched twelve-armed fullerenodendrimer **61** with C_3-spacer elongated dendrons

3.2.2
Dendrimers with C_{60} Multiple Adduct Cores with Two Different Types of Addends

By a defined sequence of regioselective cyclopropanations multiple adducts with two different types of addends are available. For example, the easily-accessible addition products of diethoxymalonate to C_{60}, namely the monoadduct **62**, the *e*-bisadduct **63**, the C_3-symmetrical *e,e,e*-trisadduct **64**, the C_S-symmetrical tetrakisadduct **65**, and the C_{2v}-symmetrical pentakisadduct **66**, represent C_{60} tectons with an incomplete octahedral addition pattern and serve as starting materials for mixed adducts with a complete octahedral addition pattern (Fig. 9) [9, 13, 14, 46, 48]. These fullerene derivatives still possess five, four, three, two, or one intact [6,6]-double bonds in octahedral positions, ready to undergo subsequent nucleophilic additions of dendritic substituted malonates.

By fivefold nucleophilic cyclopropanation of the monoadduct precursor **62** with the first generation Frechet-type bromomalonate dendron BrCH(COO-[G1])$_2$ we obtained the first-generation [1:5]-mixed hexaadduct **67** with a complete octahedral hexaaddition pattern (Scheme 10) [5]. Similar cyclopropanations of bis-**63** to pentakisadduct precursors **66** with the first, second, and third generation bromomalonates BrCH(COO-[Gx])$_2$ (x = 1 – 3), respectively, afforded second-generation [2:4]-hexakisadduct **68** (Scheme 11) [10], second-generation [3:3]-hexakisadduct **69** (Scheme 12) [10], third-generation [3:3]-hexakisadduct **70** (Scheme 13) [10], third-generation [4:2]-hexakisadduct **71** (Scheme 14) [10], first-generation [5:1]-hexakisadduct **72** (Scheme 15) [10], second-generation [5:1]-hexakisadduct **73** (Scheme 16) [10], and third-generation [5:1]-hexa-

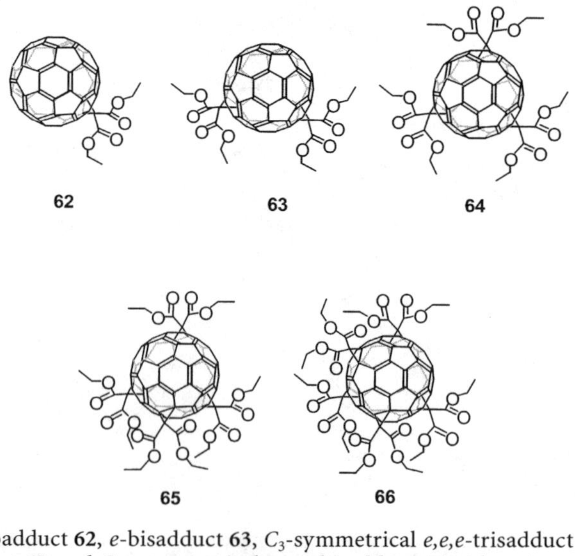

62 63 64

65 66

Fig. 9. C_{60} monoadduct **62**, *e*-bisadduct **63**, C_3-symmetrical *e,e,e*-trisadduct **64**, C_s-symmetrical tetrakisadduct **65**, and C_{2v}-symmetrical pentakisadduct **66** with an incomplete octahedral addition pattern, to be used as core tectons for the preparation of mixed hexakisadduct fullerenodendrimers

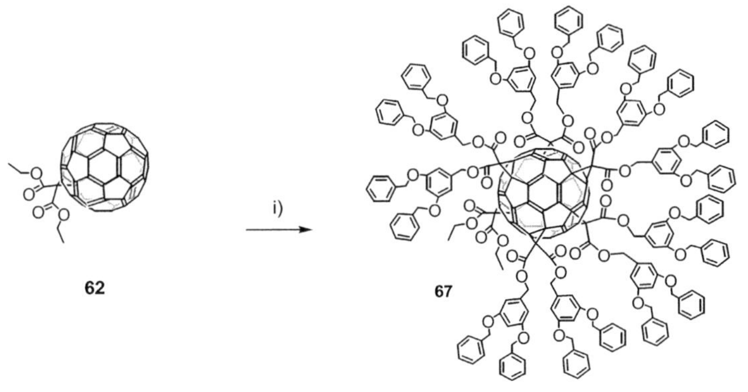

Scheme 10. Synthesis of [1:5]-mixed hexakisadduct **67** by template-mediated cyclopropanation of **62**: i) DMA, $BrCH(COO\text{-}[G1])_2$, DBU, toluene

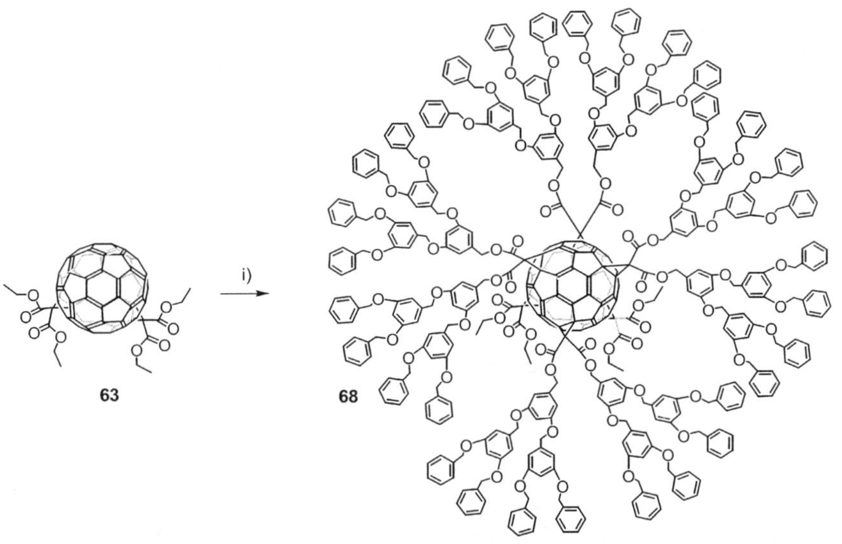

Scheme 11. Synthesis of [2:4] **68** by template-mediated cyclopropanation of **63**: i) DMA, $BrCH(COO\text{-}[G2])_2$, DBU, toluene/CH_2Cl_2, 3d, rt

kisadduct **74** (Scheme 17) [10] with a mixed octahedral addition pattern [5, 10]. The two mixed [3:3]-dendrimers $C_{66}(CO_2\text{-}Et)_6(CO_2\text{-}[G2])_6$ **69** and $C_{66}(CO_2\text{-}Et)_6(CO_2\text{-}[G3])_6$ **70** have C_3 symmetry and are inherently chiral due to the nature of their addition pattern. Both of them were obtained as racemic mixtures from the racemic starting trisadduct **64** [10].

The success of this concept is based on the regioselectivity of attacks on the [6,6]-double bonds located in equatorial positions relative to the addends already bound [16]. The spectroscopic characterization of the dendrimers was facilitated

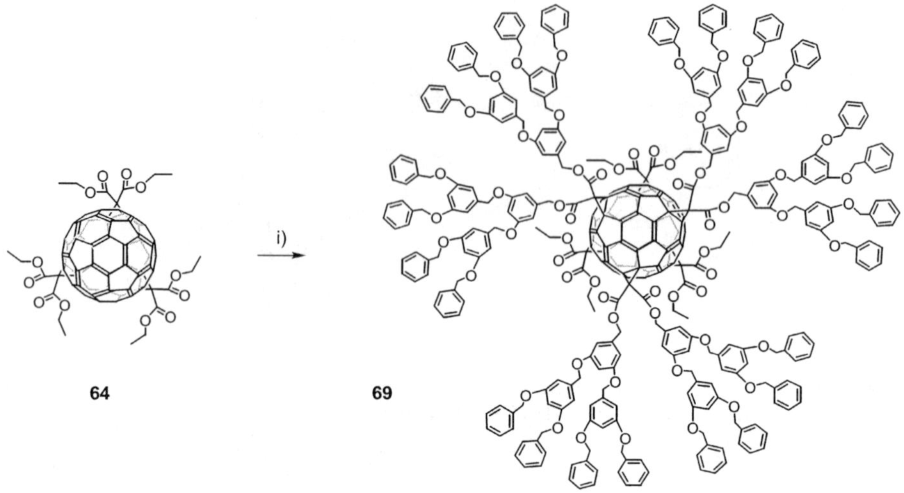

64

69

Scheme 12. Synthesis of [3:3] second-generation hexakisadduct dendrimer **69** by template-mediated cyclopropanation of **64**: i) DMA, $BrCH(COO\text{-}[G2])_2$, DBU, toluene/CH_2Cl_2, 3d, rt

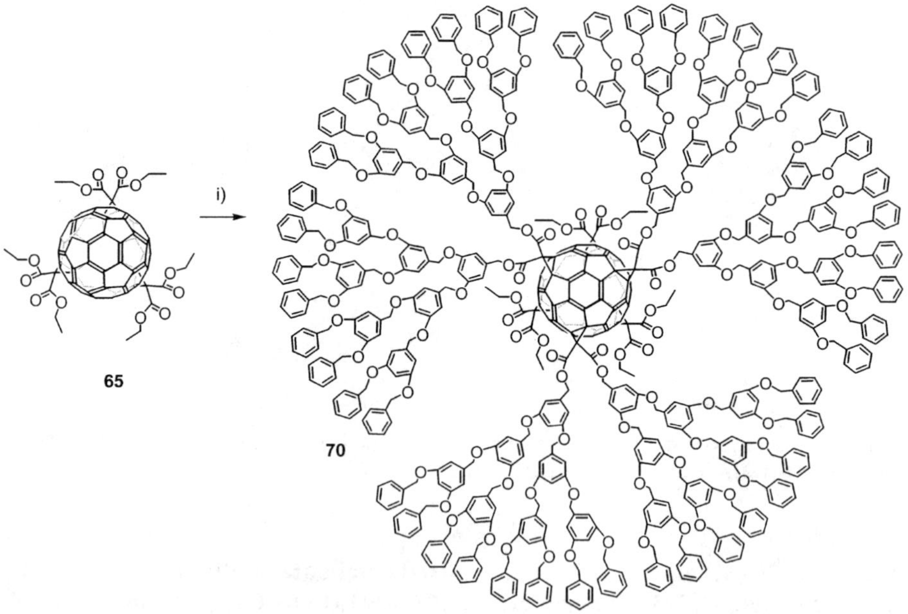

65

70

Scheme 13. Preparation of the racemate of inherently chiral third-generation Frechet-type fullerenodendrimer **70** with ($1 \rightarrow 2$) *aryl*-branching: i) DMA, $BrCH(COO\text{-}[G3])_2$, DBU, toluene/CH_2Cl_2, 3d, rt

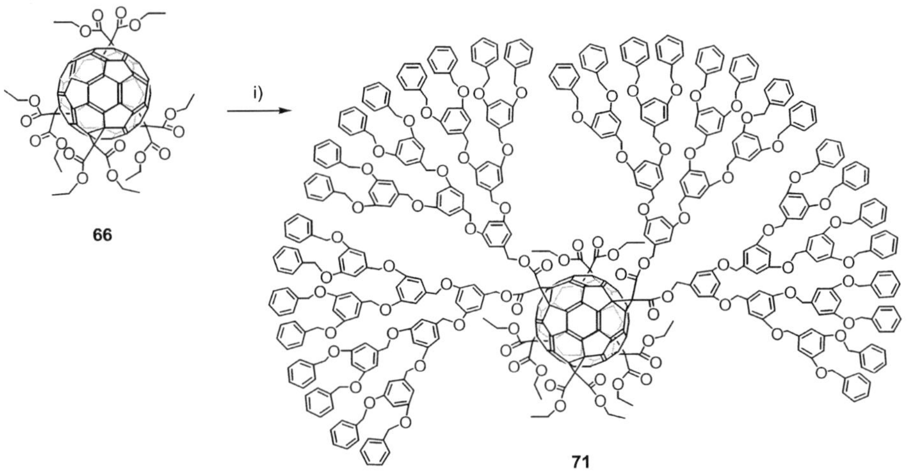

66 71

Scheme 14. Preparation of four-armed [4:2]-mixed hexakisadduct fullerenodendrimer **71** by cyclopropanation of tetrakisadduct **66**: i) DMA, $BrCH(COO\text{-}[G3])_2$, DBU, toluene/CH_2Cl_2, 3d, rt

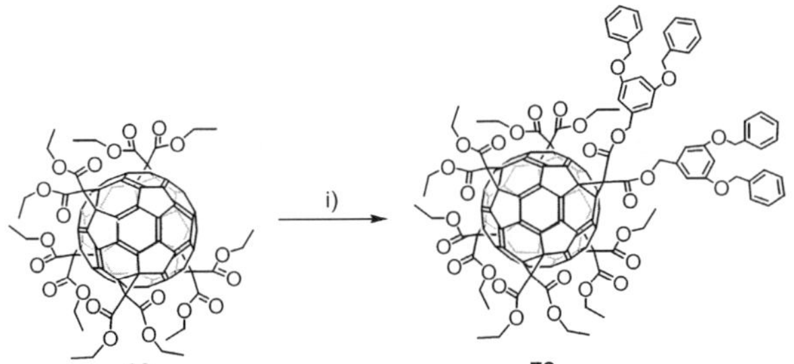

66 72

Scheme 15. Preparation of first-generation [5:1]-mixed hexakisadduct fullerenodendrimer **72** by further cyclopropanation of pentakisadduct **66**: i) DMA, $BrCH(COO\text{-}[G1])_2$, DBU, toluene/CH_2Cl_2, 3d, rt

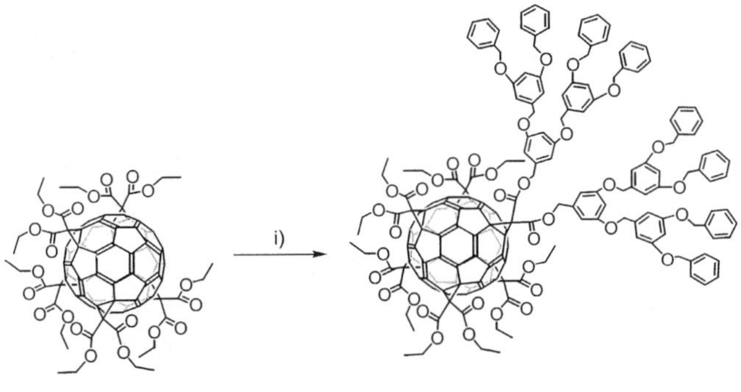

66 73

Scheme 16. Preparation of second-generation [5:1]-mixed hexakisadduct fullerenodendrimer **73** by further cyclopropanation of pentakisadduct **66**: i) DMA, $BrCH(COO\text{-}[G2])_2$, DBU, toluene/CH_2Cl_2, 3d, rt

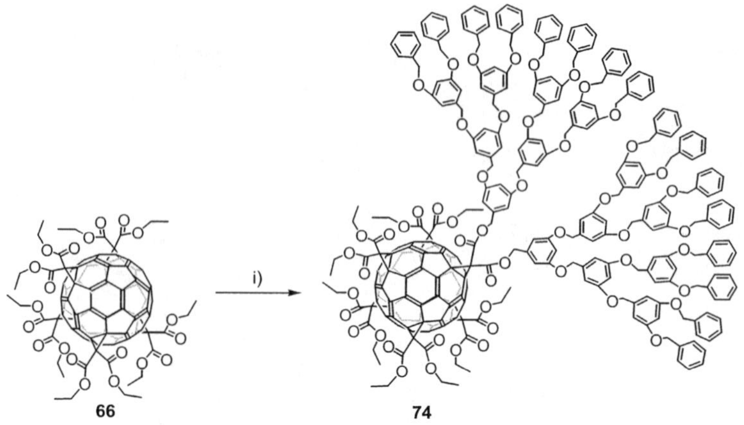

Scheme 17. Preparation of third-generation [5:1]-mixed hexakisadduct fullerenodendrimer **74** by further cyclopropanation of pentakisadduct **66**: i) DMA, $BrCH(COO\text{-}[G3])_2$, DBU, toluene/CH_2Cl_2, 3d, rt

by the high symmetry of the addition pattern leading to characteristic fingerprints in their NMR and electronic spectra. The UV/Vis spectra revealed the characteristics of fullerene hexakisadducts, i.e., two doublet absorption bands at 271 nm and 281 nm, and at 315 nm and 335 nm, respectively [4–6, 8, 13, 14]. Going from lower to higher generations causes the short wavelength doublet band to broaden until the twin peak maximum collapses to a single, broad absorption peak. In the ^{13}C-NMR spectra of **72–74**, the expected lines of the sp^2 carbons of these C_{2v}-symmetrical hexakisadducts are positioned in two groups at $\delta = 141$ and 146. The ^{13}C-NMR spectra of **70, 71**, and **74** impressively demonstrate the strong influence of the local high symmetry. In each spectrum three groups of signals appear for the fullerene C-atoms reminiscent of those of the three magnetically inequivalent C-atoms of the hexakisadducts carrying just one type of addend (overall T_h-symmetry). No influence of the overall symmetry of the three third-generation dendrimers **70, 71**, and **74**, which is C_3, C_s, and C_{2v}, respectively, is apparent. With increasing dendron density (G3-**74** → G3-**71** → G3-**70**), the three signal groups of the fullerene sp^2 and sp^3 C-atoms at $\delta = 146$, 141, and 69 become less intense relative to those of the C-atoms of the dendrons [10].

The synthesis of enantiomerically pure mixed [3:3]-hexakisadduct dendrimers with an inherently chiral C_3-symmetrical core pattern was achieved with the four diastereomeric and enantiomeric *all-R-*^fC*-*(+)-, *all-S-*^fA*-*(–)-, *all-R-*^fA*-*(+)-, and *all-S-*^fC*-*(–)-tris[bis(4-phenyl-2-oxazolinemethano)]-fullerene precursor adducts **75** (^fC = fullerene *C*lockwise, ^fA = fullerene *A*nticlockwise) [49], with known absolute configurations [18, 48] (Fig. 10). Dendritic second generation 3,5-dihydroxybenzyl bromomalonate [5] was used to complete the octahedral addition pattern to give the mixed hexakisadducts **76** [11]. The absolute configuration of the precursor trisadducts is retained within the inherently chiral C_3 symmetric hexakisadducts. Isomeric purity was unambiguously established using a combination of NMR spectroscopy and other techniques.

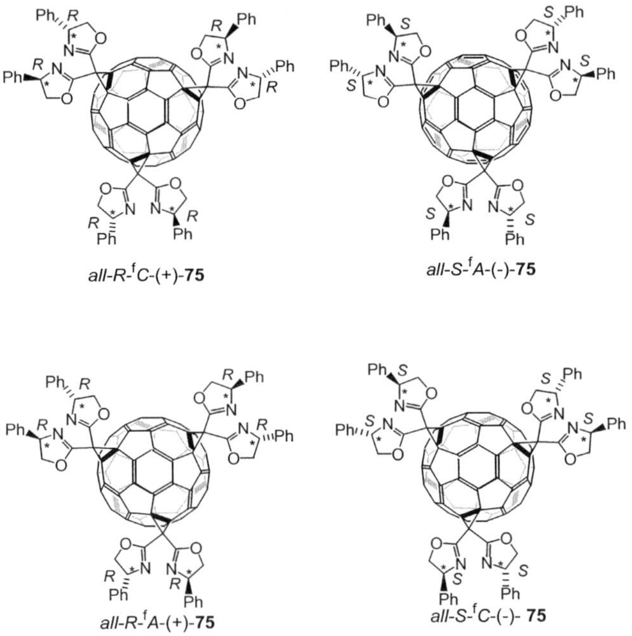

Fig. 10. The four diastereomeric and enantiomeric *all-R-*^{f}*C*-(+)-, *all-S-*^{f}*A*-(–)-, *all-R-*^{f}*A*-(+)- and *all-S-*^{f}*C*-(–)-tris[bis(4-phenyl-2-oxazolinemethano)]-fullerene precursor adducts **75** (^{f}C = fullerene *C*lockwise, ^{f}A = fullerene *A*nticlockwise) [47] with known absolute configurations

Examples for the corresponding products are the enantiomeric dendrimers *all-R-*^{f}*C*-**76** and *all-S*^{f}*A*-**76** (Fig. 11) [11], which because of their enantiomeric relationship revealed identical spectroscopic data.

While the chiral precursor trisadducts show very large Cotton effects and $[\alpha]_D$ values, the chiral hexakisadducts **76** exhibit weak chiroptical properties only. In the CD spectra, almost flat lines are observed between 300 nm and 800 nm and the small $[\alpha]_D$ values of **76** are difficult to determine [11].

These functionalized fullerenes lead to a new class of "chemzymes" [48], artificial enzymes consisting of inherently chiral mixed hexakisadducts functionalized with catalytically active sites. The C_2-symmetrical bisoxazoline unit as a bidentate ligand is known to form complexes with metals such as Fe, Mg, and Cu. These complexes can be used as catalysts for stereoselective syntheses, e.g., the cyclopropanation of styrene with ethyldiazoacetate in the presence of the Cu-bisoxazoline catalyst [50]. Preliminary experiments with the chiral fullerenodendrimers indeed showed that the globular macromolecular ligands exhibited catalytic properties. However, the observed stereoselectivities due to these first prototypes of fullerene-based *dendrizymes* were very low compared to those obtained with other bisoxazoline catalysts [11].

In 1999 we proposed a new concept of functional dendrimers (Fig. 12), which is based on fullerene-porphyrin dyads and the regioselective formation of mixed hexakisadducts [8]. Highly compact globular porphyrinodendrimers can

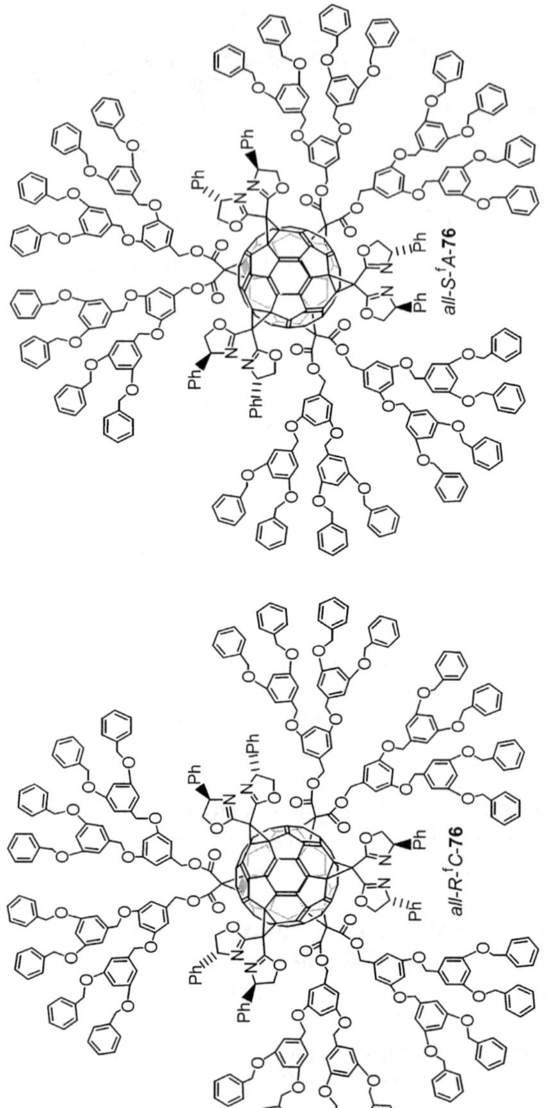

Fig. 11. Inherently chiral enantiomeric fullerenodendrimer *all-R-*^{f}*C*-**76** and *all-S-*^{f}*A*-**76**, mixed [3:3] C_{60} hexakisadducts, prototypes of catalytically active "dendrizymes"

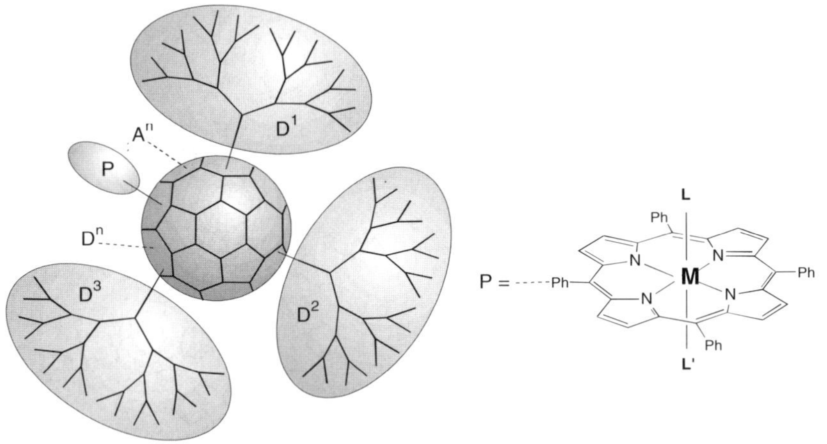

Fig. 12. Schematic presentation of functional dendrimers involving the C_{60} core as a structure-determining building block and a tetraphenyl porphyrin as a functional unit. $D^1, D^2, D^3, \ldots D^n$ denote identical or dissimilar dendrons. The A moieties are additional addends which may also be attached to the core. They may have another function or they may be just positional blockers, which enable the easy construction of a given addition pattern within the fullerene core

be obtained by completing C_{60} monoadducts to hexakisadducts with a complete octahedral T_h-symmetrical addition pattern using dendritic malonate addends. The dendrons provide a specific and widely-adjustable microenvironment around the porphyrin; the dendritic coverage will influence the redox potentials of the fullerene and porphyrin moieties and modulate their electrochemical and coordination properties. Accessibility to these functional dendrimers may be useful when designing catalysts to mimic heme proteins such as cytochrome P450.

Exhaustive cyclopropanation of Zn-porphyrin-fullerene dyad **77** with the Frechet-type dendritic bromomalonate $BrCH(COO\text{-}[G1])_2$ via template activation with DMA in the presence of DBU afforded the first-generation dendrimer-encapsulated Zn-porphyrin fullerene dyad **78** in 2% yield (Scheme 18) [8]. Dendrimer **78** with $(1 \rightarrow 2)$ *aryl*-branching and *ether* connectivity was completely characterized.

To avoid the steric hindrance and to increase the yield of the convergently synthesized functional dendrimers, the new type of "elongated" dendrons with C_3-spacers was taken for exhaustive cyclopropanation of the dyad **77** [8]. The first-generation dendrimer **79** was prepared by using bromomalonate **21** by means of template activation and was obtained in 13% yield after HPLC isolation. Second-generation porphyrin dendrimer **80** was obtained directly from dendritic malonate **20** taking advantage of the modified Bingel reaction in 2% yield after repeated HPLC purification and isolation (Fig. 13) [8].

In order to get an idea about the size and conformation of such macromolecular hexakisadducts and to explore the availability of empty space, cavities and clefts, a series of molecular modeling studies with the dendrimers **79** and **80** was

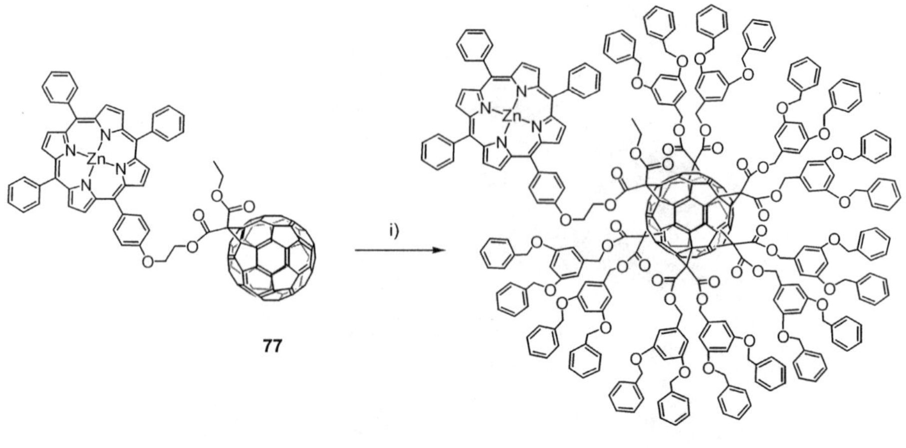

Scheme 18. Synthesis of the mixed hexakisadduct Zn-porphyrin-functionalized fullerenodendrimer **78** by fivefold cyclopropanation of dyad **77** with first-generation Frechet-type bromomalonate dendrons: i) 10 equiv. DMA, 8 equiv. $BrCH(COO\text{-}[G1])_2$, DBU, toluene

carried out [8, 31]. The conformers obtained exhibited a globular structure, even for the first-generation adduct **79**. However, while the porphyrin chromophore still largely penetrates through the dendritic branches in **79**, in the second-generation representative **80** it is partly covered (Fig. 14). The calculated diameter of **80** is in the range of 5–6 nm [8, 31].

We investigated the interactions between the porphyrin group and the fullerene core and the influence of the dendritic coverage on the redox potentials of the porphyrin and fullerene moieties in the dyad **77** and the porphyrin functionalized fullerodendrimers **79** and **80** by cyclic voltammetric (CV) studies [8]. As a result, the first reductions observed for **79** and **80** and assigned to fullerene-based processes revealed a pronounced cathodic shift with increasing generation number. The second reduction step, assigned to a porphyrin process, is also cathodically shifted with increasing generation and demonstrates the influence of the dendritic cover on the electronic properties of these dyad dendrimers [8].

At the interface between biology and material sciences, the development of artificial nanostructures as mimics of natural systems or as prototypes of novel functional aggregates is playing an important role. With our concept of regioselective mono- or multiple-funtionalization of C_{60} to yield an incomplete octahedral substitution pattern followed by completion of the addition pattern we were able to synthesize a new class of artificial dendrimeric lipids [9]. These membrane and vesicle forming amphiphiles have a globular structure with the spherical C_{60} being used to define the molecular geometry. Five pairs of dodecyl chains and one pair of polyamide dendrons are bound to C_{60} in an octahedral [1:5]-addition pattern by methylene bridges. The five pairs of lipophilic chains, all attached to octahedral positions relative to the primary addend, were inserted in one step by cyclopropanation of **26** with didodecylmalonate, using reversible template activation with DMA (Scheme 19) [4, 46]. Chromatographic

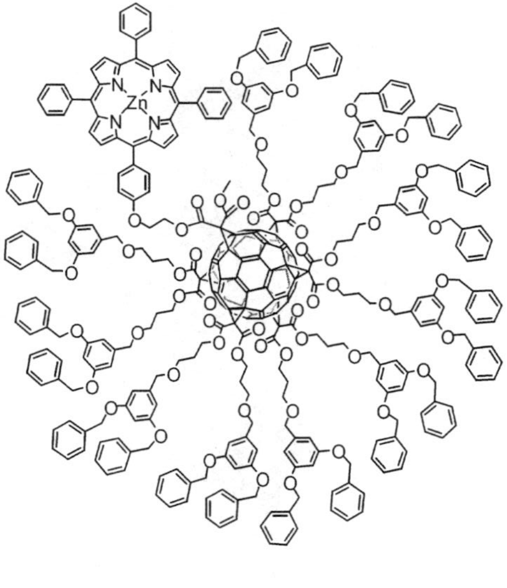

79

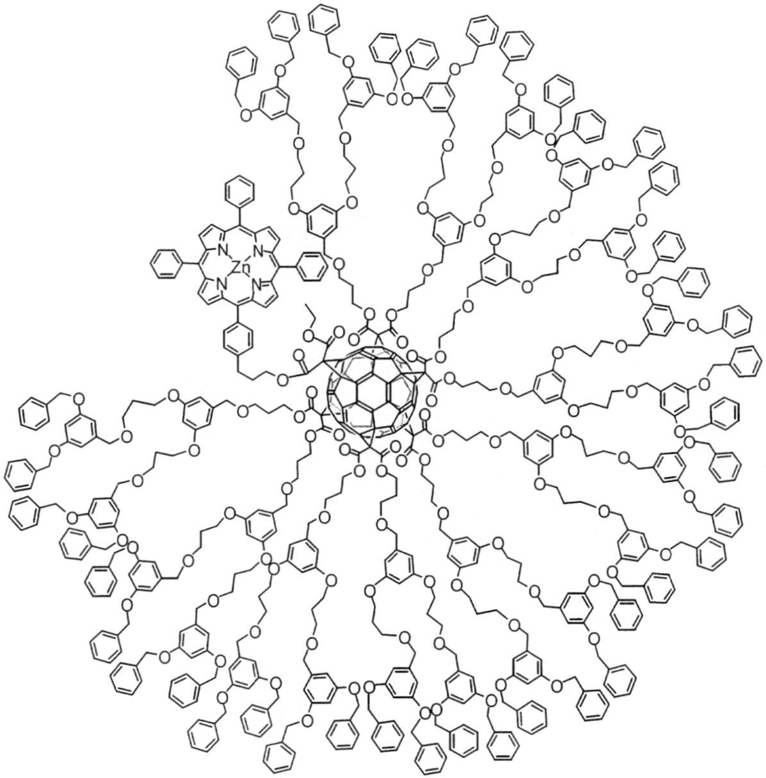

80

Fig. 13. First-generation **79** and second-generation porphyrin-fullerenodendrimer **80** obtained from modified Bingel cyclopropanation with the "elongated" malonate dendrons, DBU, CBr_4 and HPLC purification

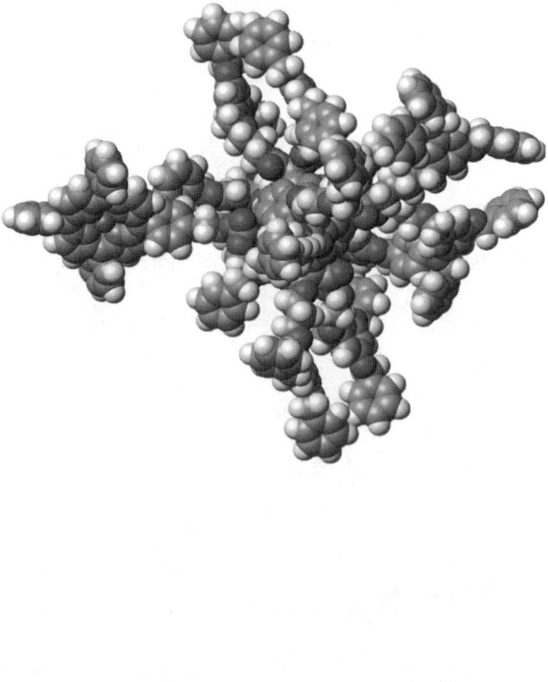

79

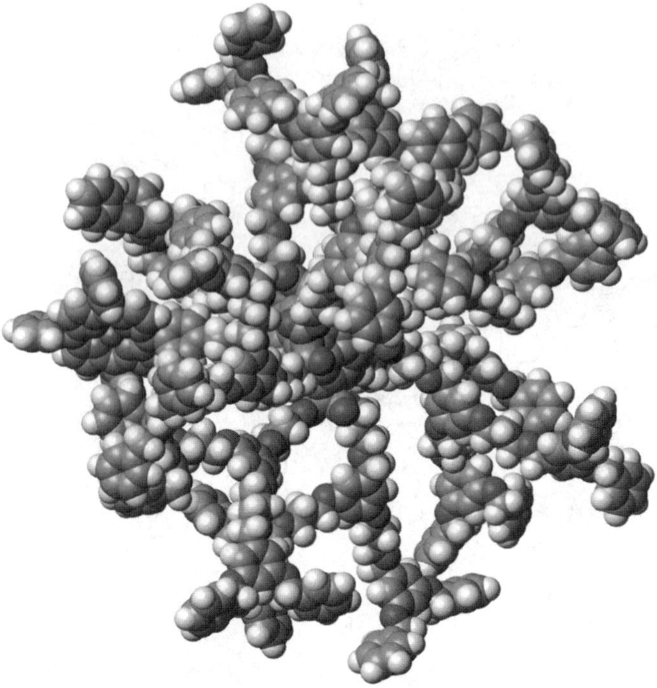

80

Fig. 14. Optimized structures of the porphyrin-dendrimer-fullerenes **79** and **80.** All the structures were minimized with the MM+ force field implemented in the program package HYPERCHEM [45] purification yielded 23% of the mixed hexaadduct **81** as a viscous liquid, which was deprotected to **82** with trifluoroacetic acid (TFA). Amide dendron **31** [7,34], which was generated convergently, was coupled using *N,N′*-dicyclohexylcarbodiimide (DCC) and 1-hydroxybenzotriazol (1-HOBT) to give dendrimer **83** in a 30% yield. The *tert*-butyl groups were removed in 96% yield using TFA and the target molecule **84** obtained as a yellow powder [9], with a melting point of 186 °C (Scheme 19). The corresponding first-generation *tert*-butyl-protected dendrimer **85** and the deprotected hexaacid **86** were obtained using the same reaction sequence in 47% and 96% yield, respectively, by amide formation using amide dendron **28** [32].

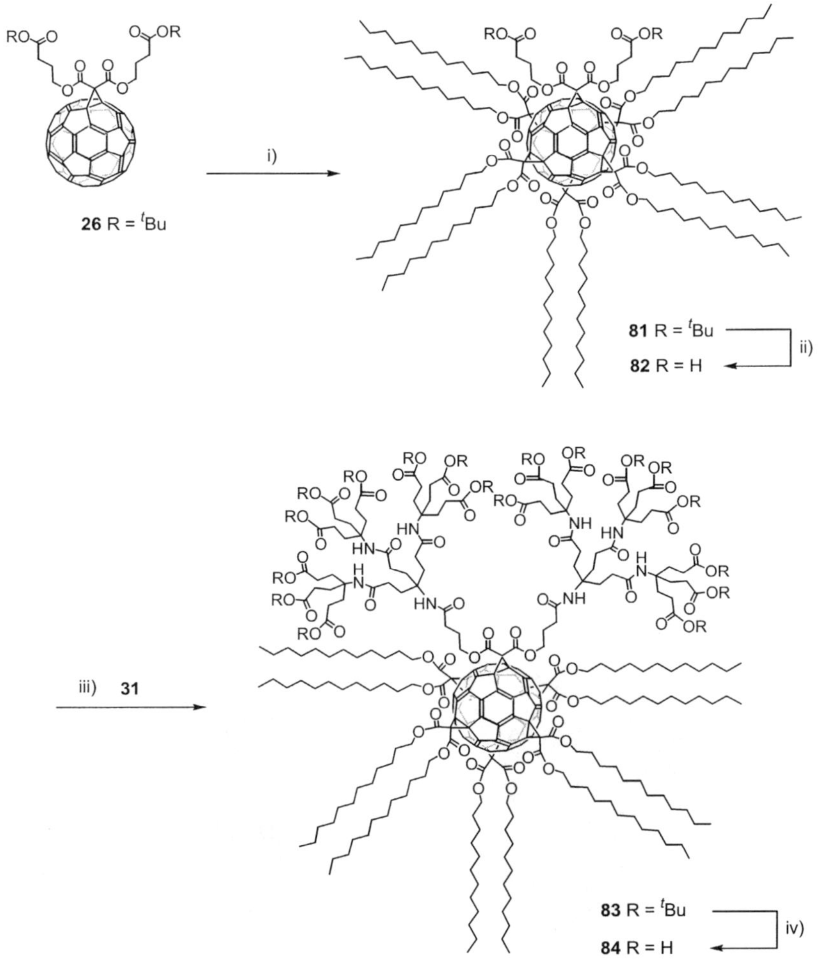

Scheme 19. Synthesis of the *tert*-butyl protected [1:5]-mixed second-generation fullerenodendrimer **83** and the deprotected globular dendritic amphiphile **84**: i) (a) DMA, (b) CBr_4, DBU, didodecyl malonate, toluene; ii) TFA, toluene; iii) "elongated" second-generation Newkome-type [G2]-NH_2 dendron, DCC, 1-HOBT, THF; iv) TFA, toluene

In order to visualize the structure of the globular amphiphile **84**, we performed molecular dynamics simulations [51]. Figure 15 shows the result of this computation. The two types of addends are located in two areas which are, to a large extent, separated from each other. The densely packed hydrophilic and hydrophobic regions almost completely enclose the C_{60} core. The average dimension of about 3.5 nm along the main polar axis is similar to that of naturally occurring phospho- or glycolipids. In contrast, the typical diameters of 1.4–2.3 nm found in directions perpendicular to this axis are considerable larger than those found for natural double-chain lipids (diameters of about 0.7 nm) [9].

Fullerenodendrimer **84** is easily dissolved in neutral water (pH 7.4, phosphate buffer) forming a turbid opalescent solution. With higher concentrations a gel-phase precipitates. The amphiphile is also soluble in most organic solvents, except hexane. In aqueous solutions stable rod-like aggregates were observed by laser light scattering experiments and the corresponding CMC determined with 3.6×10^{-7} M [32]. Furthermore, the aggregation behavior of amphiphile **84** was investigated by freeze-fracture electron and by cryo-electron microscopy (Fig. 16) [9]. Freeze-fracture electron microscopy revealed distinctly vesicular structures. The diameters of the vesicles range from 100 nm to 400 nm. No multilamellar structures were observed, which indicated a preference for the formation of single-layer (unilamellar) vesicles. The propensity to form unilamellar structures seems to be assisted by the high density of negative charges (at pH 7.4) in the hydrophilic part of **84**. Cryo-electron microscopy (cryo-TEM) also revealed the vesicular structures, and this method provided an even more precise picture. Thin layers (100 nm to 200 nm) of the amorphously-vitrified aqueous samples can be directly imaged, giving high-resolution projection images of the uncontrasted sample in the native environment of the solvent. We could observe a wide variety of aggregates, the fine structure of which were invariably based upon the formation of double-layer membranes. The dominant structures were vesicles of dimensions ranging between 50 nm and 400 nm. The

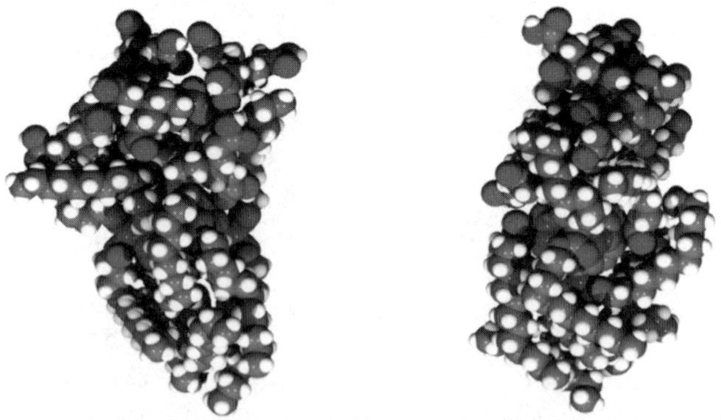

Fig. 15. Molecular dynamics simulation of **84** (orthogonal views). Conditions: heating time 5 ps, run time 15 ps, cooling time 50 ps, step size 0.001 ps, starting temperature 0 K, simulation temperature 1000 K, final temperature 0 K, vacuum [42]

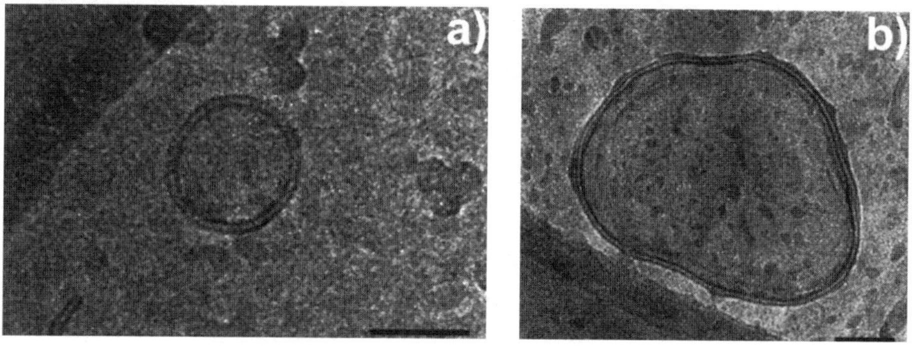

Fig. 16. **a** Cryo-TEM image of a vesicle of **84** (diameter 80 nm). The bilayer structure (approximately 7 nm) of the membrane is clearly resolved, and the outer dark (electron-dense) regions, with a thickness of approximately 2 nm, conform well to the predicted dimensions of the hydrophilic head. The brighter inner seam with a width of approximately 3 nm is, therefore, considered to be the hydrophobic alkyl chain. **b** Cryo-TEM image of a deformed vesicle of **83** (diameter 400 nm). Several regions point to a multilayered growth of the membrane [9]

density profile of the membrane was clearly resolved and the overall thickness determined as about 7 nm (Fig. 16) [9].

Monolayers of the new amphiphilic molecule **84** at the air/water interface were studied at different lateral pressures by a combination of film balance techniques, neutron reflection, and infrared reflection-absorption spectroscopy [52]. The monolayers could be compressed and expanded without significant hysteresis and the alkyl chains remained fluid at all pressures. The p*K* value was determined as 7.5 and pH-dependent measurements allowed a variation of the negative carboxylic headgroup charge by about 18 charges. The thickness of the amphiphile's monolayer at high lateral pressure was 30 Å, similar to that of phospholipids in the condensed state. In contrast, the molecular area of **84** was about sixfold higher than that of phospholipids at high pressure. The negatively-charged monolayer showed strong coupling to water-soluble cytochrom c from the subphase, leading to the formation of a 30 Å thick protein layer underneath the amphiphile layer. The protein content varied drastically with the pH value [52].

When the methanofullerene monoadduct **26** is subjected to a second cyclopropanation, the C_{60}-*e*-bisadduct **87** is obtained, which is also a minor byproduct of the synthesis of **26** itself [53]. Reaction with didodecylmalonate leads to the [2:4]-mixed hexakisadduct **88** [53]. Deprotection with TFA to **89** and coupling with the first generation dendron **28** under DCC conditions gives the *tert*-butyl protected fullerenodendrimer **90** in 25% yield (Scheme 20). Deprotection proceeded almost quantitatively to yield the amphiphilic dodecacarboxylic acid **91** (Scheme 20) [53]. Compound **91** is very soluble in chloroform, whilst its solubility in water at pH 7 (phosphate buffer) and 25 °C is <2 mg/ml.

Coupling of **89** with the second-generation dendron **31** (45% yield) and subsequent deprotection with TFA afforded the dendrimeric hexatriacontacar-

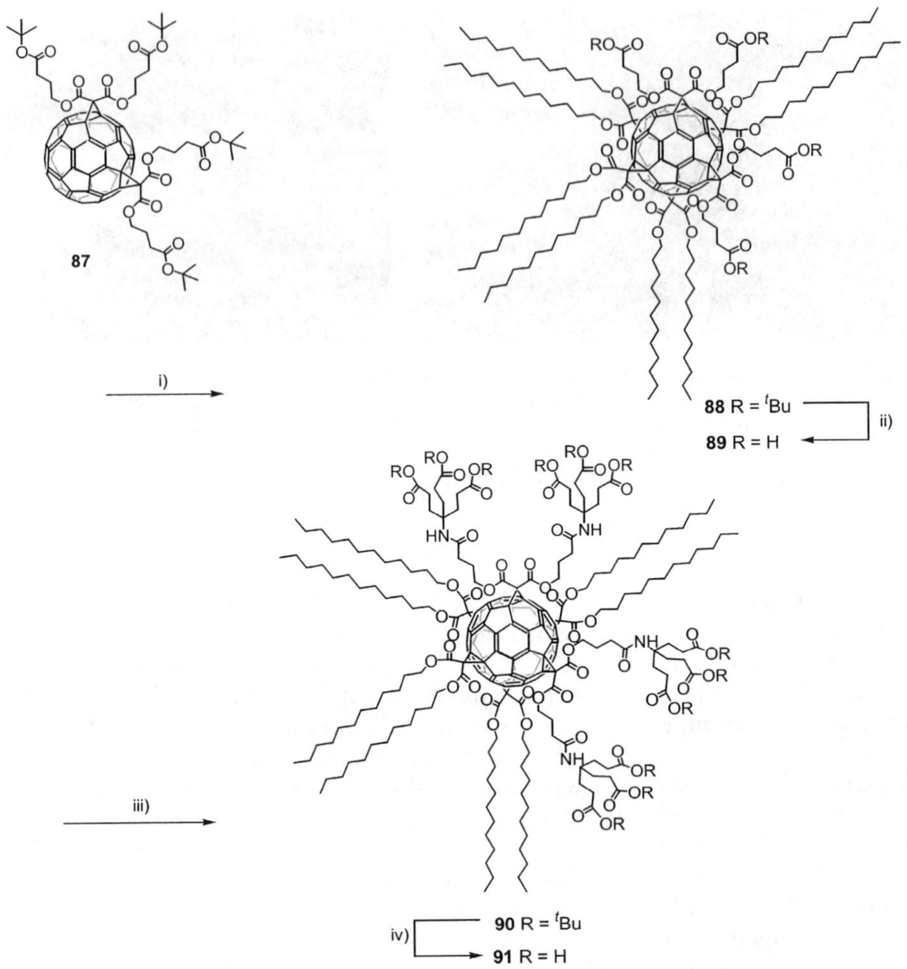

Scheme 20. Synthesis of the *tert*-butyl protected [2:4]-mixed first-generation fullerenodendrimer **90** and the globular dendritic amphiphile **91**: i) (a) DMA, (b) CBr_4, DBU, didodecyl malonate, toluene; ii) TFA, toluene; iii) "elongated" first-generation Newkome-type [G1]-NH_2 dendron, DCC, 1-HOBT, THF; iv) TFA, toluene

boxylic acid **92** (Fig. 17) [32]. The polyacid **92** is not soluble in $CHCl_3$, but readily soluble in THF. The solubility in water at 25 °C is >5 mg/ml, resulting in a solution that scatters light strongly.

With this novel concept we are able to produce brand new monolayer- and bilayer-forming amphiphiles that form vesicles and membrane mimics with novel physical properties [52]. These amphiphilic molecules exhibit surface properties such as high hydration and protein-binding capacity, a potentially high surface charge density, and electrostatic potential modulation over a wide pH range, which can be useful in advanced biomimetic surface applica-

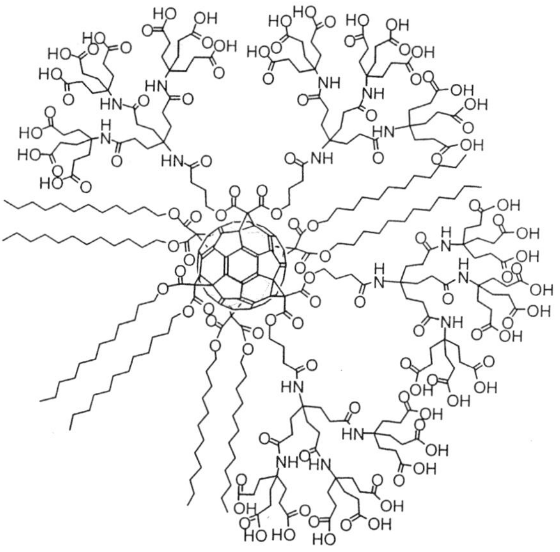

92

Fig. 17. Water soluble four-armed amphiphilic fullerenodendrimer **92**, representing a dendritic hexatriacontacarboxylic acid with a C_{60} core decorated with a [2:4]-mixed addition pattern

tions. The amphiphiles can be specifically functionalized and tuned with respect to spherical topology, solubility, functionality, reactivity, electronic and spectroscopic properties, etc. An interesting aspect of these amphiphiles for biological surface functionalization is the potentially high number of negative charges per molecule. Slight pH variations should result in drastic changes of the surface charge and make the assembly and disassembly of the molecules pH-switchable [52].

4
Oligoynes as Dendrimers Cores

End-capped polyynes, consisting of conjugated triple bonds, are one-dimensional rod-shaped molecules and may be considered as model compounds for the hypothetical sp-carbon allotrope *carbyne* sp-C_∞ (Fig. 1b) [54]. Their stability should be substantially affected by polymerization, decreasing as the chain length increases [55]. The distance between two sp-carbon chains is the critical factor for bringing about the polymerization process [55, 56]. Bulky and sterically demanding end-caps keep the carbon chains apart from one another [57], preventing translational approach and facile polymerization.

Recently we prepared oligoynes with dendritic end-caps on both ends of the sp-carbon rods [58]. Terminal dendritic branches can be adjusted specifically with respect to their size and may be considered as covalently-bound solvate coverage. Since the dendrimeric ends will not significantly interfere with the electronic properties of the polyyne, dendritic termini can be used as protective groups to avoid polymerization of the polyunsaturated axis [58]. The polyyne chain was built up starting from the dendritic terminus. First-, second-, and third-generation Frechet-type [27] benzyl ether dendrons were used to build dendronized acetylenes. From chain elongation by heterocoupling with short-chained oligoynes and final homocoupling according to established acetylene coupling techniques [59] we obtained several series of dendrimers such as **93**–**95** (see Fig. 18) with (1 → 2) *aryl*-branching and *ether* connectivity [58].

In solution the dendritic oligoynes **93**–**95** are stable but they still decompose in the solid state [58]. This indicates that even the third-generation dendritic protective groups are unable to separate sufficiently the alkyne chains from one another to prevent polymerization. One explanation is that, because of the "flat" dendritic protective groups, the molecules favor a stacking of the unsaturated chains resulting in enhanced polymerization.

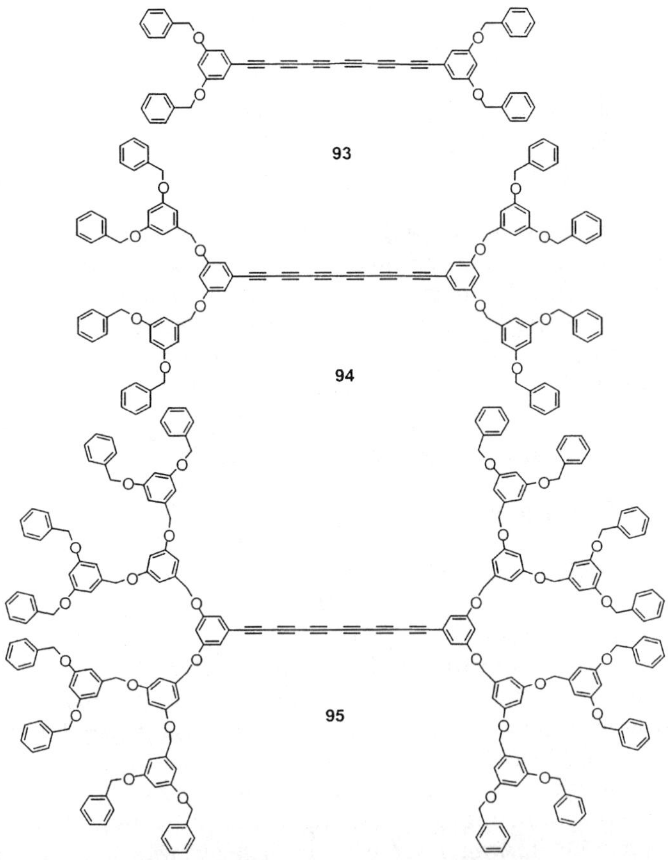

Fig. 18. Series of "dendron-protected" hexaynes **93**, **94**, and **95** decorated with first-, second-, and third-generation Frechet-dendrons, (1 → 2) *aryl*-branching and *ether* connectivity

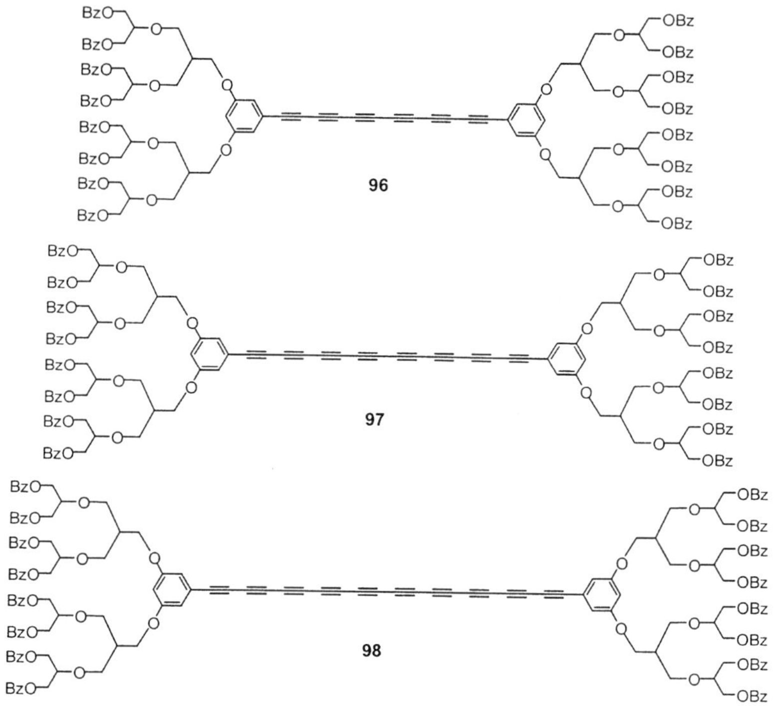

Fig. 19. Oligoyne-core dendrimers **96**, **97**, and **98** with flexible aliphatic dendrons

To address the problem of stability, we decided to modify the dendrons by exchanging the "flat" (planar) aromatic termini to more bulky aliphatic dendritic end-caps [58], introduced by Frechet et al. into dendrimer chemistry [60]. 3,5-Dihydroxybenzyl acetylene derivatives were decorated with these sterically-demanding alkyl branches. Chain elongation from heterocoupling with short oligoynes and subsequent homocoupling afforded the symmetrically end-capped dodecahexayne **96**, hexadecaoctayne **97**, and eicosadecayne **98** (Fig. 19). The oligoynes were completely characterized and appeared to be sufficiently stable in solution and in the solid state [58].

5 Carbon Nanotube Cores

Since their discovery in 1991 [61], interest in carbon nanotubes (Fig. 1c) has been steadily increasing. Nanotubes can be considered as a new carbon allotrope and exhibit an enormous range of potential technical and material sciences applications. Being formed by graphite vaporization, however, the *as prepared* carbon nanostructures consist of inhomogeneous bundles and ropes of insoluble single-walled (SWNTs) and multi-walled nanotubes (MWNTs), varying in structure, diameter, and length. Successful end-group functionalization of short nanotubes will allow the preparation of stable or-

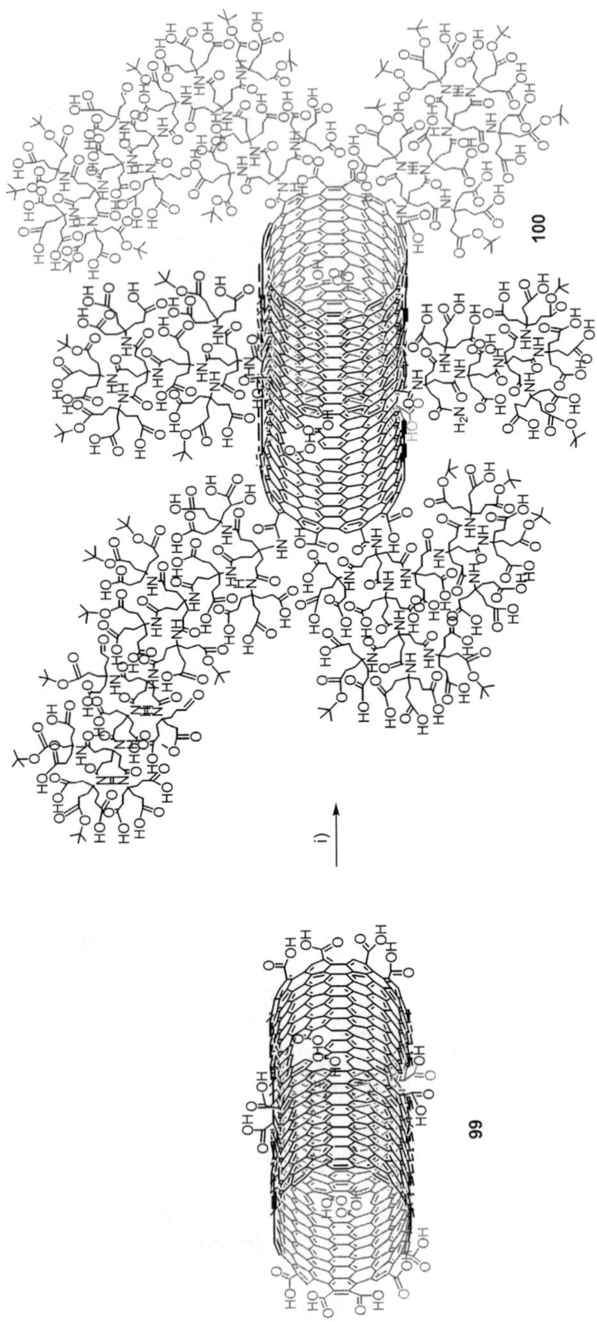

Scheme 21. Preparation of nanotube dendrimer **100** by amide coupling of carboxyl-functionalized carbon nanotube **99** with Newkome-type second-generation dendron and EDC coupling reagent: i) EDC, H_2O, 5d, rt

ganic solutions of these materials [62], permitting the study of their chemical and physical properties and making them accessible for subsequent chemical modifications.

We developed a simple and effective purification method which allows us to use nanotubes as a starting material for chemical modification. The process involves three steps: (i) shortening by oxidation with nitric acid, (ii) ultrasonication to reduce the size of the nanoparticles, and (iii) size exclusion chromatography [63]. By Raman spectroscopy and AFM microscopy on surface-modified silicon wafers, the carbon nanotube fraction appeared to consist mainly of multiply carboxylated single walled tube bundles and a very small quantity of amorphous carbon impurities [63].

The purification process terminates the open ends of the SWNTs with carboxylic acid groups [63, 64], susceptible to reactions with amines to give carboxylic amides. This is illustrated in Scheme 21 with a (10,10)-SWNT-COOH **99**. The treatment of this "nanotube acid" mixture with *tert*-butyl protected second-generation Newkome dendrons [G2]-NH_2 under peptide synthesis conditions afforded a crude mixture of polyamides [64]. After a series of washings, ultrasonication, and centrifugation steps we finally obtained a supernatant solution containing ethanol-soluble nanotube polyamide dendrimers **100** (Scheme 21) [64]. In the course of the amide formation a simultaneous deprotection was observed, which led to a partial cross-linkage of tubes and polymerization of the dendrons.

Despite the exhaustive and tedious purification and chromatographic processes, the dendronized nanotubes still represent a mixture with respect to structure, dimension, and extent of functionalization. Considering, that a SWNT of 1 μm length comprises about 170,000 carbon atoms in more than 80,000 benzenoid rings and has a molecular weight of more than 2,000,000 D, it is obvious that a detailed spectroscopic characterization is not likely. With these attributes, however, the ethanol-soluble dendronized carbon nanotubes are big enough to be resolved by atomic force microscopy (AFM). From these AMF images which, of course, represent a "snapshot" of one single molecule only out of an oligodisperse mixture, we could learn that, in addition to terminal carboxylation, to some extent the tube-walls have also been oxidized and subsequently dendronized. Furthermore, indications of multiple amide formation and cross-linking between different nanotubes were found, which can be attributed to the waste amount of dendritic reagent.

6 References

1. Newkome GR, Moorefield CN, Vögtle F (1996) Dendritic molecules: concepts, synthesis, perspectives. VCH, Weinheim
2. Tomalia DA, Kirchhoff PM. US Pat 4,694,064 (1987); Eur Pat Appl EP 234408 (1987)
3. Schlüter A-D (1998) Dendrimers with polymeric core: towards nanocylinders. In: Vögtle F (ed) Dendrimers. Top Curr Chem 197:165
4. Lamparth I, Maichle-Mössmer C, Hirsch A (1995) Angew Chem 107:1755; Angew Chem Int Ed Engl 34:1607
5. Camps X, Schönberger H, Hirsch A (1997) Chem Eur J 3:561

6. Camps X, Hirsch A (1997) J Chem Soc, Perkin Trans 1 1595
7. Brettreich M, Hirsch A (1998) Tetrahedron Lett 39:2731
8. Camps X, Dietel E, Hirsch A, Pyo S, Echegoyen L, Hackbarth S, Röder B (1999) Chem Eur J 5:2362
9. Brettreich M, Burghardt S, Böttcher C, Bayerl T, Bayerl S, Hirsch A (2000) Angew Chem 112:1915; Angew Chem Int Ed Engl 39:1845
10. Herzog A, Hirsch A, Vostrowsky O (2000) Eur J Org Chem 171
11. Djojo F, Ravanelli E, Hirsch A, Vostrowsky O (2000) Eur J Org Chem 1051–1059
12. Nierengarten J-F (2000) Chem Eur J 6:3667
13. Hirsch A, Lamparth I, Grösser T, Karfunkel HR (1994) J Am Chem Soc 116:9385
14. Hirsch A, Lamparth I, Karfunkel HR (1994) Angew Chem 106:453; Angew Chem Int Ed Engl 33:437
15. Hirsch A, Lamparth I, Grösser T, Prato M, Lucchini V, Wudl F (1996) Regioselective multiple additions to buckminsterfullerene. In: Andreoni W (ed) The chemical physics of fullerenes 10 (and 5) years later. Kluwer Academic Publishers, Netherlands, p 267
16. Hirsch A (1998) Top Curr Chem 199:1
17. Djojo F, Herzog A, Lamparth I, Hampel F, Hirsch A (1996) Chem Eur J 2:1537
18. Djojo F, Grimme S, Hirsch A (1999) Eur J Org Chem 3027
19. Diederich F, Kessinger R (1999) Acc Chem Res 32:537
20. Smith DK, Diederich F (1998) Chem Eur J 4:1353
21. (a) Kroto HW, Heath JR, O'Brien SC, Curl RF, Smalley RE (1985) Nature 318:162; (b) Krätschmer W, Lamb LD, Fostiropoulos K, Huffman DR (1990) Nature 347:354
22. Hirsch A (1994) The chemistry of the fullerenes. Thieme, Stuttgart New York
23. Diederich F, Thilgen C (1996) Science 271:317
24. Hirsch A (1995) Synthesis 895
25. Hirsch A, Vostrowsky O (2001) Eur J Org Chem 829
26. Wooley KL, Hawker CJ, Frechet JMJ, Wudl F, Srdanov G, Shi S, Li C, Kao M (1993) J Am Chem Soc 115:9836
27. Hawker CJ, Frechet JMJ (1990) J Chem Soc, Chem Commun 1010; Wooley KL, Hawker CJ, Frechet JMJ (1991) J Chem Soc, Perkin Trans 1 1059
28. Hawker CJ, Wooley KL, Frechet JMJ (1994) J Chem Soc, Chem Commun 925
29. Bingel C (1993) Chem Ber 126:1957
30. Hermann A, Rüttimann M, Thilgen C, Diederich F (1995) Helv Chim Acta 78:1673
31. Schönberger H, Schwab CH, Hirsch A, Gasteiger J (2000) J Mol Model 6:379
32. Brettreich M (2000) Dissertation. University of Erlangen-Nürnberg
33. Newkome GR, Nayak A, Behera RK, Moorefield CN, Baker GR (1992) J Org Chem 57:358
34. Brettreich M, Hirsch A (1998) Synlett 1369
35. Schuster DI, Wilson SR, Kirschner AN, Schinazi RF, Schlueter-Wirtz S, Tharnisch P, Barnett T, Ermolieff J, Tang J, Brettreich M, Hirsch A (2000) In: Martin N, Maggini M, Guldi DM (eds) Fullerenes 2000. Electrochemical Society Proceedings 9:267
36. Gharbi N, Pressac M, Tomberly V, Da Ros T, Brettreich M, Hadchouel M, Arbeille B, Trivin F, Ceolin R, Hirsch A, Prato M, Szwarc H, Bensasson RV, Moussa F (2000) In: Martin N, Maggini M, Guldi DM (eds) Fullerenes 2000. Electrochemical Society Proceedings 9:240
37. Yu G, Gao J, Hummelen JC, Wudl F, Heeger AJ (1995) Science 270 (#5243):1789
38. Feldrapp K, Brütting W, Schwoerer M, Brettreich M, Hirsch A (1999) Synthetic Metals 101:156
39. Nierengarten J-F, Habicher T, Kessinger R, Cardullo F, Diederich F, Gramlich V, Gisselbrecht JP, Boudon C, Gross M (1997) Helv Chim Acta 80:2238
40. Newkome GR, Lin X, Weis CD (1991) Tetrahedron: Asymmetry 2:957
41. Young JK, Baker GR, Newkome GR, Morris KF, Johnson CS Jr (1994) Macromolecules 27:3464
42. Cardullo F, Diederich F, Echegoyen L, Habicher T, Jayaraman N, Leblanc RM, Stoddart JF, Wang S (1998) Langmuir 14:1955
43. Chuard T, Deschenaux R (1996) Helv Chim Acta 79:736

44. Dardel B, Deschenaux R, Even M, Serrano E (1999) Macromolecules 32:5193
45. Nierengarten J-F, Felder D, Nicaud J-F (2000) Tetrahedron Lett 41:41
46. Lamparth I, Herzog A, Hirsch A (1995) Tetrahedron 52:5065
47. DeGennes PG, Hervet H (1983) J Physiques-Lettres 44:L-351
48. Djojo F, Hirsch A (1998) Chem Eur J 4:344
49. Thilgen C, Herrmann A, Diederich F (1997) Helv Chim Acta 80:183
50. Fritschi H, Leutenegger U, Pfaltz A (1988) Helv Chim Acta 71:1553
51. HyperChem (1996) Release 5, Hypercube Inc., Ontario, Canada
52. Maierhofer AP, Brettreich M, Burghardt S, Vostrowsky O, Hirsch A, Langridge S, Bayerl T (2000) Langmuir 16:8884
53. Vostrowsky O, Brettreich M, Burghardt S, Hirsch A, Böttcher C, Maierhofer AP, Hetzer M, Bayerl S, Bayerl T (2000) In: Kuzmany H, Fink J, Mehring M, Roth S (eds) Electronic properties of novel materials-molecular nanostructures. AIP Conference Proceedings, Melville, New York, p 50
54. Schermann G, Grösser T, Hampel F, Hirsch A (1997) Chem Eur J 3:1105
55. Wegner G (1981) Angew Chem 93:352; Angew Chem Int Ed Engl 20:361
56. Wegner G (1972) Makromol Chem 154:35; Wegner G (1984) Makromol Chem Suppl 6:347; Wegner G (1974) Chimia 28:475
57. Eastmond R, Walton DRM (1972) Tetrahedron 28:4591; Eastmond R, Johnson TR, Walton DRM (1972) Tetrahedron 28:4601; Johnson TR, Walton DRM (1972) Tetrahedron 28:5221
58. Gibtner T (2001) Dissertation. University Erlangen-Nürnberg
59. Siemsen P, Livingston RC, Diederich F (2000) Angew Chem 112:2741; Angew Chem Int Ed Engl 39:2632
60. Grayson SM, Frechet JMJ (2000) J Am Chem Soc 122:10,335
61. Iijima S (1991) Nature 354:56
62. Chen J, Hamon MA, Hu H, Chen Y, Rao AM, Eklund PC, Haddon RC (1998) Science 282:95
63. Holzinger M, Hirsch A, Bernier P, Duesberg GS, Burghard M (2000) Appl Phys A 70:599
64. Holzinger M, Hirsch A, Bernier P, Duesberg GS, Burghard M (2000) In: Kuzmany H, Fink J, Mehring M, Roth S (eds) Electronic properties of novel materials-molecular nanostructures. AIP Conference Proceedings, Melville, New York, p 246

Supramolecular Chemistry of Dendrimers

Steven C. Zimmerman, Laurence J. Lawless

Department of Chemistry, University of Illinois, 600 South Mathews Ave, Urbana, Illinois 61801, USA
E-mail: sczimmer@uiuc.edu

This review will focus on recent progress in supramolecular dendrimer chemistry. We have chosen to present several representative examples that illustrate the diverse ways in which dendrimers can be used to create supramolecular systems. The early focus is on host-guest chemistry where molecular recognition may occur within the dendrimer interior or at its surface. Interior binding may be directed, for example, by a specific group at the dendrimer core, or it may be a nonspecific hydrophobic effect (e.g., dendrimer as unimolecular micelle). Molecular recognition at the "surface" is distinguished by the large number of end-groups and the potential for multivalent interactions.

The nanoscopic size and recognition abilities of dendrimers make them ideal building blocks for self-assembly and self-organization systems. The review will focus on ways in which dendrimers may be formed by self-assembly and ways in which preformed dendrimers may interact with one another. Two types of self-organizing systems will be illustrated: liquid crystalline dendrimers and dendrimers organized at interfaces.

Keywords. Dendrimer, Complexation, Binding, Encapsulation, Nanosphere, Self-Assembly, Hydrogen Bonding

1 Introduction

1.1 Definitions and Scope

The field of dendrimer chemistry is rapidly advancing, and there continues to be a need for literature reviews. Our laboratory published two reviews on the supramolecular chemistry of dendrimers just four years ago [1, 2]. In the interim, numerous important reports have appeared, and therefore this is an appropriate time to update our earlier review. Thus, this chapter will focus primarily on work reported in 1999 and the first half of 2000. Because several specialized reviews on the topic of supramolecular dendrimer chemistry have appeared recently (see below), this review will present a broad overview of the field. The concepts will occasionally be illustrated with selected examples from earlier literature. This chapter will not cover the history of the field, methods of synthesis, or the structure and properties of dendrimers except as it is relevant to their supramolecular chemistry. Readers who are interested in these or more general aspects of dendrimer chemistry are directed to the outstanding book by Newkome et al. [3]. General reviews and those dealing with specific aspects of supramolecular dendrimer chemistry also have been published recently by Astruc et al. [4], Smith and Diederich [5], Emrick and Fréchet [6], Frey and Schlenk [7], Hawker [8], Inoue [9], Majoral and Caminade [10], Baars and Meijer [11], Moore [12], Müllen et al. [13], Newkome et al. [14], Schlüter and Rabe [15], Stoddart et al. [16], Tomalia et al. [17], Vögtle et al. [18], and many others.

Many of the terms that are used in this review are not well defined in the literature and their usage varies among authors. We use the term "self-assembly" to denote the process by which collections of molecules are formed [19]. These collections may contain a very small or very large number of molecules, but order should exist due to a "pre-programmed" atomic level recognition process. Thus, the chemist determines the ultimate structure. Self-organization refers to an identical process but where order arises spontaneously due to the inherent desire for molecules to order themselves into the lowest thermodynamic state, for example in the formation of liquid or molecular crystals and the formation of micelles and liposomes.

There continues to be debate about the exact structure of dendrimers, in particular whether they are fully extended with maximum density at the surface or whether the end-groups fold back into a densely packed interior [1, 2]. Recently, experimental evidence has been obtained in support of both compact folded and extended structures. For example, Amis et al. has reported [20] the synthesis of seventh generation poly(amidoamine) (PAMAM) dendrimers with partial deuteration of the peripheral layer ($CD_2CDHCONHCH_2CH_2NH_2$). Contrast matching experiments (in CD_3OD) using small angle neutron scattering allowed the radius of gyration to be determined which was similar to that of the whole dendrimer. This finding is consistent with localization of the terminal groups near the surface of the dendrimer. Wooley et al. have synthesized two fifth-generation Fréchet-type dendrimers with ^{19}F at the core, one with a ^{13}C label in the third generation layer, the other in the fifth generation (peripheral) layer [21]. Solid state rotational-echo double-resonance (REDOR) NMR experiments indicate a similar distance between the core and the third- and fifth-generation labels consistent with a fold-back of peripheral groups.

As seen in Fig. 1, the structure of some dendrimer repeat units, for example, the 1,3-diphenylacetylene unit developed by Moore [22], must by their very nature fold back on themselves. Parquette and coworkers [23, 24] have designed and synthesized a new class of dendrimers, which are designed to fold back via hydrogen bonding and adopt defined chiral ordered structures. With many dendrimers it is likely that no single structure is adopted but rather different structures depending on the nature of branching units and its environment. Thus, in referring to surface and internal recognition events, we note that the "surface" refers to the end-groups and the interaction being discussed might actually occur on the inside of the dendrimer. Likewise, "internal" refers to the core or the subunits that interconnect the core and end-groups, and this recognition could occur at a solvent exposed surface if the end-groups fold back.

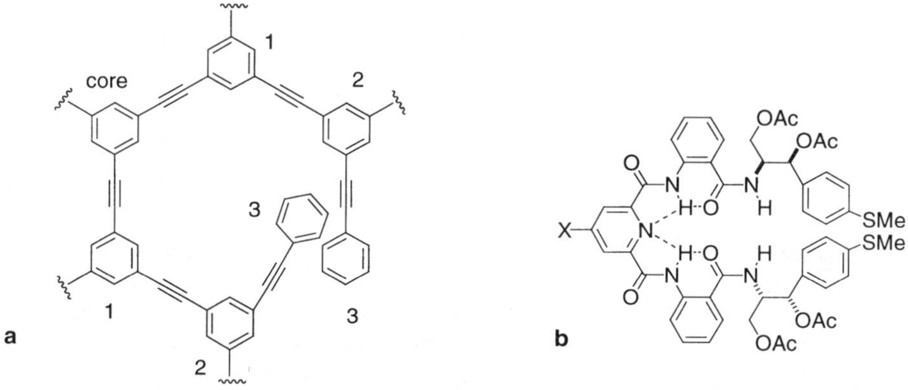

Fig. 1. **a** Moore-type dendrimers consist of phenyl acetylene subunits. At the third generation different arms may occupy the same space and the fourth generation layer potential overlaps with the second generation layer. **b** Parquette-type dendrons are chiral, non-racemic, with intramolecular folding driven by hydrogen bonding

2 Host-Guest Chemistry Involving Dendrimers

2.1 Unique Structures for Surface and Internal Complexation

The unique structure of dendrimers provides special opportunities for host-guest chemistry (Fig. 2). The multiple end-groups allow multiple complexation events to occur simultaneously at these sites, which can lead to special types of interfacial molecular recognition. For example, dendrimers are especially well equipped to engage in multivalent interactions. At the same time, one of the earliest proposed applications of dendrimers was as container compounds wherein small substrates are bound within the internal voids of the dendrimer [25]. Experimental evidence for unimolecular micelle properties was established many years ago both in hyperbranched polymers [26] and dendrimers [27, 28].

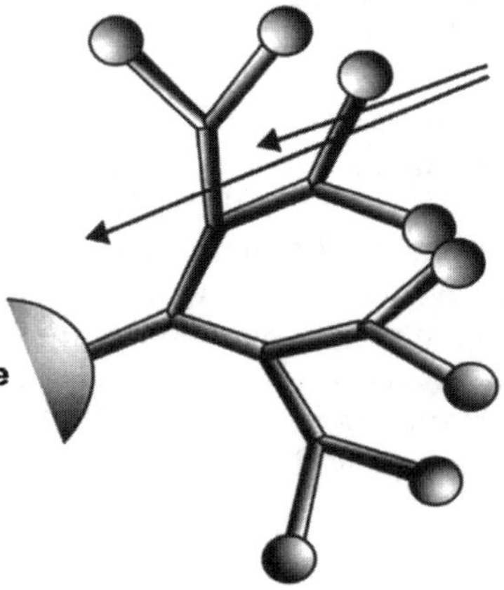

Fig. 2. Schematic showing the three main parts of a dendrimer, the core, end-groups, and subunits linking the two

2.2 Nonspecific Internal Binding

This nonspecific approach to binding is nicely illustrated by the coating of poly(propylene imine) (PPI) dendrimers with a hydrophilic outer layer by Meijer and coworkers (see dendrimer **1**) [29]. With basic amines and a somewhat hydrophobic interior, dendrimer **1** dissolves in water and binds rose Bengal (**2**) and 4,5,6,7-tetrachlorofluorescein (**3**), with association constants (K_{assoc}) of $5 \times 10^5\ M^{-1}$ and $3 \times 10^4\ M^{-1}$, respectively. The importance of the acid-base association was supported by the pH effect on binding. Finally, SAX measurements showed localization of the guest molecules on the dendrimer interior.

A dendrimer-like inverted unimolecular micelle was recently described by Sun and coworkers [30]. Using the Bingel-Hirsch type addition reaction to C60

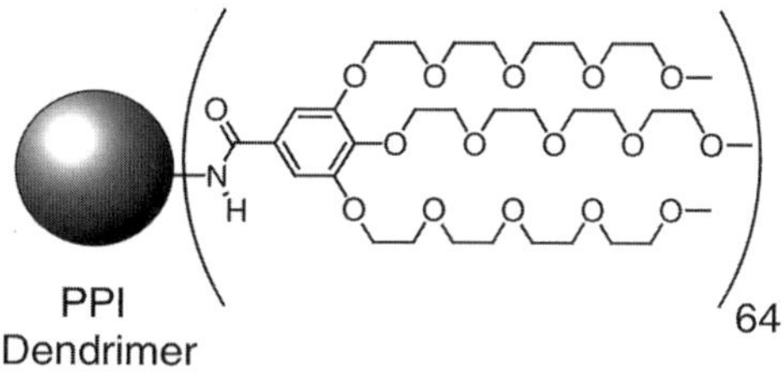

1

(sixfold), the straightforward synthesis of **4a**–**c** was achieved. Sonicating this compound in dodecane with an aqueous lithium chloride solution led to incorporation of a portion of both water and metal ions on the inside of **4**. As would be expected, the amount of aqueous ion incorporated was dependent on the length of the hydrophilic block (i.e., **4a** → **b** → **c**). The authors also showed that the micellar structures could be used as "nanoreactors" to produce silver nanoparticles of relatively uniform sizes.

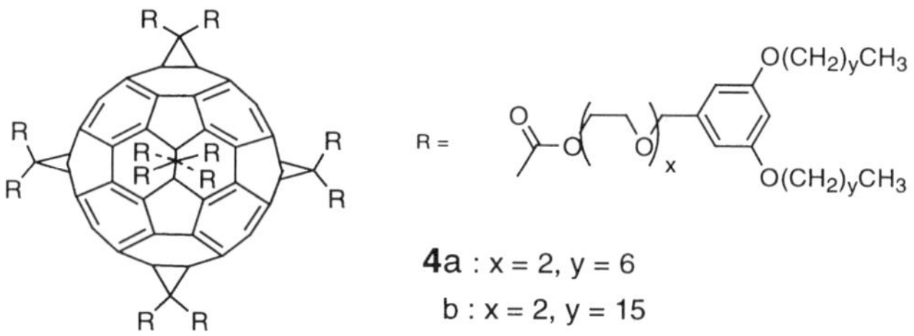

4a : x = 2, y = 6
b : x = 2, y = 15
c : x = 4, y = 15

In a related study, Crooks and coworkers [31] showed that inverted micelles could be produced by a self-assembly process. Thus, a fourth generation PAMAM dendrimer was shown to readily dissolve in 1% dodecanoic acid-toluene to a degree that suggested nearly complete formation of surface ion pairs (i.e., ammonium ion-carboxylate pairings; see **5** in Fig. 3). The IR was consistent with this suggestion. Similar structures have been prepared by covalent modification and shown to encapsulate guest molecules. Beyond avoiding the need for covalent chemistry and its attendant purification difficulties, the self-assembly approach is reversible. Thus, addition of acid leads to protonation of the PAMAM dendrimer, which in turn causes it to migrate to an aqueous layer. The authors not only demonstrate the reversible transport and encapsulation of methyl orange (**6**) into the self-assembled inverted micelles – they also show that catalytically active Pd nanoparticles can be prepared within the micelles.

There is considerable interest in the use of dendrimers as unimolecular micellar carriers of water insoluble drugs or for targeted delivery of drugs using the peripheral groups for tissue or cellular specificity. The simple binding experiments that have been reported to date strongly support the utility of dendrimers as unimolecular micelles. Little effort has focused on the capacity of dendrimers. It is likely that the capacity will be considerably lower than that of liposomes. Wendland and Zimmerman have shown that dendrimers may be

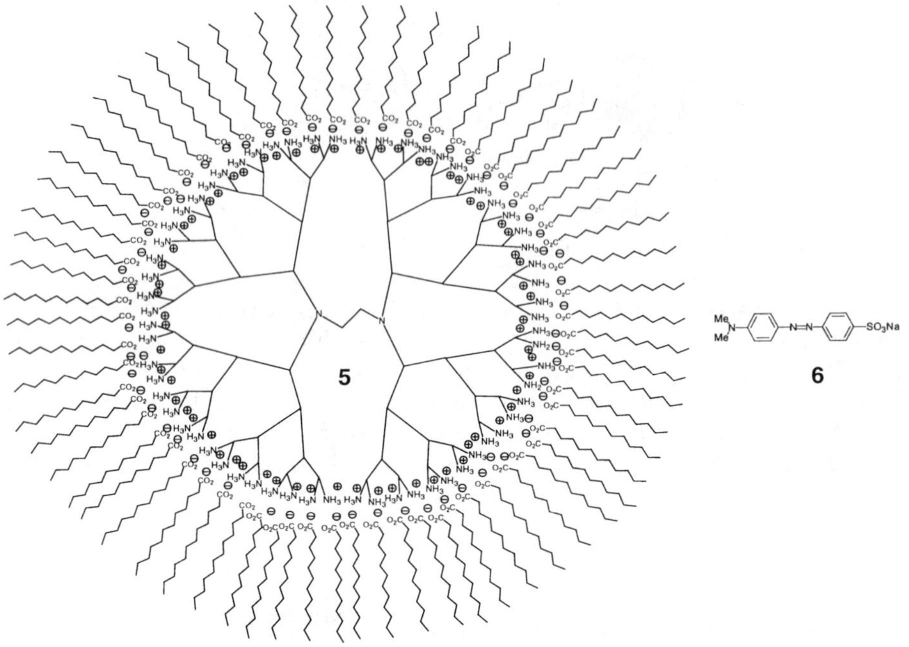

Fig. 3. Ionic assembly of PAMAM dendrimer and decanoic acid (**5**) studied by Crooks and co-workers. In water the assembly is capable of complexing methyl orange (**6**)

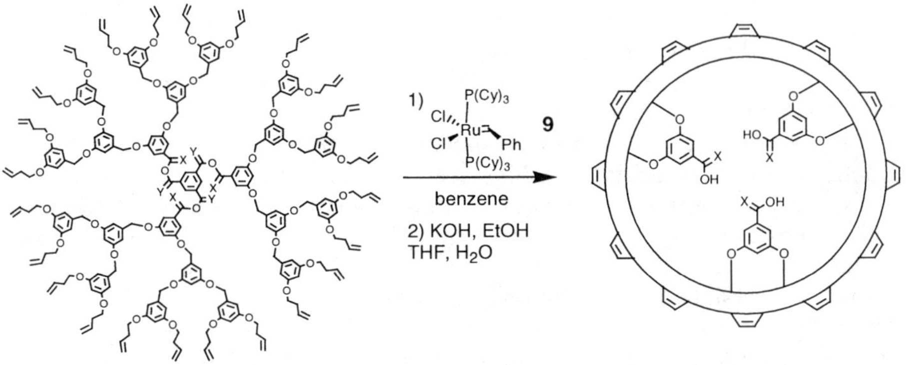

7 : X = H,H; Y = O
8 : X = O; Y = H,H

10 : X = H,H
11 : X = O

Scheme 1. Wendland and Zimmerman process for "coring" dendrimers. Cross-linking with the ring closing metathesis reaction is followed by basic hydrolysis/alcoholysis which removes the core unit "cored," which may open the way to increasing their carrying capacity [32]. As shown in Scheme 1, the ring closing metathesis (RCM) reaction of dendrimers **7** and **8** occurs with the commercially available Grubbs' catalyst **9**, giving nearly full cross-linking of the peripheral homoallyl groups. The core is then removed under basic conditions to give "cored" dendrimers **10** and **11**. The coring process can leave different functional groups behind.

An interesting example of metal ion sensing by a multi-chromophoric dendrimer was reported by Balzani et al. [33]. The dendrimer studied, **12**, contains a trimesic acid core, a bis(ethylamino) spacer, then two lysine layers, and 24 dansyl units as the end groups. In 5:1 acetonitrile-dichloromethane solution containing tributylamine, **12** showed strong fluorescence quenching upon addition of Co^{2+} and Ni^{2+} whereas no change was seen when a control compound (*N*-butyl 5-dimethylamino-1-naphthalene sulfonamide) was subjected to the same conditions. This result, combined with the results of other experiments, suggests that two or more sulfamide anions cooperate in the metal ion binding. Significantly, at stoichiometric metal ion concentrations, a single ion is found to quench the fluorescence of nine chromophores. This type of signal amplification is particularly useful for sensing applications.

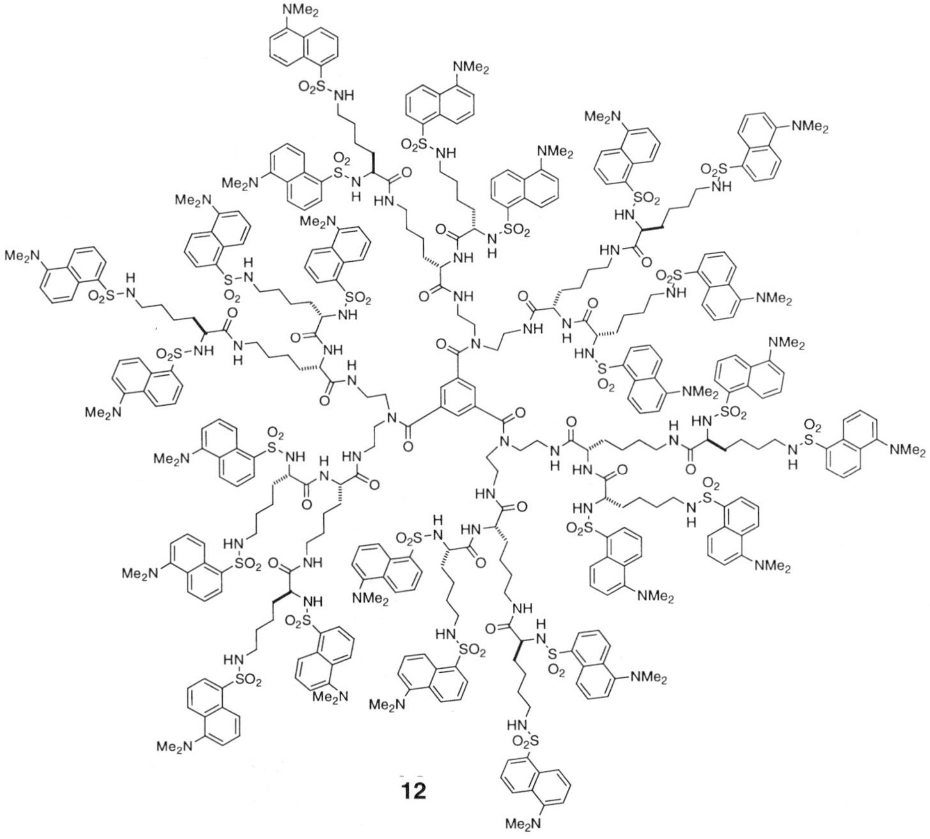

12

2.3
Directed Internal Binding

The incorporation of host or guest molecules at the core of a dendrimer allows the binding to be directed specifically at the core. Early examples showed that hosts that use either hydrogen bonding interactions or hydrophobic complexation led to specific guest binding. Of course the host-guest designation is arbitrary, but compounds traditionally considered guests have recently been attached to dendrimer cores.

2.3.1
Hydrogen Bonding

Remarkably few reports have appeared wherein specific recognition sites on the interior of dendrimers are used to direct internalization of guest molecules. Early work by Newkome et al. [34] on glutarimide complexation and studies by Zimmerman et al. [35] on amidinium binding showed that hydrogen bonding could occur on dendrimer interiors with similar binding constants to those observed in free solution. In the latter case, the dendrimer type and generation number did not affect the ability to complex a small guest, and the K_{assoc} values were fully responsive to the solvent polarity. The results suggest that even large dendrimers can be filled with solvent and this controls the microenvironment at the core. Diederich has reported chiral, non-racemic "dendroclefts" where the dendrimer diminishes the degree of enantioselective binding of α-glucosides but increases the diastereoselective binding [36]. In this example, the dendrimer plays an integral role likely due to additional hydrogen bonding interactions possible between host and guest.

Newkome and coworkers have synthesized a series of dendritic monomers, **13**, **14**, and **15** containing one, three, and six 2,6-diamidopyridine units, respectively [37]. These were subsequently covalently linked to epichlorohydrin-activated agarose and the surface-modified gel's ability to bind amital (**16**) deter-

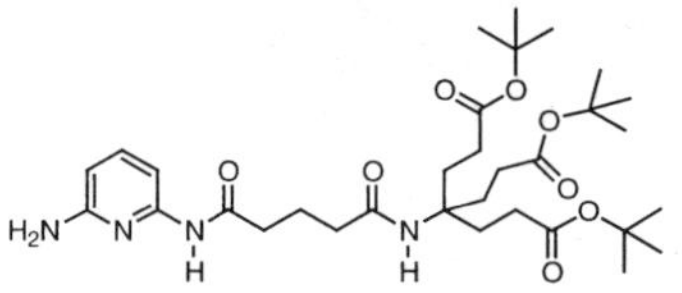

13

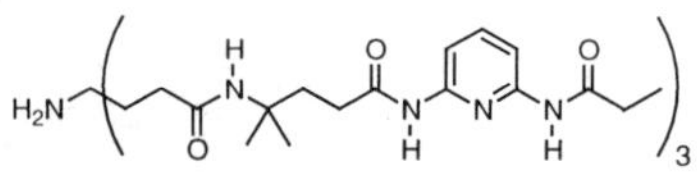

14

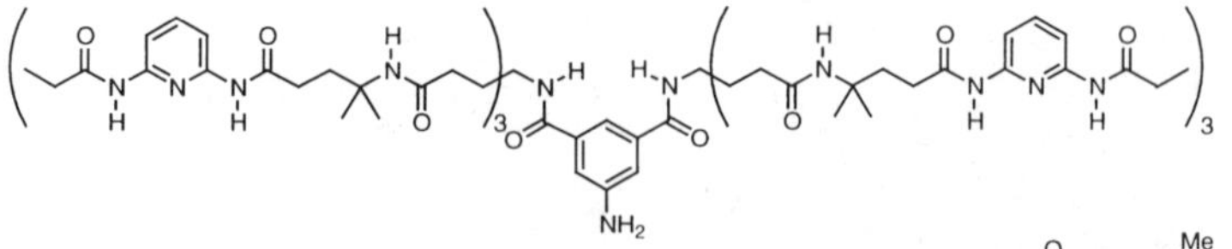

15

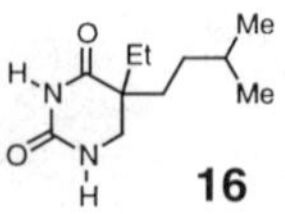

16

mined. In this case, the dendritic structure diminished the extent of amital uptake. Indeed, a 13-fold increase in uptake was observed for dendron (**13**) and a linear analog relative to the gel derivatized with **15**. Although two proximal arms of **15** (and **14**) might simultaneously complex amital, thereby increasing its binding efficiency, the authors propose that two adjacent arms in **14** and **15** self-complex. Thus, an energy price must be paid prior to binding.

2.3.2
Apolar Binding

Kaifer and coworkers have extensively studied ferrocene-based dendrimers as macromolecular redox agents [38a]. Recently, these workers have synthesized Newkome-type dendrimers (e.g., **17** and **18**) containing a single ferrocene unit at the focal point (Fig. 4) [38b]. Because ferrocene is known to be an excellent guest for β-cyclodextrin (**19**), the electrochemical potentials were determined in water with and without β-cyclodextrin present, and K_{assoc} values for cyclodextrin binding were measured as a function of generation number of the dendrimer. It was found that both the dendrimer and its binding to cyclodextrin affected the electrochemical properties of the ferrocene. Also, the affinity of the cyclodextrin for the ferrocene was reduced with the third generation dendrimer **18** showing the largest effect ($K_{assoc} = 50\ M^{-1}$) and the first generation ferrocene, **17**, ($K_{assoc} = 950\ M^{-1}$) at the low end of the normal range for ferrocene-cyclodextrin complexes. The electrochemical redox potentials of the ferrocene are clearly affected by both the dendrimer and its complexation to cyclodextrin.

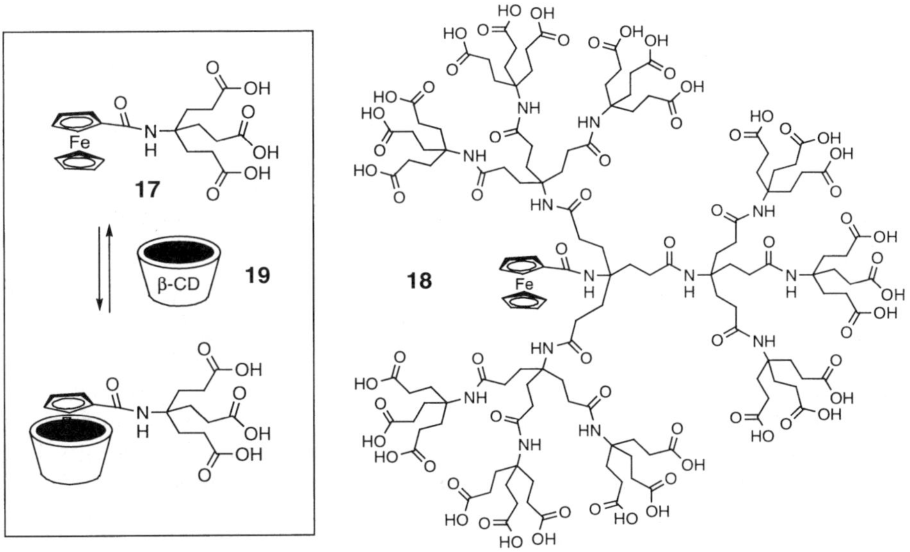

Fig. 4. Kaifer's third-generation Newkome-type dendrimer with a ferrocene core (**18**). Equation showing ferrocene complexation (first generation, **17**) into the secondary side of β-cyclodextrin (**19**)

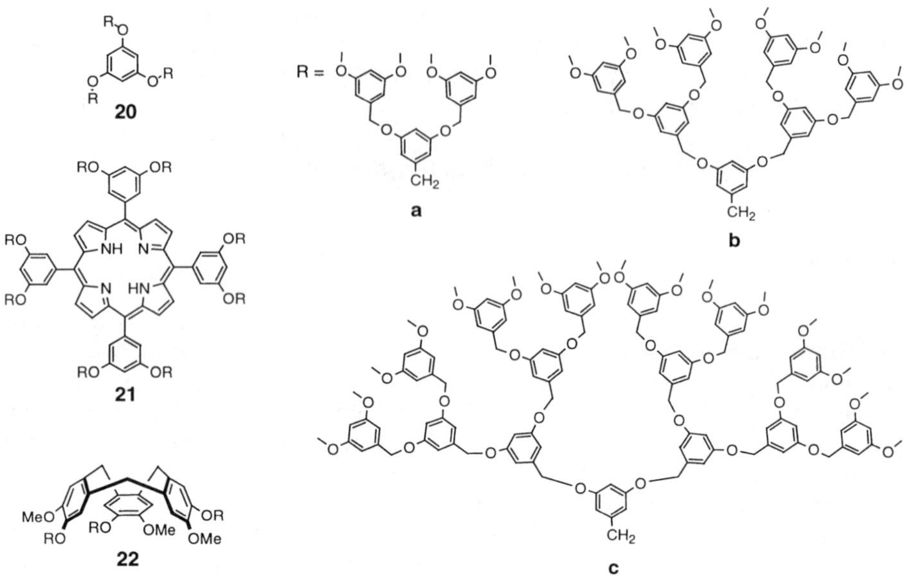

Fig. 5. Shinkai's dendritic hosts for C60. Three generations of Fréchet-type dendrons (**a**–**c**) attached to phloroglucinol (**20**), porphyrin (**21**), and cyclotriveratylene (**22** – lacking OMe groups) core units

Shinkai and coworkers found [39] that Fréchet-type dendrimers with phloroglucinol (**20**), porphyrin (**21**), and cyclotriveratrylene (**22**) cores (Fig. 5) all bound C60 in apolar organic solvents. In each case, the K_{assoc} values increased with generation number. For example, in toluene with hosts **20a–c**, the K_{assoc} values were 5 (**20a**), 12 (**20b**), and 68 M^{-1} (**20c**), respectively. Spectroscopic evidence was presented indicating complexation at the core. For the cyclotriveratylene-based hosts **22a–c**, the K_{assoc} values in methylene chloride were 130, 190, and 300 M^{-1}, respectively. A Job plot indicates 1:1 stoichiometry. The results indicate that the electron-rich dendrons increase the binding to the core element, presumably by classical electron donor-acceptor interactions (i.e., electrostatic, polarization, and dispersion forces).

2.4 Topological Complexation

There are several ways in which one could imagine topological complexation of molecules by dendrimers. One of the earliest proposals was that dendrimers with extremely densely packed end-groups might permanently encapsulate guests. Meijer et al. realized this process [40] in work that was previously reviewed [1]. Mechanical complexation could also occur by catenane or rotaxane formation (see below). Pseudorotaxane formation has been used to self-assemble dendrimers and this work is discussed in Sect. 3.3.

Vögtle et al. have described two types of chiral dendrimeric assemblies based on rotaxane and catenane topologies (Fig. 6) [41]. Both types of structures were made by chemoselective alkylation of the preexisting rotaxane (**23**, R = H) or

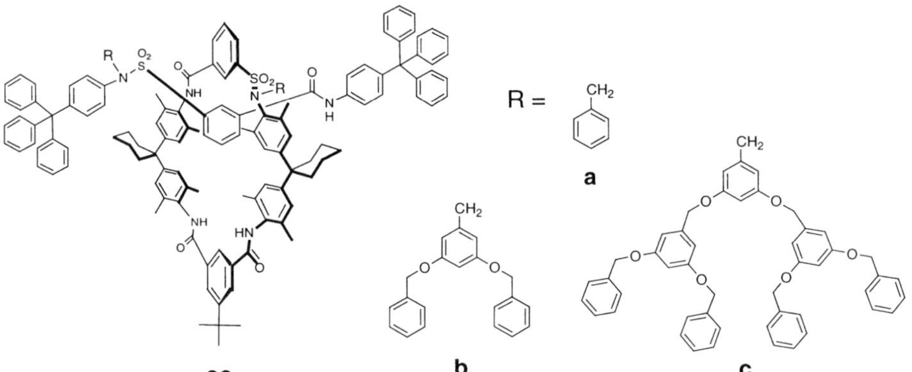

Fig. 6. Rotaxane 23 with dendritic stoppers developed by Vögtle and coworkers

catenane core sulfonamide with Fréchet-type dendrimers. The zero-generation and first generation chiral rotaxanes (**23a** and **23b**) could be resolved by chiral HPLC (baseline resolution), but the second-generation compound (**23c**) eluted as a single peak. The first through third generation catenanes (structures not shown) could also be resolved with baseline resolution using chiral HPLC. All of the resolved compounds exhibited distinct CD spectra with a larger peak amplitude observed for **23a** and **23b** relative to the unalkylated rotaxane (**23**, R = H).

2.5 Surface Binding

Quite a number of examples have appeared recently where multiple and simultaneous complexation events occur on the surface (peripheral groups) of a dendrimer. An excellent example of this approach is the interaction of adamantyl dendrimers **24a–e** with β-cyclodextrin (**19**), reported by Reinhoudt and coworkers (Fig. 7) [42]. These polypropylene imine based dendrimers containing between 4 and 64 adamantyl end-groups are insoluble in water but readily dissolve in the presence of β-cyclodextrin. The number of cyclodextrin molecules bound per dendrimer was determined by NMR. Each generation (**24a–d**) was fully occupied except for the last (**24e**) wherein about 40 out of 64 possible sites were complexed, presumably due to steric effects. The supramolecular assemblies were all shown to bind 8-anilinonaphthalene-1-sulfonate (ANS) on their interior. A related surface complexation study was also reported by Kaifer using cobaltocenyl-terminated PPI dendrimers [43].

Newkome and Hill have reported the reaction of Newkome-type polyol dendrimers with polyoxometalates (POMs, e.g., $(H(CH_3O)_3P_2V_3W_{15}O_{59})^{-5}$) to produce dendritic tetra(POM) structures [44]. Because of the chelate effect, these compounds were significantly more stable than the methanol adducts. The compounds were shown to catalyze the oxidation of tetrahydrothiophene in the presence of *tert*-butylperoxide and an acid catalyst, and to be easily precipitated following the reaction. The stability of these compounds in the presence of water makes them appealing as "green" oxidation catalysts.

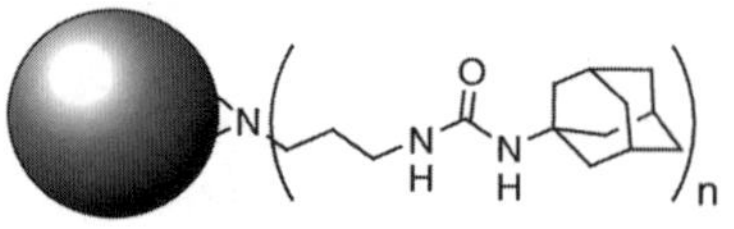

n	compd no.
4	**24a**
8	**24b**
16	**24c**
32	**24d**
64	**24e**

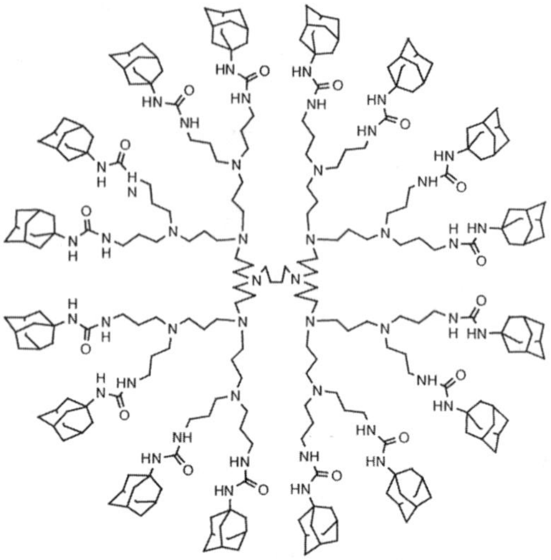

24c

Fig. 7. Adamantane-tipped PPI dendrimers developed by Meijer and Reinhoudt, complex multiple β-cyclodextrin units in water

3 Self-Assembly of Dendrimers

3.1 Concept and Definitions

Because dendrimers contain three distinct structural parts – the core, end-groups, and branched units connecting core and periphery – there are three strategies for self-assembling dendrimers. The first is to create dendrons with a core unit that is capable of recognizing itself or a ditopic or polytopic core structure, thus leading to spontaneous formation of a dendrimer [45–48]. The second approach is to add layers or generations to the end-groups non-covalently [49–51]. These two strategies are analogous to the convergent and divergent approaches to dendrimer synthesis. Finally, dendrimers can be self-assembled by adding layers or generations via recognition units on the branched monomers inside the dendrimer. This would be equivalent to grafting dendrons onto reactive sites within a dendrimer.

3.2 Hydrogen Bond Mediated Self-Assembly

Second-, third-, and fourth-generation Fréchet-type dendrimers have been assembled into stable hexameric aggregates using bis(isophthalic acid) units at the core [45]. The largest such aggregate has a molecular weight of 34 kD and is the size of a small protein. More stable hexamers with a first-generation substituent were obtained with a heterocycle containing two self-complementary hydrogen bonding sites [47]. A dendritic trimeric assembly was also reported as a discrete constituent of a self-assembling discotic liquid crystal [52].

3.3
Self-Assembly Using Pseudorotaxane Formation

A self-assembling dendrimer using pseudorotaxane formation as the organizing force was reported by Gibson and coworkers (Fig. 8) [53]. Triammonium salt 25 was found to be insoluble in chloroform-*d* but became soluble upon addition

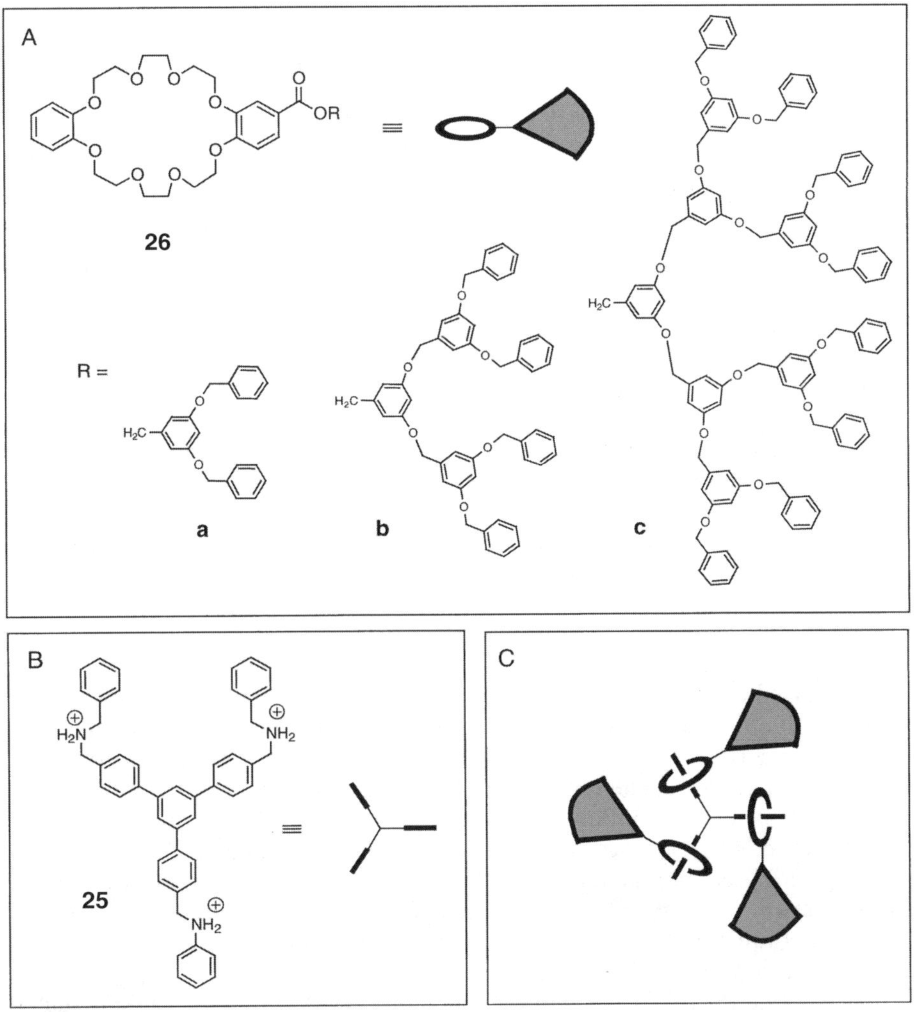

Fig. 8A–C. Gibson's self-assembling dendrimers using pseudorotaxanes formation: **A** crown-ethers with dendritic substituents; **B** triammonium ion core; **C** schematic of tridendron formed by triple pseudorotaxane self-assembly

of dendron **26** containing a crown at the focal point. Stoddart has extensively investigated related catenanes, rotaxanes, and pseudorotaxanes and showed that such crowns will complex ammonium ions [54]. In the case of **26**, complexation was evidenced by the NMR spectrum indicating a 3:1 complex (Fig. 8c). The assembled dendritic structure was estimated by molecular modeling to have an 8–12 nm "diameter." The smaller generation crowns **26a** and **26b** similarly formed self-assembled dendrimers, with the time to reach equilibrium being generation-dependent: **26a** (ca. 36 h), **26b** (ca. 48 h), and **26c** (ca. 72 h).

3.4 Metal Mediated Self-Assembly

There continues to be considerable activity in the area of metal-mediated self-assembled dendrimers. Newkome has continued an extensive investigation of self-assembled dendrimers based on the Ru(II)-terpyridine system [14]. These elegant systems allow various subunits to be rapidly connected into dendritic "networks." The latest system involved the synthesis and study of two isomeric "methane"-type dendritic metallo-macromolecules (**27** and **28**, Fig. 9) [55]. Both compounds were made with metal ligation as the final assembly step. Both compounds were characterized both spectrophotometrically and electrochemically, as well as by MALDI-TOF. A related set of compounds was synthesized containing eight internal carboxylate groups, which provided intramolecular counter ions. These compounds gave clean MALDI-TOF spectra at lower laser power than their ester counterparts containing Cl^- ions.

Gorman and Smith have extended the study of iron-sulfur core dendrimers (**29a–c**, Fig. 10) to films and compared their electrochemical properties to those previously obtained in solution [56]. The observation that the dendrimer generation number affects the redox potential in films but not in solution is attributed to a dilution effect of the metal centers. Thus, from the second to fourth generation iron-sulfur core dendrimer, a 500 mV drop in the $E_{1/2}$ is seen. Despite the significant change in thermodynamics, the redox kinetics in the film reveal no effect of dendrimer size. In solution the generation two to four comparison showed a two order of magnitude change in rate.

Van Veggel and coworkers reported the attachment of their previously described palladium-based dendrimer to a gold decane thiol monolayer (Fig. 11) [57]. The first and second-generation dendrimers were synthesized with a long chain dialkyl sulfide at the focal point. The time dependence of the attachment to the gold surface was followed by AFM. An increase in absorption over time was observed with a surface coverage of about 1% reached after 20 h. The AFM imaging indicated heights for **30a** and **30b** that are consistent with the dimensions determined from CPK models. The widths for **30a** and **30b** were measured to be 20 ± 4 nm and 23 ± 4 nm, respectively.

Aida has continued his extensive studies of biomimetic self-assembled dendrimers with a timely report on the "dendrimer effect" in a non-heme metalloprotein mimic [58]. The work is based on the report by Tolman et al. [59] that copper complexes such as **31a** react with molecular oxygen to form complex **32a** (Fig. 12) which may serve as mimics of the bis(μ-oxo)bridged bimetallic found in

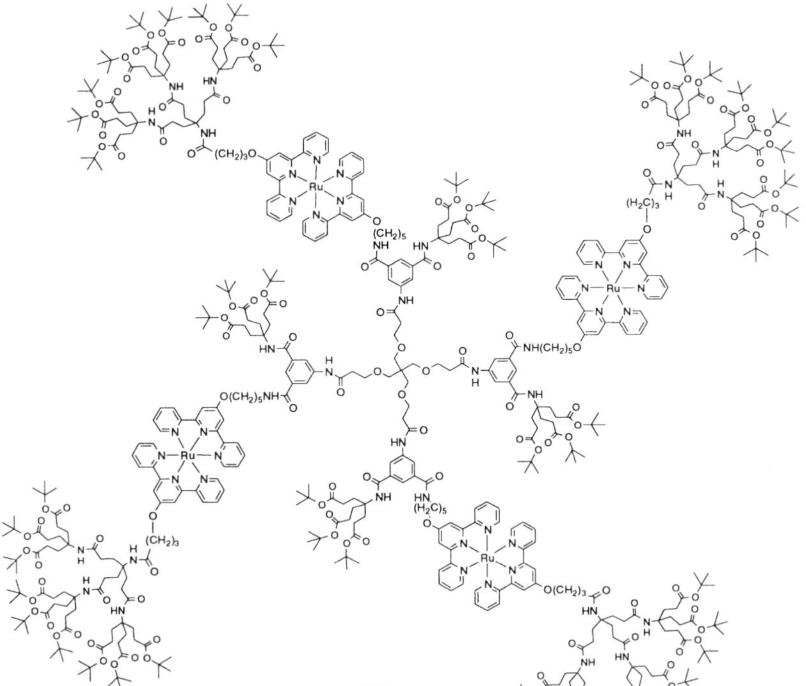

27

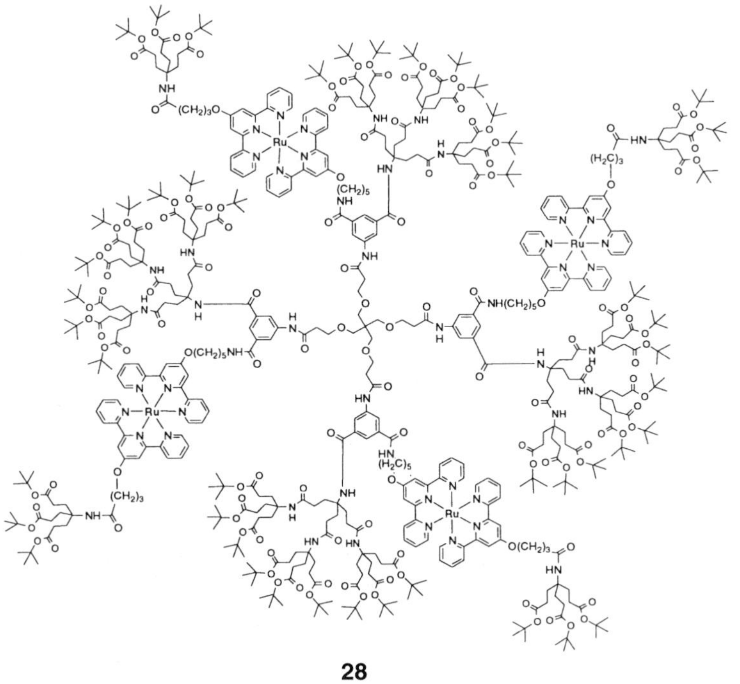

28

Fig. 9. Newkome's isomeric Ru-trpy self-assembled dendrimers **27** and **28**

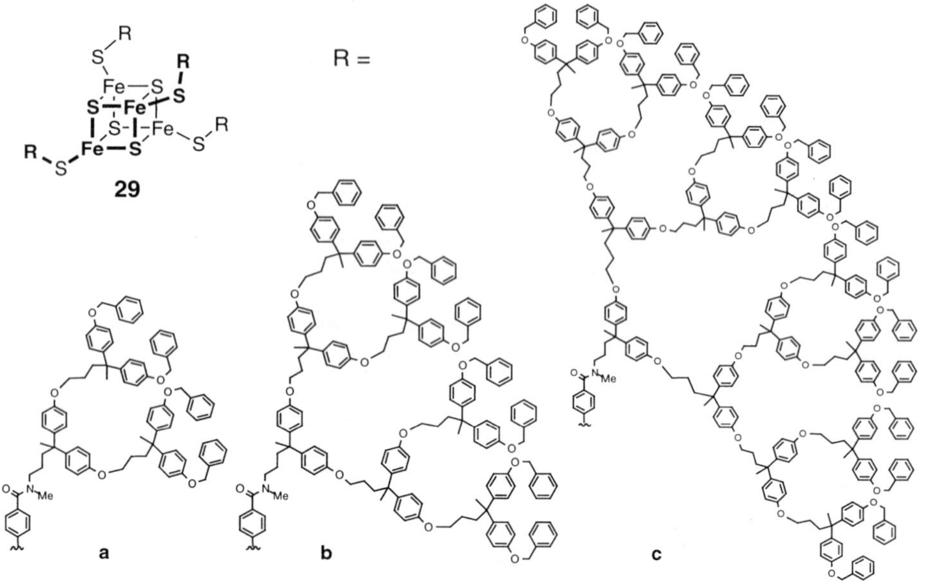

Fig. 10. Second, third, and fourth-generation dendrons (**a**–**c**) attached to iron-sulfur core **29**, synthesized and studied by Gorman and coworkers

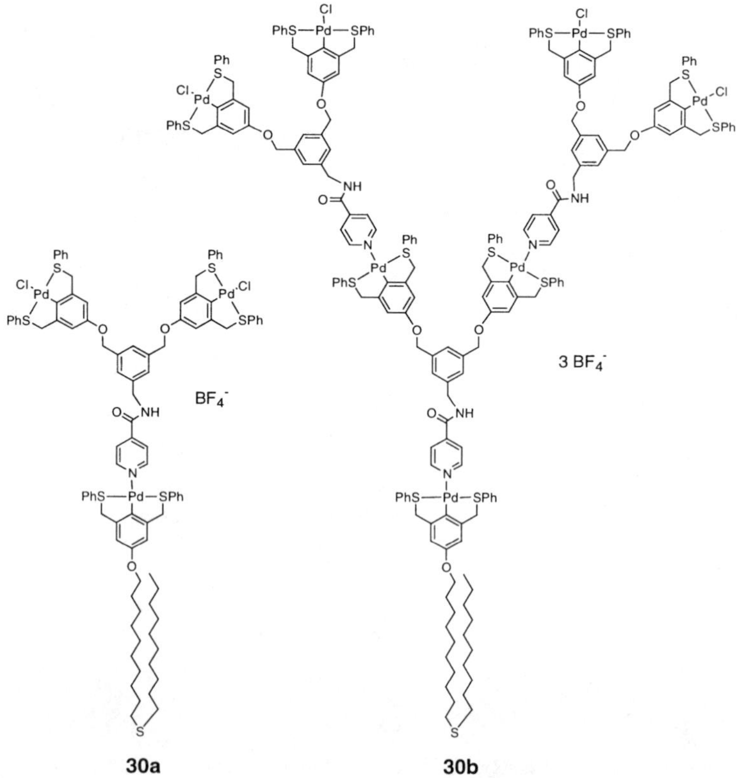

30a **30b**

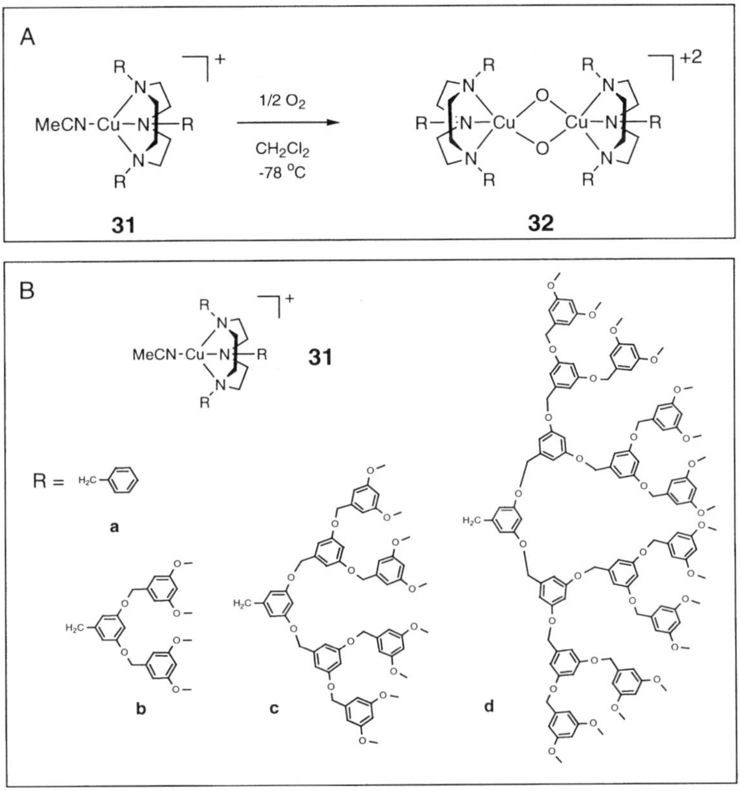

Fig. 12. **A** Tolman's synthesis of bis(μ-oxo)bridged copper complex and, **B** Aida's dendritic analog. A dendritic effect results in stabilization of the complex which is a protein mimetic

the active site of methane monooxygenase and ribonucleotide reductase. Whereas Tolman's complex had a very short half-life of 7 s at –10 °C, Aida and coworkers report a dramatic stabilizing effect by replacement of the benzyl groups in **31 a** with Fréchet type dendrons. The greatest stabilizing effect was found for **31 c** where the half-life increased to 3075 s. The $(Cu_2(\mu\text{-}O)_2)^{2+}$ core even showed significant stability at room temperature. Kinetics of decomposition indicated that the increased stability was entropic in origin, and the authors suggested that the tightly packed dendrons prevent the benzylic methylene groups attached to the triazaundecane core from orienting toward the oxygenated core. The largest dendrimer, **31 d**, failed to undergo conversion to **32 d** presumably due to steric hindrance.

Fig. 11. Two generations of palladium-containing self-assembled dendrons reported by van Veggel and Reinhoudt. The long-chain sulfide group at the core is used to anchor the dendron to a gold monolayer for imaging

4 Self-Organization of Dendrimers

4.1 Concept and Definitions

As previously indicated, self-organizing dendritic systems for the purpose of this review are those where long range ordering occurs due to forces that are less specific and directional than those in self-assembly and where the organization is guided more by nature than the investigator. As will be seen, Percec has shown remarkable control over the structure of mesophases by changes in molecular structure, thus blurring the distinction between self-assembly and self-organization. Perhaps somewhat remarkable is the fact that very large dendrimers are capable of forming liquid crystalline phases at all. The other main type of dendritic self-organization studied recently occurs at interfaces, and dendrimers have proven to be excellent building blocks for constructing mono- and multilayers.

4.2 Liquid Crystalline Phases

Over the past several years, Percec and coworkers have shown that a diverse array of dendrimers can exhibit liquid crystalline (LC) phases [60–64]. In an extensive series of structure-property investigations, the nature of the LC phases were characterized and it was shown how the molecular structure of the dendrons relates to self-assembled structures that ultimately organize into the specific LC phase (Scheme 2). For example, the smaller generation dendrons, or those less substituted, are proposed to adopt fan-shaped structures that assemble into washers and then into 2-D hexagonal columns. The larger generation and more highly functionalized dendrons either form larger fans, in which case fewer molecules are required to form the washer, or they adopt conical shapes, in which case a spherical assembly is produced leading to a 3-D cubic thermotropic liquid crystalline phase. The repeat units in the dendrimers are Fréchet-type (3,5-benzyloxybenzyl) and one main structural variant (the 3,4,5-benzyloxybenzyl motif). Strictly speaking, the dendrimers are dendrons because they have a functional group at the focal point (core). The focal groups are generally simple groups such as carboxylic acids, alcohols, and ester groups.

In a first, Percec and coworkers show [62] that a spherical dendrimer alone can form a cubic phase, thus bypassing the need for self-assembly by smaller conical or pie-shaped subunits. Building upon the earlier results with gallic acid-derived phenyl-benzyl ether dendrons the authors examined a related series containing the same internal repeat unit but 3,4-alkyloxyphenyl units on the surface. The rationale for this small change is that the fifth generation dendrimer could not be prepared due to steric hindrance at the surface. With the surface 3,4-alkyloxyphenyl units, dendrimers of the first generation (G1) to the fifth generation (G5) were prepared and studied by differential scanning

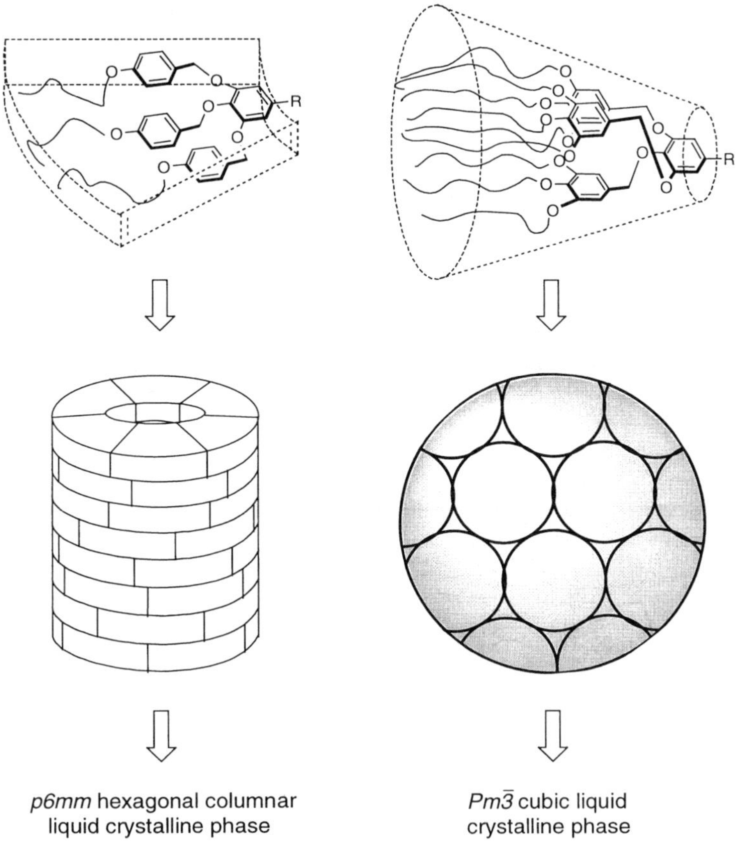

Scheme 2

calorimetry (DSC), thermal optical polarized microscopy (TOPM), X-ray diffraction (XRD), and scanning force microscopy (SFM). Each dendrimer G2-G5 formed a $\mathrm{Pm}\bar{3}n$ cubic lattice with the XRD analysis of the fifth generation dendrimer (MW 42,731) indicating a single isolated spherical structure (lattice dimensions 57 Å) repeated within the cubic lattice. The lattice dimensions are similar to the molecular dimensions measured in a SFM experiment performed on a monolayer prepared on mica.

4.3 Interfacial Organization

Müllen and coworkers have reported the synthesis of a series of relatively rigid polyphenylene dendrimers with nanoscopic sizes based on a clever divergent, Diels-Alder strategy [13]. With two members of this class, **33** and **34** (Fig. 13), whose diameters are 5.5 and 3.8 nm, respectively, they have examined their ability to form self-assembled monolayers on highly oriented pyrolytic

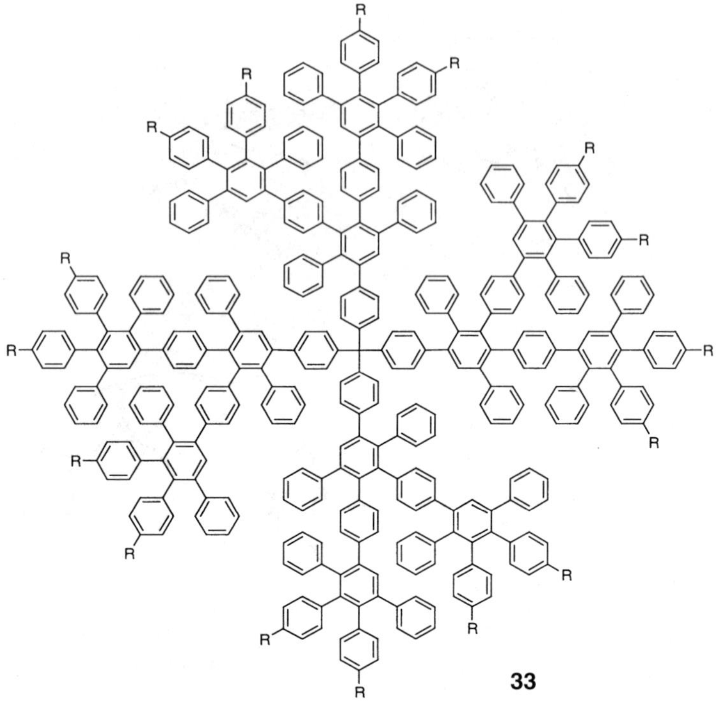

33

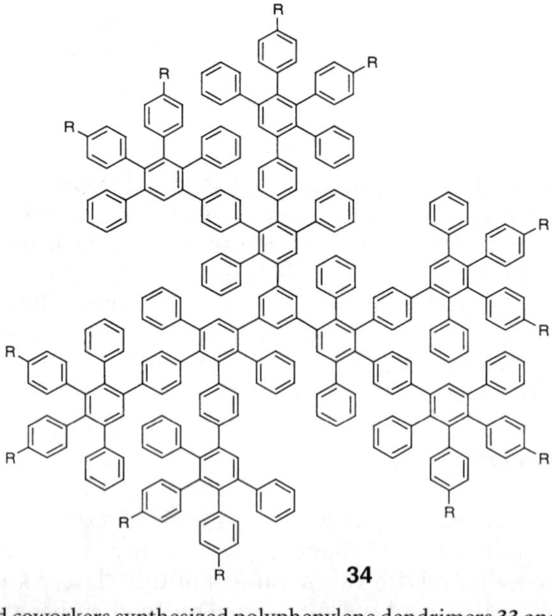

34

Fig. 13. Müllen and coworkers synthesized polyphenylene dendrimers **33** and **34** and examined their self-organization in monolayers on highly oriented pyrolytic graphite (HOPG)

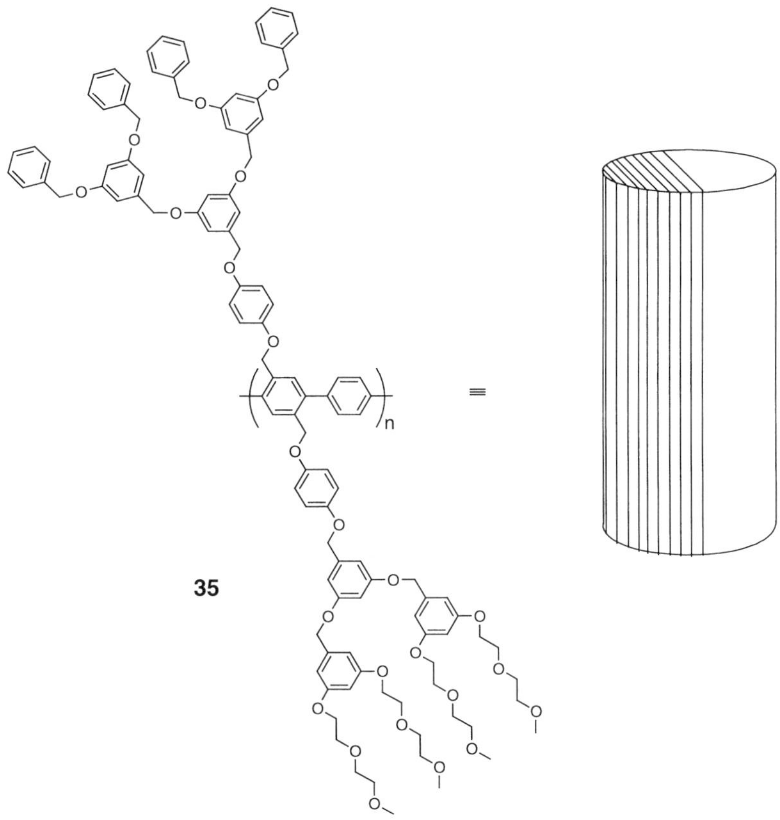

Fig. 14. Schlüter and coworkers synthesized dendritic rods (e.g., 35) and reported the self-organization of amphiphilic structures with a dendritic PEG half (hydrophilic) and a Féchet-type dendritic segment (hydrophobic)

graphite (HOPG) [65]. Dendrimer **33** was found to form regions (50–100 nm × 70–800 nm) consisting of parallel rows with a 5.9 ± 0.7-nm spacing. In most cases, only a few percent of the surface was covered, but in 2 out of 11 tries the entire surface was covered by a rod-like structure. With dendrimer **34**, 2-D crystals were obtained with each unit cell containing at least two dendrimers. The rod-like structures formed from **33** are the first such structures formed by self-assembly of non-rod-shaped dendrimers. With both **33** and **34**, the underlying graphite controlled the array orientation of the monolayer.

Schlüter and coworkers have continued their extensive investigations of polyphenylene polymers with dendritic substituents that adopt cylindrical structures (Fig. 14) [15]. By analogy to amphiphilic helical peptides, they have synthesized a new class of cylinders in which roughly half the surface contains hydrophobic Fréchet-type dendrons and the other hydrophilic oligoethylene glycol arms [66]. Some of these compounds (e.g., **35**) form stable monolayers in Langmuir troughs. In the case of **35**, it could be efficiently transferred to mica,

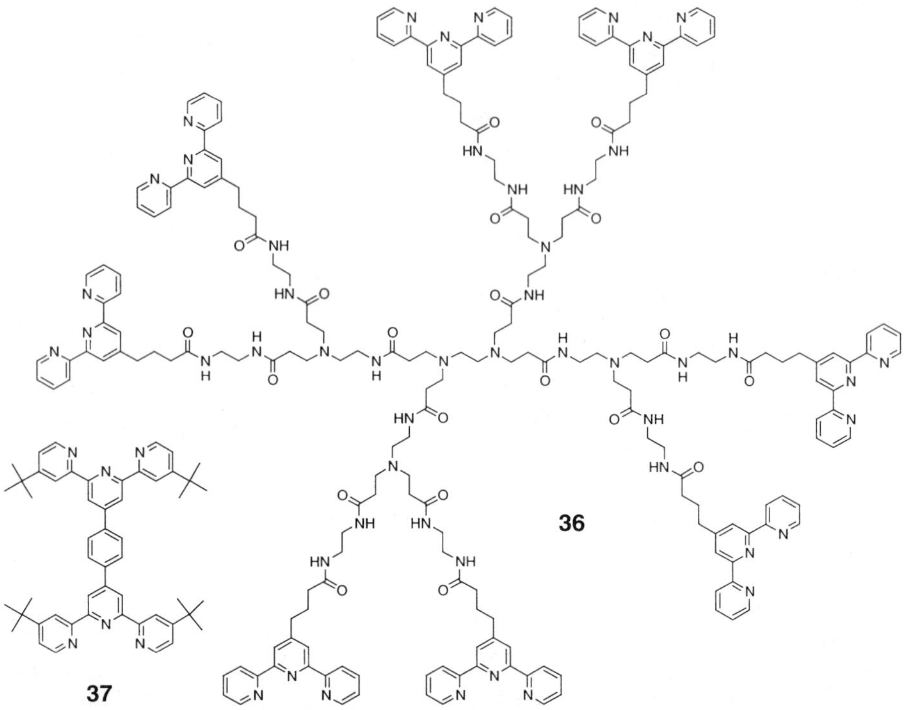

Fig. 15. Abruña and coworkers self-assembled a dendritic network on highly oriented pyrolytic graphite (HOPG) surfaces by treating **36** with Fe(II), with and without the linking unit **37**

forming a monolayer that is stable for at least a week. SFM on the transferred films from **35** indicated linear features with a 3.7-nm repeat consistent with alignment of the cylinders and orientation of the hydrophilic arms toward the polar phase, and the hydrophobic dendrons toward the air.

Abruña and coworkers reported [67] a very interesting system wherein self-assembly is mediated by metal ions at an interface leading to ordered arrays of dendrimers. The assembly of three dendrimers with peripheral terpyridine groups (den-n-tpy; n = 4, 8, 32) were examined on highly oriented pyrolytic graphite (HOPG) surfaces. The majority of the studies were performed on **36** (n = 8) with Fe(II) (Fig. 15). Films prepared on HOPG were characterized using STM, which revealed patterns consistent with a 2-D packing of 1-D arrays whose structure was proposed to be (**tpy**-dend-$(\text{tpy}')_6$-**tpy**-Fe(II)$)_n$ where **tpy** is a coordinatively bridged terpyridine linking dendrimers and tpy′ is uncoordinated. When the assembly is carried out in the presence of **37**, similar arrays are observed but with larger spacings. This result suggests that **37** is incorporated into the assembly most likely as a group linking the dendrimers. This approach gives especially stable arrays of dendrimers, and the authors note that the result is some of the highest resolution images of dendrimers obtained to date.

Dendrimers containing one or more azobenzene units as photoresponsive units have been synthesized and studied by Jiang and Aida [68], Junge and

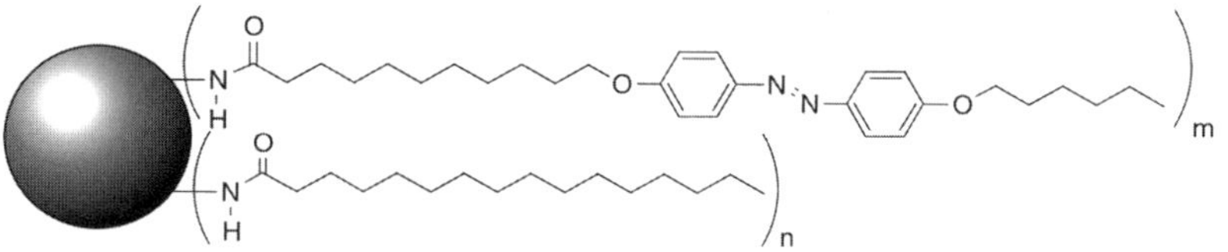

38 **a**: m,n = 32
b: m = 64, n = 0

Fig. 16. Weener and Meijer's PPI-based dendrimers **38** containing a variable number of surface azobenzene units

McGrath [69], and Vögtle et al. [70]. These compounds are each prepared via stepwise syntheses, which place the azobenzene units in precise locations within the dendrimers. An alternative approach was reported by Weener and Meijer [71] in which a fifth-generation polypropylene imine (PPI) dendrimer reacts with an equal amount of the pentafluoro-phenyl esters of palmitoic acid and 11-(4-(4-hexyloxyphenylazo)-phenyloxy)undecanoic acid. The result is a random shell-functionalized copolymer with an average of 32 azobenzene and 32 palmitoyl units per dendrimer (see **38a**, m, n = 32) (Fig. 16). A second dendrimer was analogously prepared with 64 azobenzene units (**38b**, m = 64, n = 0). Both dendrimers formed stable Langmuir monolayers at an air-water interface, and these could be transferred with good efficiency to glass. Alternating UV and IR irradiation of the two Langmuir-Blodgett films at constant pressure showed only **38a** to undergo a reversible change in area. The authors conclude that the area change is due to *cis-trans* isomerization, which is possible because the surrounding alkyl chains prevent dye aggregation, known to inhibit the isomerization.

A number of investigators have examined dendrimer monolayers on gold. Early examples feature dendrimers with a single thiol group at the focal point [72, 73]. To increase the packing efficiency, Chechik and Crooks converted [74] a fourth-generation PAMAM dendrimer into thiol-coated dendrimers by reaction with the bis(*N*-hydroxysuccinimide ester) of 3,3′-dithiodipropionic acid followed by reduction with zinc metal. Three different dendrimers were prepared, one with nearly full surface reaction, and two others with an average 10% and 20% coverage (presumably random). X-ray photoelectron spectroscopy studies on gold monolayers prepared from the latter dendrimer indicate essentially complete reaction of the thiol groups with the gold surface. This finding is consistent with the dendrimer rearranging itself for maximum surface-dendrimer reaction. Gold nanoclusters were also shown to react with the thiol dendrimers leading to a stable nanocomposite.

5
Summary and Outlook

Dendrimers have captured the interest of an extraordinarily broad range of scientists. No doubt much of the fascination arises from their unique architecture wherein internal voids and multiple end-groups are displayed on a nano-sized macromolecule. Organic chemists are equally attracted to the fact that dendrimers synthesized by the convergent approach are homogeneous (i.e., ultrapure) and in many respects have properties similar to those shown by compounds with low molecular weights. Thus, they are compounds that are relatively easy to prepare, characterize, and manipulate.

As noted at the outset, the goal of this review was a selective presentation of some of the supramolecular dendrimer chemistry that has appeared in the past year or two. The examples presented herein certainly show that the dendrimer shell can create a microenvironment that is capable of directed or nonspecific internal complexation. The dendritic shell is also capable of affecting the properties of internal organic metallic complexes, with obvious implications for catalysis. Notably, dendrimers serve as outstanding building blocks for self-assembly and self-organization. The richness of the past research in supramolecular dendrimer chemistry only reinforces the view that remarkable new discoveries and useful application will emerge in the next several years.

Acknowledgements. The authors thank the members of the Zimmerman group for their many discussions of supramolecular dendrimer chemistry, which facilitated the writing of this review. We gratefully acknowledge funding from the National Institutes of Health, which supported the writing of this review.

6 References

1. Zimmerman SC (1997) Curr Opin Colloid Interfac Sci 2:89
2. Zeng FW, Zimmerman SC (1997) Chem Rev 97:1681
3. Newkome GR, Moorefield CN, Vögtle F (1996) Dendritic macromolecules: concepts, syntheses, perspectives. VCH, Weinheim
4. Astruc D, Blais JC, Cloutet E, Djakovitch L, Rigaut S, Ruiz J, Sartor V, Valerio C (2000) Top Curr Chem 210:229
5. Smith DK, Diederich F (2000) Top Curr Chem 210:183
6. Emrick T, Fréchet JMJ (1999) Curr Opin Colloid Interfac Sci 4:15
7. Frey H, Schlenk C (2000) Top Curr Chem 210:69
8. Hawker CJ (1999) Curr Opin Colloid Interfac Sci 4:117
9. Inoue K (2000) Prog Polym Sci 25:453
10. Majoral JP, Caminade AM (1999) Chem Rev 99:845
11. Baars MWPL, Meijer EW (2000) Top Curr Chem 210:131
12. Moore JS (1999) Curr Opin Colloid Interfac Sci 4:108
13. Berresheim AJ, Müller M, Müllen K (1999) Chem Rev 99:1747
14. Newkome GR, He EF, Moorefield CN (1999) Chem Rev 99:1689
15. Schlüter AD, Rabe JP (2000) Angew Chem, Int Ed Engl 39:865
16. Matthews OA, Shipway AN, Stoddart JF (1998) Prog Polym Sci 23:1
17. Tomalia DA, Wang ZG, Tirrell M (1999) Curr Opin Colloid Interfac Sci 4:3
18. Vögtle F, Gestermann S, Hesse R, Schwierz H, Windisch B (2000) Prog Polym Sci 25:987
19. Whitesides GM, Mathias JP, Seto CT (1991) Science (Washington, DC) 254:1312
20. Topp A, Bauer BJ, Klimash JW, Spindler R, Tomalia DA, Amis EJ (1999) Macromolecules 32:7226
21. Kao HM, Stefanescu AD, Wooley KL, Schaefer J (2000) Macromolecules 33:6214
22. Moore JS (1997) Acc Chem Res 30:402
23. Recker J, Tomcik DJ, Parquette JR (2000) J Am Chem Soc 122:10,298
24. Huang BH, Parquette JR (2000) Org Lett 2:239
25. Maciejewski M (1982) J Macromol Sci Chem A17:689

26. Kim YH, Webster OW (1990) J Am Chem Soc 112:4592
27. Newkome GRM, Baker GR, Saunders MJ, Grossman SH (1991) Angew Chem, Int Ed Engl 30:1178
28. Hawker CJ, Wooley KL, Fréchet JMJ (1993) J Chem Soc, Perkin Trans 1 21:300
29. Baars M, Kleppinger R, Koch MHJ, Yeu SL, Meijer EW (2000) Angew Chem, Int Ed Engl 39:1285
30. Fu K, Kitaygorodskiy A, Sun Y-P (2000) Chem Mater 12:2073
31. Chechik V, Zhao MQ, Crooks RM (1999) J Am Chem Soc 121:4910
32. Wendland MS, Zimmerman SC (1999) J Am Chem Soc 121:1389
33. Balzani V, Ceroni P, Gestermann S, Gorka M, Kauffmann C, Vögtle F (2000) J Chem Soc, Dalton Trans 3765
34. Newkome GR, Woosley BD, He E, Moorefield CN, Güther R, Baker GR, Escamilla GH, Merrill J, Luftmann H (1996) Chem Commun 2737
35. Zimmerman SC, Wang Y, Bharathi P, Moore JS (1998) J Am Chem Soc 120:2172
36. Smith DK, Diederich F (1998) Chem Commun 2501
37. Strumia MC, Halabi A, Pucci PA, Newkome GR, Moorefield CN, Epperson JD (2000) J Polym Sci, Part A Polym Chem 38:2779
38. a. Cardona CM, Mendoza S, Kaifer AE (2000) Chem Soc Rev 29:37; b. Cardona CM, McCarley TD, Kaifer AE (2000) J Org Chem 65:1857
39. Eckert JF, Byrne D, Nicoud JF, Oswald L, Nierengarten JF, Numata M, Ikeda A, Shinkai S, Armaroli A (2000) New J Chem 24:749
40. Jansen JFGA, Debrabander van den Berg EMM, Meijer EW (1994) Science (Washington, DC) 266:1226
41. Reuter C, Pawlitzki G, Wörsdörfer U, Plevoets M, Mohry A, Kubota T, Okamoto Y, Vögtle F (2000) Eur J Org Chem 3059
42. Michels JJ, Baars M, Meijer EW, Huskens J, Reinhoudt DN (2000) J Chem Soc, Perkin Trans 2:1914
43. Gonzalez B, Casado CM, Alonso B, Cuadrado I, Moran M, Wang Y, Kaifer AE (1998) Chem Commun 2569
44. Zeng HD, Newkome GR, Hill CL (2000) Angew Chem, Int Ed Engl 39:1772
45. Zimmerman SC, Zeng FW, Reichert DEC, Kolotuchin SV (1996) Science (Washington, DC) 271:1095
46. Wang Y, Zeng FW, Zimmerman SC (1997) Tetrahedron Lett 38:5459
47. Kolotuchin SV, Zimmerman SC (1998) J Am Chem Soc 120:9092
48. Issberner J, Vögtle F, Decola L, Balzani V (1997) Chem Eur J 3:706
49. Huck WTS, Vanveggel FCJM, Reinhoudt DN (1996) Angew Chem, Int Ed Engl 35:1213
50. Balzani V, Campagna S, Denti G, Juris A, Serroni S, Venturi M (1994) Coord Chem Rev 132:1
51. Newkome GR, Cardullo F, Constable EC, Moorefield CN, Thompson AMWC (1993) J Chem Soc, Chem Commun 925
52. Suarez M, Lehn JM, Zimmerman SC, Skoulios A, Heinrich B (1998) J Am Chem Soc 120:9526
53. Gibson HW, Hamilton L, Yamaguchi N (2000) Polym Adv Technol 11:791
54. Amabilino DB, Stoddart JF (1995) Chem Rev 95:2725
55. Newkome GR, He E, Godinez LA, Baker GR (2000) J Am Chem Soc 122:9993
56. Gorman CB, Smith JC (2000) J Am Chem Soc 122:9342
57. Huisman BH, Schonherr H, Huck WTS, Friggeri A, van Manen HJ, Menozzi E, Vancso GJ, van Veggel F, Reinhoudt DN (1999) Angew Chem, Int Ed Engl 38:2248
58. Enomoto M, Aida T (1999) J Am Chem Soc 121:874
59. Halfen JA, Mahapatra S, Wilkinson EC, Kaderli S, Young VG, Que L, Zuberbuhler AD, Tolman WB (1996) Science (Washington, DC) 271:1397
60. Ungar G, Percec V, Holerca MN, Johansson G, Heck JA (2000) Chem Eur J 6:1258
61. Percec V, Cho WD, Ungar G, Yeardley DJP (2000) Angew Chem, Int Ed Engl 39:1598
62. Percec V, Cho WD, Moller M, Prokhorova SA, Ungar G, Yeardley DJP (2000) J Am Chem Soc 122:4249
63. Yeardley DJP, Ungar G, Percec V, Holerca MN, Johansson G (2000) J Am Chem Soc 122:1684

64. Percec V, Cho WD, Ungar G (2000) J Am Chem Soc 122:10,273
65. Loi S, Wiesler UM, Butt HJ, Müllen K (2000) Chem Commun 1169
66. Bo ZS, Rabe JP, Schlüter AD (1999) Angew Chem, Int Ed Engl 38:2370
67. Diaz DJ, Storrier GD, Bernhard S, Takada K, Abruña HD (1999) Langmuir 15:7351
68. Jiang DL, Aida T (1997) Nature (London) 388:454
69. Junge DM, McGrath DV (1997) Chem Commun 857
70. Archut A, Azzellini GC, Balzani V, De Cola L, Vögtle F (1998) J Am Chem Soc 120:12,187
71. Weener JW, Meijer EW (2000) Adv Mater 12:741
72. Gorman CB, Miller RL, Chen KY, Bishop AR, Haasch RT, Nuzzo RG (1998) Langmuir 14:3312
73. Bo ZS, Zhang L, Zhao B, Zhang X, Shen JC, Hoppener S, Chi LF, Fuchs H (1998) Chem Lett:1197
74. Chechik V, Crooks RM (1999) Langmuir 15:6364

Non-Covalent Synthesis of Metallodendrimers

Henk-Jan van Manen, Frank C. J. M. van Veggel, David N. Reinhoudt

Laboratory of Supramolecular Chemistry and Technology and MESA+ Research Institute, University of Twente, P. O. Box 217, 7500 AE, Enschede, The Netherlands
E-mail: smct@ct.utwente.nl

The field of dendrimers has undergone an explosive growth since the first dendritic structures were reported two decades ago. These three-dimensional, highly branched macromolecules have attracted interest from such diverse areas as polymeric, organic, inorganic, biomedical, theoretical, and physical chemistry. Future applications were already hypothesized from the early days of dendrimer research. From an application point of view, the incorporation of a range of metals into the dendritic framework is of particular interest. The resulting metallodendrimers might be applied in fields such as catalysis, sensors, medical diagnosis, light-harvesting devices, and nanoparticles. In this chapter, metallodendrimers are discussed in which the metals are essential for maintaining the dendritic structure. This means that all the dendrimers described have been assembled non-covalently using coordination chemistry. Although this restriction narrows the metallodendrimer field significantly, there is still an enormous variety in the architecture of reported non-covalent metallodendrimers. Where possible, emphasis is placed on the characterization methods and specific behavior of the dendrimers, because characterization is of utmost importance in establishing their often complicated three-dimensional structure. Finally, we have emphasized properties that may lead to future applications.

Keywords: Dendrimers, Metals, Non-covalent synthesis, Coordination chemistry

1 Introduction

After the early theoretical and experimental work by Flory in the 1940s on three-dimensional branched macromolecules [1], Vögtle et al. [2] described in 1978 the first example of an iterative synthetic methodology towards well-defined branched architectures, now commonly referred to as *dendrimers*. In the mid-1980s, research on these fascinating fractal-like polymers was initiated by the groups of Tomalia [3] and Newkome [4].

A common definition of dendrimers is that they are globular, monodisperse, highly branched macromolecules of well-defined size and shape that are constructed via an iterative sequence of reaction steps. Two main approaches are usually distinguished in their synthesis, i.e., the *divergent* (from the core to the outside) and *convergent* (from the periphery inwards) dendritic growth [5].

The combination of dendrimer chemistry with the specific properties of (transition) metals is very interesting from the point of view of possible applications. Materials with new catalytic, optical, magnetic, electro- and photochemical, and biomedical properties might be created from the combination of dendritic structures and metals. This combines the fields of organic, inorganic, supramolecular, and polymer chemistry into a highly interdisciplinary field of research.

Virtually all positions in the dendritic framework are amenable to metal incorporation. This review will focus on metallodendrimers in which the metals are essential for creating the dendritic structure, i.e., the metals serve as the structural units "gluing" the organic building blocks together. This narrows down the metallodendrimer field to three different classes:

- Dendrimers containing metals as cores
- Dendrimers containing metals as branching units
- Dendrimers containing metals as building block connectors

A consequence of this restriction is that metallodendrimers in which the metals have been positioned at the periphery (e.g., ferrocenes [6], catalytic sites [7], etc.) or have been incorporated as structural auxiliaries (site-specific [8] or random [9] inclusion) will not be covered. The reader is referred to some excellent reviews and monographs about (metallo)dendrimers for in-depth information about these and other topics [5, 6, 10]. There are also review articles devoted to other specific classes of dendrimers, e.g., heteroatom-containing (Si, P, B, Ge, or Bi) dendrimers [11], chiral dendrimers [12], carbohydrate-containing dendrimers [13], dendronized polymers [14], and dendrimers in diagnostics [15].

In this review, most dendritic structures are depicted with one dendron arm fully drawn, whereas the other arms, identical to the one drawn, are represented by cones.

2 Metals as Cores

2.1 Introduction

The incorporation of metals at the core of dendrimers may serve a number of purposes. In general, research in this area is aimed at tailoring the properties of the core metal as a function of the type and generation (size) of the organic dendrons folded around it. The shielding of the dendritic core from the environment is a well-known phenomenon and has been under investigation since the first report of core isolation by Fréchet et al. in 1993 [16].

2.2 Encapsulated Metal Clusters

The encapsulation of electroactive moieties is important in understanding electron transfer found in biological systems [17] (e.g., cytochromes) and also in the construction of molecular electronic devices [18]. By attaching dendrons to a porphyrin core, enzyme mimics have been prepared in which electron transfer to or from the porphyrin is affected by the bulkiness of the dendrons and the microenvironment they create around the porphyrin [19].

Iron-sulfur clusters serve in nature as electron transfer and storage sites, and as structure-enforcing units in enzymes [17]. Gorman and coworkers have placed dendritic wedges (G1–G4) around electroactive iron-sulfur clusters of the form $[Fe_4S_4(SR)_4]^{2-}$, resulting in new hybrid inorganic/organic dendritic architectures [20]. The dendrimers were prepared by a ligand exchange reaction of aromatic thiol-substituted dendrons on $(n\text{-}Bu_4N)_2[Fe_4S_4(S\text{-}tert\text{-}Bu)_4]$. The second generation dendrimer is shown in Fig. 1.

In the ^1H NMR spectrum, a substantial broadening and change in chemical shifts of the thiolate aromatic ring protons close to the paramagnetic iron-sulfur cluster was observed. The small longitudinal relaxation time constants (T_1) of these protons confirmed their fast relaxation, which is expected for protons in close proximity to a paramagnetic group. In a subsequent study [21], a dramatic shortening of T_1 values for the protons in the dendrimers containing a paramagnetic iron-sulfur core was found in comparison with similar dendrimers that contained a diamagnetic tetraphenylmethane core. This observation led the

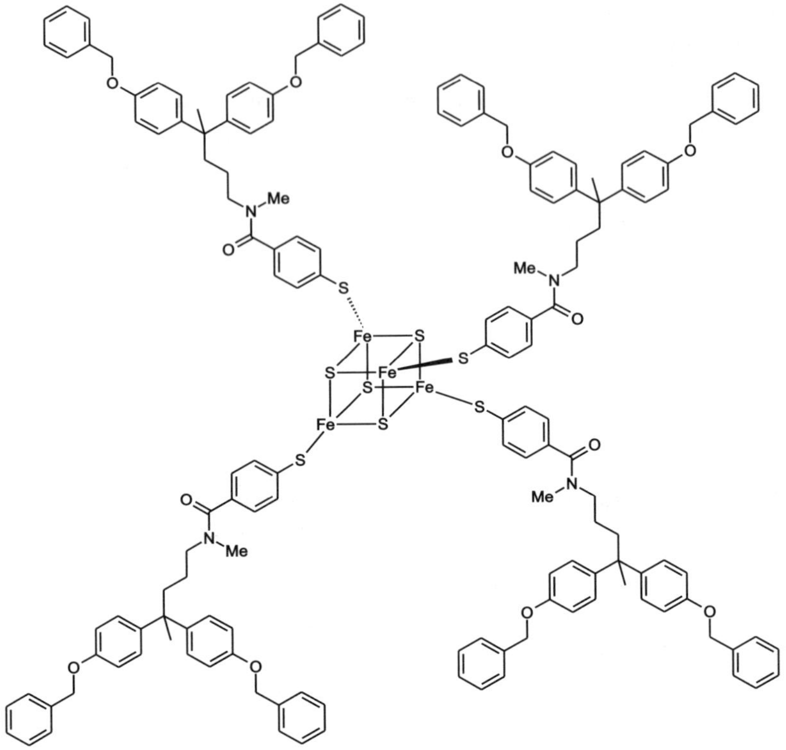

Fig. 1. Second generation dendrimer containing an iron-sulfur cluster as a core

authors to conclude that protons in each generation of the dendrimers must approach the core of the molecule closely in order to experience the attenuation of T_1 values. This conclusion is consistent with radial density distributions of different dendritic generations, as calculated from molecular dynamics simulations.

The insulating effects of increasing steric bulk of the dendritic ligands on the electrochemical properties of the iron-sulfur core were demonstrated by cyclic voltammetry. Going from G0 to G4, increasingly more negative reduction potentials $E_{1/2}$ were observed. Moreover, G0–G3 displayed increasingly larger voltage differences between the current maxima of the reduction and return oxidation waves (ΔE), which indicates an increasing kinetic difficulty of reduction/oxidation. In the dendritic zinc porphyrins reported by Diederich et al. [19a] the increasingly electron-rich microenvironment created by the dendritic branches around the core hinders reduction and facilitates oxidation of the porphyrin ring, which is reflected in more negative reduction potentials and less positive oxidation potentials, respectively, with increasing generation. Apparently, electronic effects rather than steric effects primarily determine the redox potentials of the core. Clearly, the *type* of dendrimer positioned around the core has a profound influence on its electrochemical properties. This was recently confirmed by Gorman et al. [22], who compared iron-sulfur dendrimers containing rigid phenylacetylene-type dendrons with the flexible iron-sulfur dendrimers described above. Diffusion coefficients and corresponding hydrodynamic radii of the dendrimers in dilute solution were determined by pulsed field gradient spin-echo NMR and chronoamperometry. For both the flexible and rigid series, both techniques showed that increasing the dendrimer generation results in a decrease in diffusion coefficient and thus an increase in hydrodynamic radius. Whereas the flexible series of dendrimers displayed a large dependence of the hydrodynamic radius on the solvent, the rigid dendrimers showed little change in radius as a function of solvent. This observation is consistent with the rigid and "shape persistent" [23] nature of these dendrimers. Heterogeneous electron-transfer rate constants indicate that the rigid dendrimers attenuated more effectively the electron transfer rate than the flexible ones. These results were rationalized using computational conformational searching which indicated an off-center "mobile" iron-sulfur core in the flexible series and a central and relatively immobile core in the rigid series.

Recently, other metal clusters have also served as inorganic dendritic cores. Dendrimers $Mo_6(\mu_3\text{-}Cl)_8(OR)_6$ (R = dendrons, G0–G2) were constructed by Gorman et al. from triflate- or methoxy-capped pseudo-octahedral clusters ($[Mo_6Cl_8(X)_6]^{2-}$, X = OTf or OMe) by ligand exchange reactions with dendrons containing focal phenoxide groups [24]. Similarly, exchange of labile acetonitrile ligands in the hexanuclear $[Re_6Se_8(MeCN)_6](SbF_6)_2$ cluster for pyridine-functionalized dendrons (G1) produced the corresponding dendrimers in high yields after chromatography [25]. Both these hybrid organic/inorganic dendrimers were characterized by electrospray mass spectrometry, which clearly showed the molecular ion peaks.

2.3
Dendrimers Containing Ru(II)-Bipyridine Units as Cores

Since the complexes of the $[Ru(bpy)_3]^{2+}$ family (bpy = 2,2′-bipyridine) show a unique combination of photophysical and redox properties [26], incorporation of these moieties as cores into dendritic frameworks offers the possibility of modifying their properties as a function of dendritic generation. This was investigated by Vögtle et al. [27], who synthesized dendritic ligands by a divergent strategy, starting from 4,4′-functionalized 2,2′-bipyridines. Following procedures reported by Newkome et al. [28], 4,4′-bis(bromomethyl)-2,2′-bipyridine **1** was alkylated with triethyl methanetricarboxylate to obtain the dendritic hexaester **2** shown in Fig. 2. Three of these ligands were subsequently coordinated to Ru(II) to give complex **3**, which was finally converted into the hydrophilic hydroxyl-containing Ru(II)-dendrimer **4** by amidation with tris(hydroxymethyl) aminomethane.

Unfortunately, dendritic growth beyond generation one was not possible by this synthetic method. Therefore, a slightly different methodology was pursued based on dendrimer chemistry developed by Diederich et al. [29]. The second generation Ru(II) dendrimer **5** (containing 54 peripheral esters) prepared in this manner is depicted in Fig. 3.

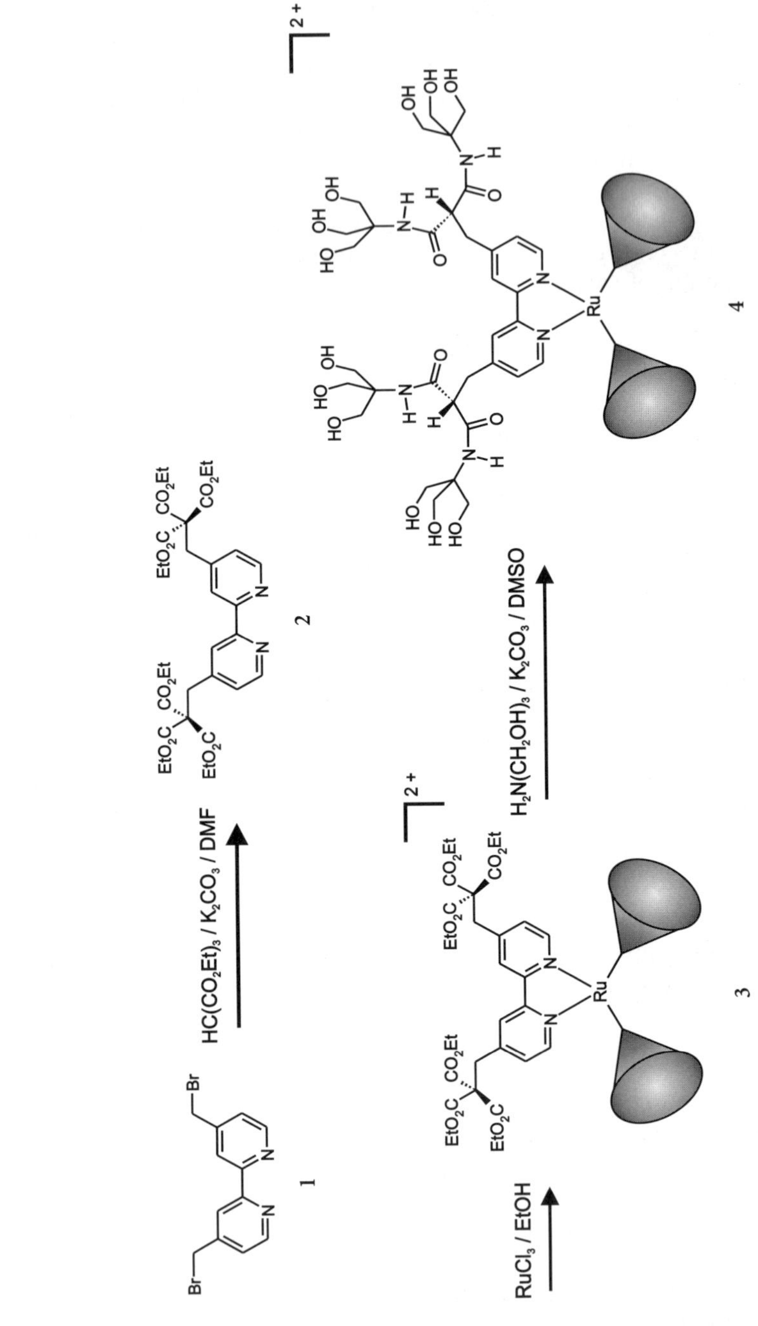

Fig. 2. Synthesis of a first generation, hydrophilic metallodendrimer based on the $[Ru(bpy)_3]^{2+}$-motif

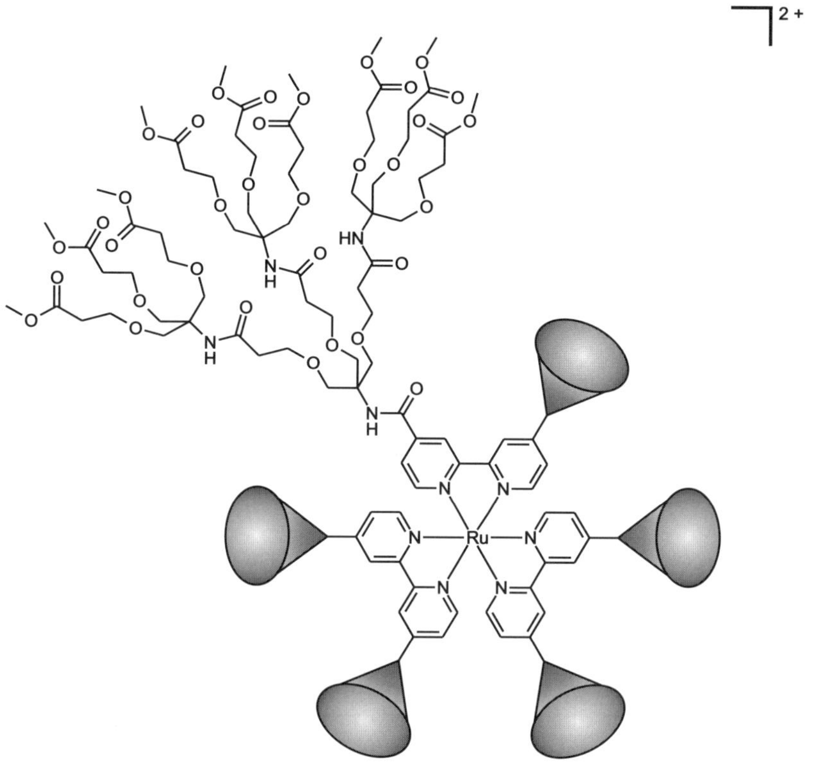

5

Fig. 3. Second generation Ru(II)-containing metallodendrimer

The absorption and emission spectra of complexes **4** and **5** are similar to those of the unsubstituted parent Ru(II)-bipyridine complexes. However, the large dendritic complexes exhibited a more intense emission and a longer excited-state lifetime than $[Ru(bpy)_3]^{2+}$ in aerated solutions. This is due to the shielding effect of the dendritic branches on the Ru(II)-bipyridine core, thereby limiting the quenching effect of dioxygen (for **5** the rate constant of dioxygen quenching is twelve times smaller than for $[Ru(bpy)_3]^{2+}$). The authors emphasized the importance of long-lived luminescent excited states in immunoassay applications, since the label signal can be read after the decay of the sample background fluorescence.

Combining the $[Ru(bpy)_3]^{2+}$ core with peripheral naphthyl units, Vögtle et al. [30] also reported that very efficient energy-transfer takes place from the potentially fluorescent excited states of the aromatic moieties of the dendritic wedges (naphthyl-functionalized Fréchet-type aryl ether dendrons were used in this case) to the Ru(II) dendritic core. This antenna effect, potentially useful in harvesting sunlight [31], was again accompanied by reduced dioxygen quenching of the core luminescence, although the effect was smaller than reported for complex **5** (Fig. 3).

Recently, Vögtle, De Cola, Balzani, and their coworkers extended the work on Ru(II)-dendrimers by decorating the periphery of Ru(II)-bipyridine aryl ether dendrimers with either benzyl or 4′-*tert*-butylphenyloxy groups [32]. All dendrimers showed the characteristic luminescence of the $[Ru(bpy)_3]^{2+}$-type core, and a similar protection of the luminescent excited state of the core from dioxygen quenching was observed as discussed above. For the compounds containing the 4′-*tert*-butylphenyloxy peripheral units, the electrochemical behavior and the excited-state electron-transfer quenching by cationic (methyl viologen dication, MV^{2+}), neutral (tetrathiafulvalene, TTF), and anionic (anthraquinone-2,6-disulfonate anion) quenchers, which quench [33] the ^{3}MLCT excited state of $[Ru(bpy)_3]^{2+}$, were investigated. The core of the largest dendrimer (24 peripheral 4′-*tert*-butylphenyloxy groups) showed an electrochemical behavior typical of encapsulated electroactive units. The quenching rate constants, obtained by Stern-Volmer kinetic analysis, decreased with increasing number and size (= generation) of the dendritic branches for each quencher. The magnitude of this effect depends on the quencher and is largest for MV^{2+} and smallest for TTF.

2.4 Dendrimers Containing Metal-Terpyridine Units as Cores

Whereas dendritic *growth* based on metal terpyridines is well-developed (see Sect. 4.3), the *encapsulation* of terpyridine complexes has received less attention, particularly in comparison with metal *bipyridine* complexes (see above). This might be due to the absence of room temperature luminescence of $[Ru(terpy)_2]^{2+}$, which renders terpyridine-based assemblies less suitable for practical applications involving light-induced processes. However, there are a few exceptions. In a first attempt to mimic redox proteins, Chow et al. synthesized aryl ether dendrons (G1 – G3) containing a focal terpyridine and they studied the redox properties of the corresponding iron (II) complexes [34]. The encapsulation was indicated by a decreasing reversibility of the metal redox centers with increasing dendritic generation. These cyclic voltammetry results were rationalized by the steric hindrance caused by the bulky dendritic shell, hindering the metal complexes from approaching the electrode surface. Similar effects due to shielding of the redox center are well known for cytochrome *c* [35] and other electron-transfer proteins [36]. The dendritic iron(II) complexes were further characterized by X-ray photoelectron spectroscopy (XPS), which showed that the solid-state coordinating environments of the iron(II) of different generations were very similar.

A methodology for preparing bis-dendrimers, developed by Newkome and coworkers [37], also utilized the metal complexation ability of terpyridines. Starting from a carboxylic acid-functionalized terpyridine, divergent dendritic growth (G1 – G4) followed by complexation of the terpyridine with $RuCl_3$ led to the first half of the bis-dendrimer (Fig. 4a). Subsequent connection of the second terpyridine dendrimer (using 4-ethylmorpholine as the reducing agent) provided the Ru(II) bis-dendrimers (Fig. 4b) in yields around 60%.

Cyclic voltammetry showed that for the low generation bis-dendrimers both the cationic and anionic scans display electrochemically and chemically re-

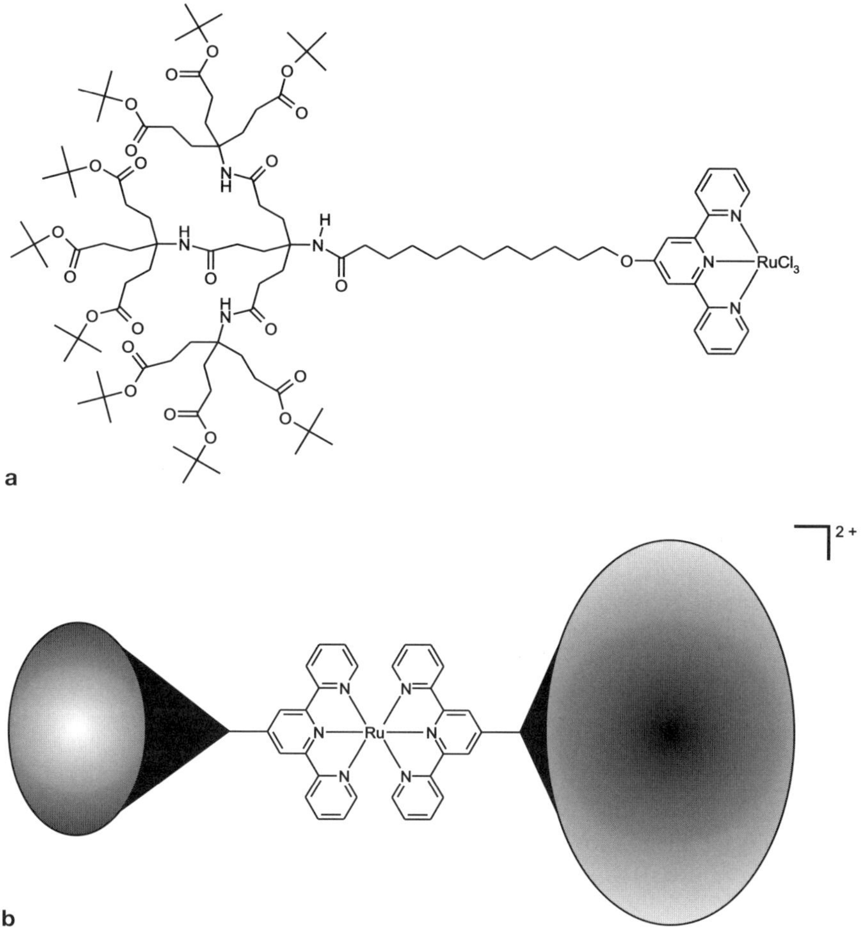

Fig. 4. **a** Half bis-dendrimers containing Ru(II)-terpyridine units. **b** Complete bis-dendrimers containing Ru(II)-terpyridine units

versible processes. However, for the sterically congested bis-dendrimers irreversible behavior (both electrochemically and chemically) was found, similar to the results of Chow et al. (see above). Moreover, hardly any potential shifts were found. Interestingly, the sequential growth of the bis-dendrimer allows the two halves to differ in size, structure, and properties. In the reported examples the two halves only differ in generation (Fig. 4 b).

2.5
Coordination of Dendrons to Metalloporphyrins

There are a few porphyrin-containing dendrimers in which coordination of dendritic ligands to the porphyrin metal is exploited in the growth of larger dendritic structures. Aida et al. used zinc porphyrins decorated with four aryl ether

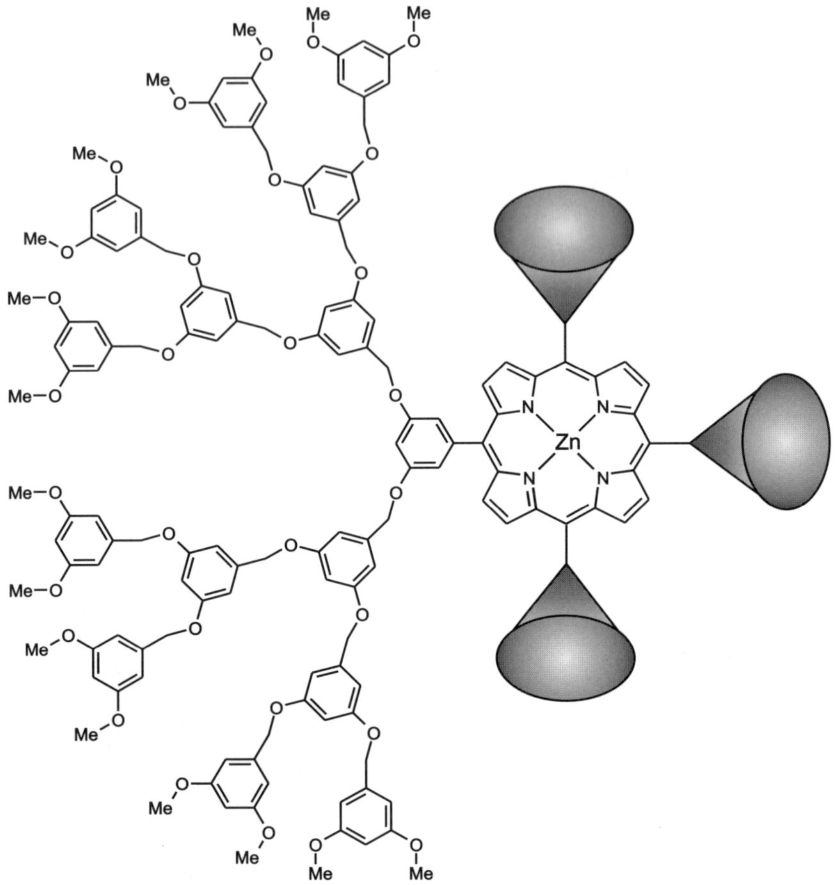

Fig. 5. Dendritic zinc porphyrin used for the interaction with imidazole-functionalized dendrons

dendrons (Fig. 5) to study the interpenetrating interaction of imidazole-functionalized dendrons with the zinc center [38].

The singlet excited state of the zinc porphyrin is not affected by the encapsulation within the dendritic framework. This was evidenced by ^{1}H NMR pulse relaxation time (T_1) measurements, which indicated that the conformational flexibility of the porphyrin skeleton is retained upon increasing the dendrimer generations. Spectrophotometric titration of dendritic imidazoles (G1, G2, and G4) to the porphyrin dendrimers (a one-to-one complexation in all cases) showed a decrease in binding constant upon increasing either the dendritic imidazoles or porphyrins, especially in going from the third to the fourth generation dendritic porphyrin. It is noteworthy that the G4-imidazole binds to the G5-porphyrin at all ($K = 2.4 \times 10^2$ M^{-1} in CH_2Cl_2 at 20 °C), suggesting significant flexibility of the aryl ether dendritic parts.

Coordination of pyridyl-functionalized porphyrins to ruthenium (II) porphyrins was exploited recently by Sanders et al. in the construction of dendritic

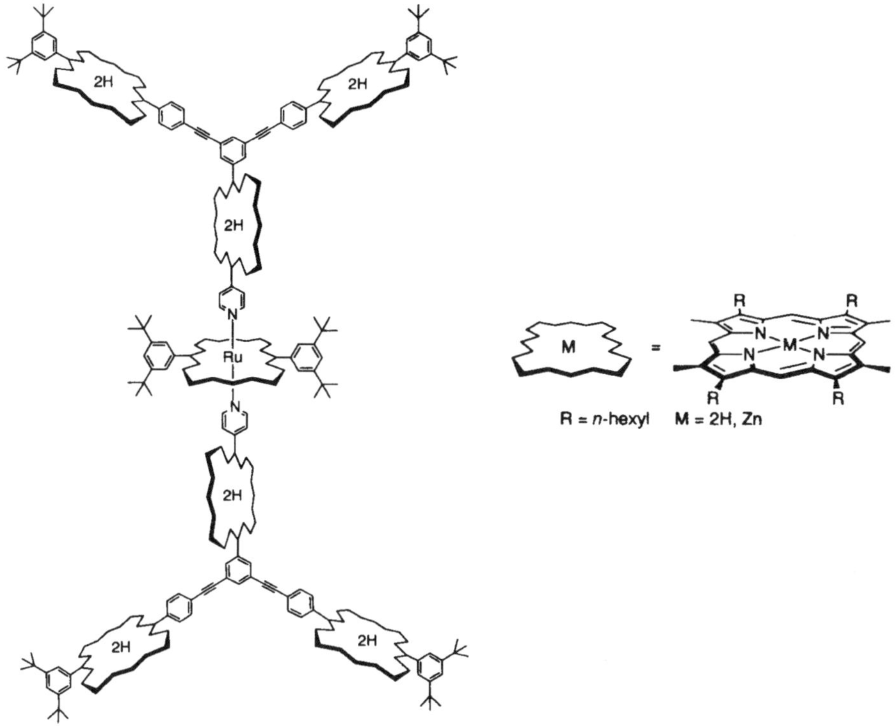

Fig. 6. Dendritic porphyrin heptamer synthesized by Sanders and coworkers

multiporphyrin arrays [39]. These are of considerable interest [40, 41] due to their excellent photophysical and redox properties which make them suitable candidates for light-harvesting devices, molecular wires, photosensitizers, and (semiconducting) liquid crystalline materials. Sanders and coworkers synthesized a porphyrin trimer dendritic wedge containing a pyridyl group at the focal point. Two of these dendrons were subsequently attached to a Ru(II) porphyrin by sequential axial displacement of the solvent and CO ligands coordinated to Ru(II) in the starting porphyrin, to give the heptamer shown in Fig. 6. The multiporphyrin arrays were characterized by NMR and UV spectroscopy.

2.6 Other "Metals as Cores" Dendrimers

Recently, Enomoto and Aida have attached three aryl ether dendrons (G2–G4) to 1,4,7-triazacyclononane and studied the oxygen-induced dimerization of the corresponding Cu(I) complexes, as a non-heme metalloprotein mimic [42]. Both the rate of formation of the bis(μ-oxo)dicopper(III) moiety and its stability towards oxidative self-decomposition are affected by the size of the attached dendrons (a "dendritic effect").

The binding of organic dendritic wedges to core (transition) metals has been undertaken in order to change the properties of the metal or to impart new functions on the resulting dendrimer-metal combination. An example of the latter has been reported by Catalano and coworkers [43, 44]. Dendritic Fréchet wedges (G1–G2) were functionalized with focal phosphines and these dendrons were coordinated to Ir(I) to produce the dendritic analogs of Vaska's compound. IR and ^{31}P NMR spectroscopy showed that C_{60} reacted reversibly with the Ir(I) dendrimers via an oxidative addition. The reversible binding was probed by examining the ^{31}P NMR line widths as a function of temperature. This gave the thermodynamic data and equilibrium constants ($K = 388\ M^{-1}$ at 257 K and $5\ M^{-1}$ at 293 K in chlorobenzene for the G2 Ir(I)-C_{60} complex). The dendritic arms do not play a major role in fullerene binding.

Fréchet and Kawa coordinated three aryl ether dendritic wedges functionalized at the focal point with carboxylates around trivalent lanthanide ions (Er, Eu, and Tb, Fig. 7) in order to shield the lanthanides from one another for possible

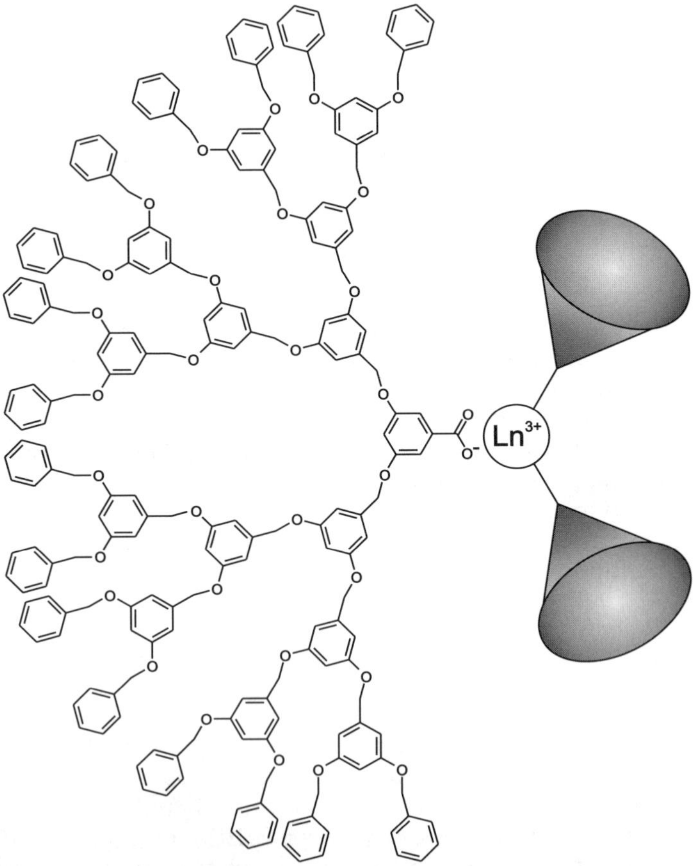

Fig. 7. Fourth generation dendrimer containing lanthanide ions in the core

use in fiber optics amplification [45]. Site isolation is important in photoluminescence as it decreases the rate of self-quenching between metal atoms.

Higher generation dendrimers indeed exhibited enhanced luminescence. The authors observed a non-conjugated "antenna effect" with energy transfer from the dendritic ligands to the lanthanide core.

2.7 Conclusions

From this section it can be concluded that positioning metals at the core of dendrimers can significantly alter their properties. More specifically, metals can be shielded from each other and from the environment by incorporation into a dendritic framework. These features might be exploited in applications such as protein mimics, catalysis, immunoassays, and energy-harvesting devices.

3 Metals as Branching Units

3.1 Introduction

A second alternative for metallodendrimer synthesis involves the coordination of more than one ligand to a (transition) metal that serves to introduce the necessary branching into the dendritic framework. Here, branching occurs *at the metal* and not in the dendritic building blocks. The majority of metallodendrimers of this class originates from the combined efforts of the groups of Balzani, Campagna, and Denti [46].

3.2 Metallodendrimers Based on Ru(II)- and Os(II)-Polypyridine Complexes

Dendrimers containing Ru(II)- and Os(II)-polypyridine complexes arose from precursors such as the hetero-tetrametallic complex reported in 1989 by Balzani and coworkers (Fig. 8) [47].

Due to their outstanding photophysical and electrochemical properties, such Ru(II)- and Os(II)-complexes had been extensively used as luminescent species and as reactants or mediators in light-induced and light-generating electron transfer processes [26]. Replacing the peripheral bipyridine units (Fig. 8) with 2,3-bis(2-pyridyl)pyrazine ligands (2,3-dpp, which in Fig. 8 are coordinated to the Os(II) core) opened the way to higher, luminescent oligomers. The synthesis, photophysical, and electrochemical properties of decanuclear homo- and heterometallic polypyridine complexes were reported in 1992 [48]. A "complexes as metals" and "complexes as ligands" strategy was followed in the assembly of the dendritic structures. This strategy relies on a protection/deprotection procedure in which one metal complex is used as a ligand equivalent (**6** in Fig. 9) and another as a metal equivalent (**7** in Fig. 9).

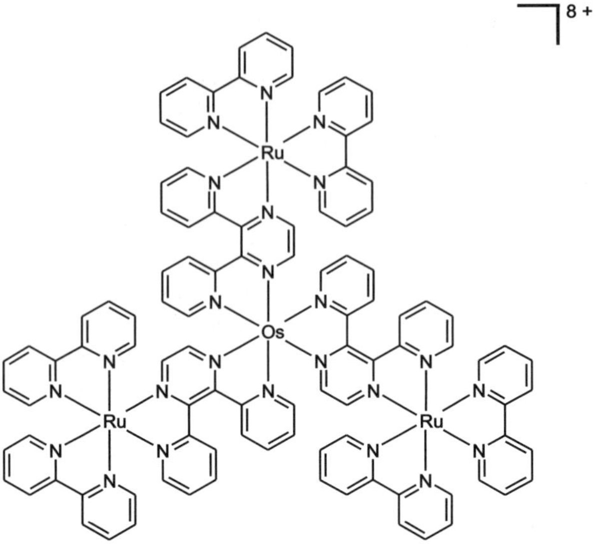

Fig. 8. Heterometallic tetrameric complex reported by Balzani and coworkers

After deprotection of the chlorides with six equivalents of Ag^+, three equivalents of the "complex metal" species were added to the "complex ligand" core to produce the decanuclear dendrimer in good yields. With this procedure, equivalent sites can only be occupied by the same type of metal ion. As reported later, protection of one of the chelating sites in the bridging ligands (2,3-dpp) allows the coordination of different metals to the same bridging ligand, leading to less symmetric dendrimers (see below). In principle, each decanuclear compound can exist as different geometrical isomers depending on the arrangement of the ligands around the metal ions. Each complex can also be a mixture of several diastereomeric species, owing to the chiral nature of each metal center. This renders structural characterization of these systems difficult. However, the chirality of similar decanuclear Ru(II) dendrimers can be controlled using substitutionally inert Ru(II) trisdiimine complexes as chiral synthons (see below) [49]. The decanuclear complexes exhibit extraordinary large molar absorption coefficients in the UV and visible spectral region. Furthermore, the excitation energy could be channeled in the desired direction by a tailored choice of the components (schematically depicted in Fig. 10).

The electrochemical data of the decanuclear complexes revealed a selectivity in the metal oxidation behavior based on its nature and location in the array, offering a fingerprint of the topological structure of the complexes. The near-infrared luminescence of several polynuclear Os(II) and Ru(II) complexes, including two decanuclear dendrimers of the type described above, was reported by Juris et al. in 1994 [50]. Furthermore, light scattering and conductivity experiments showed that the decanuclear complexes aggregate at room temperature in solution at very low concentrations. This was attributed to attractive forces between the hydrophobic assemblies [51].

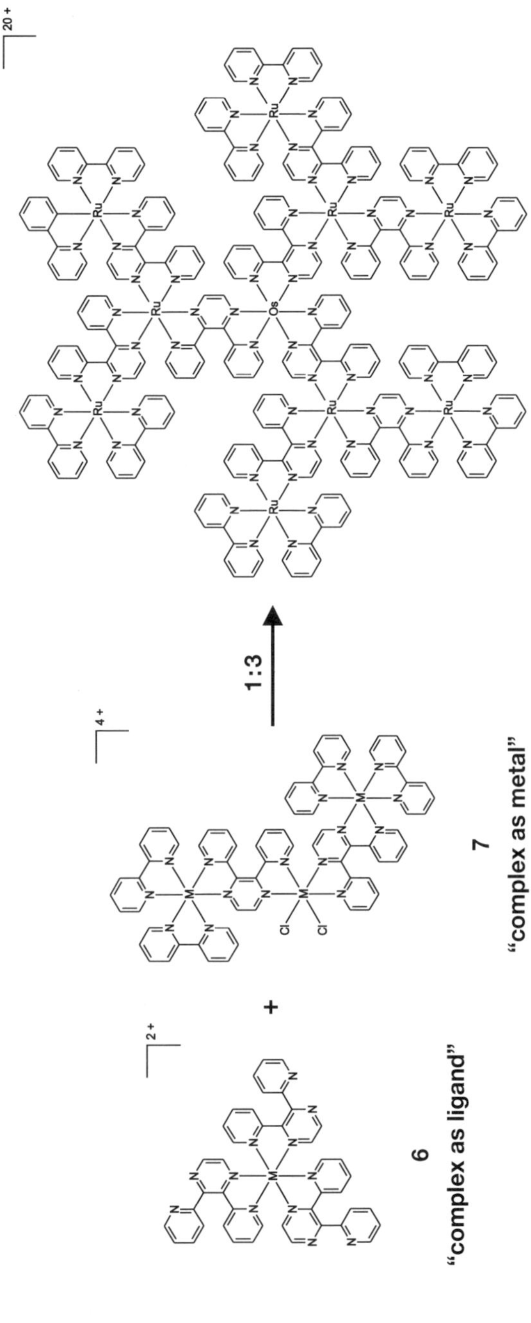

Fig. 9. The "complexes as metal"/"complexes as ligand" strategy in the assembly of Ru(II)- and Os(II)-containing metallodendrimers

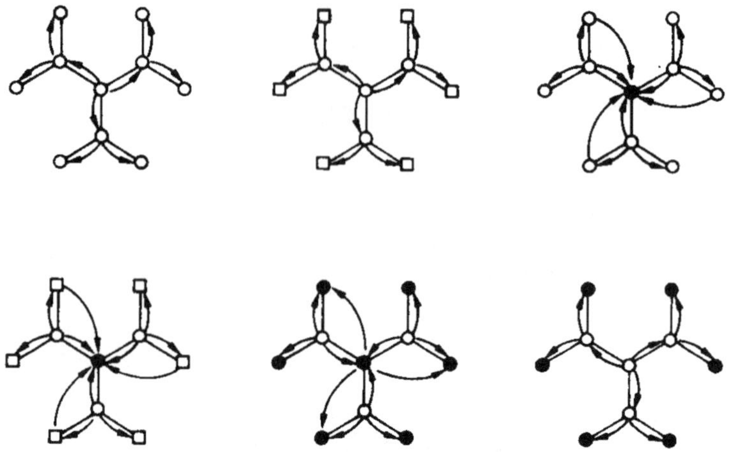

Fig. 10. Directional energy transfer (represented by the *arrows*) in Ru(II)- and Os(II)-containing metallodendrimers. *Empty and full circles* indicate Ru(II) and Os(II), respectively. In the peripheral positions, *circles and squares* indicate M $(bpy)_2$ and $M(biq)_2$ (biq = biquinoline) components, respectively

While the above discussed construction of decanuclear complexes could be regarded as *convergent* dendritic growth (although no dendritic growth was required in the preparation of the "wedges"), iterative *divergent* dendritic growth can only be performed using potentially bifunctional building blocks, where one of the two functional groups is temporarily blocked or present as a masked functionality. Such a bifunctional complex was exploited by Serroni et al. in the divergent construction of docosanuclear (22 metals) Ru(II) dendrimers [52]. Monomethylation of 2,3-dpp followed by complexation of the free chelating sites to Ru(II) resulted in the dendritic building block **8** (Fig. 11).

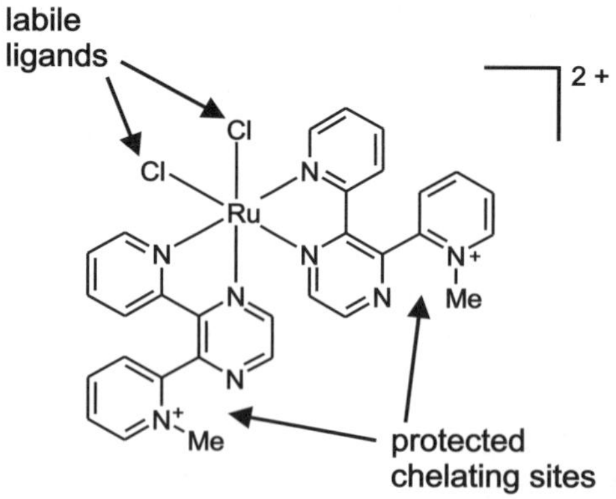

8

Fig. 11. Bifunctional Ru(II)-complex used in the divergent construction of metallodendrimers

After removal of the chloride ligands with $AgNO_3$, three of these building blocks were coupled to the core (**6** with M = Ru(II) in Fig. 9) to produce the corresponding tetranuclear complex in 70% yield. Demethylation of the six chelating sites with DABCO followed by complexation of either the mononuclear building block **8** or the trinuclear capping complex **7** (M = Ru(II), Fig. 9) provided a second generation protected decanuclear complex [53] or a docosanuclear dendrimer, respectively (both homonuclear complexes). The iterative divergent dendritic growth has so far not been extended beyond the decanuclear second generation complexes. The synthesis, luminescence, and redox properties of a heteronuclear analog of the docosanuclear dendrimer have also been reported [54].

A number of studies on the properties of similar polynuclear complexes with different metals and/or ligands has been reported [55]. However, the number of *dendrimers* of the "metals as branching units" type is quite limited. A dendritic heptanuclear Ru(II) complex containing 1,10-phenanthroline as peripheral ligands and 1,4,5,8,9,12-hexaazatriphenylene (HAT) as bridging ligands was synthesized in a divergent manner by Kirsch-De Mesmaeker and coworkers [56]. This dendritic coordination complex was characterized by electrospray mass spectrometry (ES-MS). This technique has proven its usefulness in the characterization of metallodendrimers ever since (including in our group, see below). All peaks observed in the ES spectrum correspond to a single heptametallic Ru(II) complex associated with different types of counteranions (resulting from the hydrolysis of PF_6 anions). The excited state of the heptanuclear dendritic Ru(II) complex was studied by femtosecond transient absorption spectroscopy [57]. Scanning tunneling microscopy (STM) showed ordered patterns on graphite indicating packing of the metallodendrimer molecules on the surface [58]. A distance of 27 ± 2 Å between adjacent lamellae suggest that these are rows of dendritic molecules.

As mentioned before, unequivocal characterization (e.g., by NMR) of the dendrimers reported by Balzani and coworkers has been hampered by the presence of numerous diastereoisomers in such polynuclear complexes. This problem originates from the synthetic strategy, in which the complexes are usually assembled via ligand displacement reactions with little or no direct control for the product stereochemistry. Additional stereochemical complications arise when asymmetrically substituted bidentate ligands are being used. An elegant strategy that allows the assembly of enantiomerically pure oligonuclear and dendrimeric Ru(II) complexes from chiral synthons has been developed by MacDonnell and coworkers. An irreversible covalent reaction which does not involve ligand displacement at the chiral centers is used to couple the metal centers, thereby avoiding disturbance of the metal-complex stereochemistry. The first generation dendritic D_3-symmetric tetranuclear complexes were described in 1997 [59, 60]. A representative example (Fig. 12) shows that the condensation reaction between the *o*-dione moiety of the enantiopure Δ-[Ru(1,10-phenanthroline-5,6-dione)$_3$]$^{2+}$ complex and the *o*-diamine function of enantiopure Λ-[Ru(1,10-phenanthroline)$_2$(1,10-phenanthroline-5,6-diamine)]$^{2+}$ yields the rigid D_3-symmetric trigonal-propeller-shaped complex ΔΛ$_3$-**9** containing tetrapyrido[3,2-*a*:2′,3′-*c*:3″,2″-*h*:2″,3″-*j*]phenazine bridges between stereocenters. The three other D_3 isomers were prepared similarly.

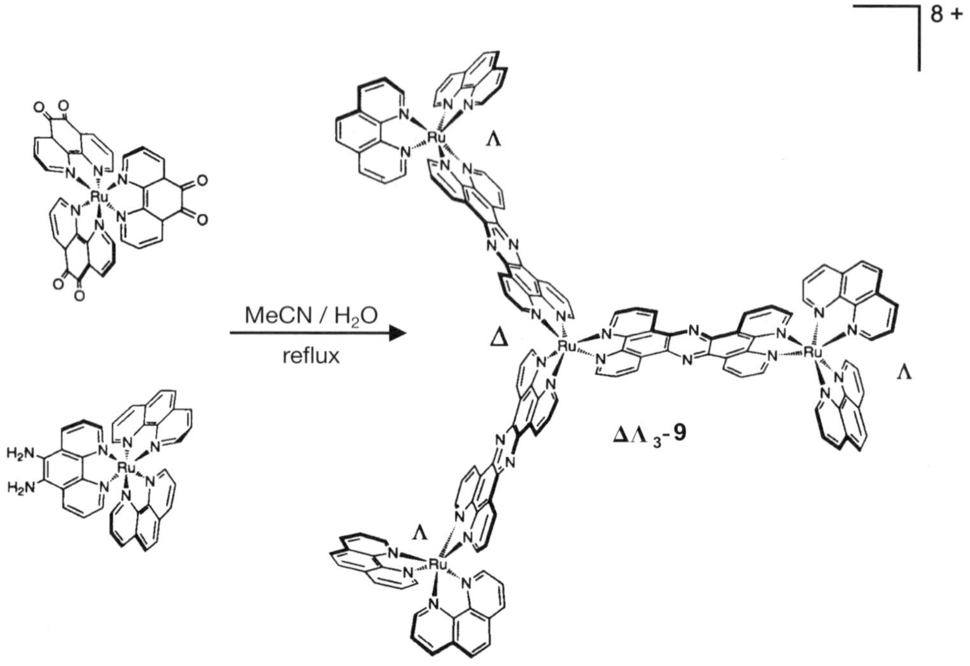

Fig. 12. Synthesis of enantiomerically pure, first generation dendritic D_3-symmetric tetranuclear Ru(II)-complexes

The high symmetry of the complexes greatly facilitated their characterization by ^1H, ^{13}C, COSY and HMQC NMR spectroscopy. MALDI-TOF mass spectrometry, a well-established technique for characterizing (metallo)dendrimers nowadays, gave the parent molecular ion signals corresponding to the species [**9** – nPF$_6$]$^+$ (n = 3–8). Finally, the absolute stereochemistry and optical purity of each of the stereoisomers was established by circular dichroism (CD) spectroscopy. Recently [61], further divergent dendritic growth was achieved from the isomers of **9** by oxidation of the terminal phenanthrolines to quinones (which proceeds with retention of the stereochemistry despite the harsh reaction conditions, H_2SO_4/HNO_3/NaBr at 100 °C), followed by condensation with Λ-[Ru(1,10-phenanthroline)$_2$(1,10-phenanthroline-5,6-diamine)]$^{2+}$. The decameric Ru(II) complexes were isolated in ≈20% yield as robust metallodendrimers (stable to heat, concentrated acids and bases, and racemization) and were characterized by ES-MS and ^1H and COSY NMR spectroscopy. Interestingly, the two diastereomeric decamers $\Lambda_6\Delta_3\Lambda$-Ru$_{10}$ and $\Lambda_6\Lambda_3\Lambda$-Ru$_{10}$, which differ in their stereochemistry at three dendritic sites, display remarkably different overall topologies, as suggested by molecular modeling. The *global* stereochemical descriptors T (achiral) and P (clockwise) were introduced to describe the flat disk-like structure of $\Lambda_6\Delta_3\Lambda$-Ru$_{10}$ (Fig. 13 left) and the right-handed propeller structure of $\Lambda_6\Lambda_3\Lambda$-Ru$_{10}$ (Fig. 13 right), respectively.

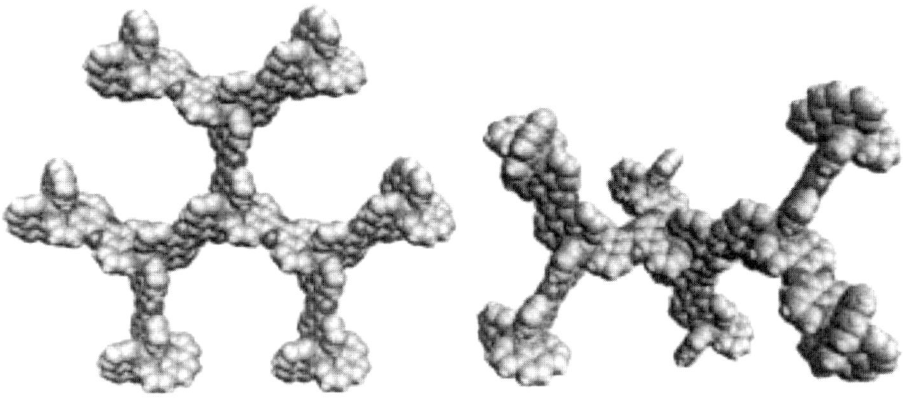

(T)-$(\Lambda_6\Delta_3\Lambda$-$\mathrm{Ru}_{10})$ (P)-$(\Lambda_6\Delta_3\Lambda$-$\mathrm{Ru}_{10})$

Fig. 13. Space-filling models of decanuclear (T)-$(\Lambda_6\Delta_3\Lambda$-$\mathrm{Ru}_{10})$ (*left*) and (P)-$(\Lambda_6\Lambda_3\Lambda$-$\mathrm{Ru}_{10})$ (*right*) showing the respective disk-like and propeller structure of these complexes

Although these differences in the global structure do not influence the CD spectra (which was attributed to weak electronic coupling between the chromophores), the colloidal properties (dynamic light scattering revealed polydisperse aggregation in acetonitrile) as measured by electric birefringence are significantly different.

3.3 Conclusions

Metallodendrimers belonging to the "metals as branching units" class were among the first metallodendrimers to be reported. Since then, various synthetic strategies have allowed the type of both the incorporated metals and ligands to be varied, thereby tailoring, e.g., the chiral, electrochemical, and photochemical properties of these metallodendrimers for future applications such as multi-electron catalysis, solar energy conversion, and chiral recognition.

4 Metals as Building Block Connectors

4.1 Introduction

In this class of metallodendrimers the necessary branching is incorporated into the dendritic building blocks instead of being located at the metal centers. As a consequence the variation is much larger for this type. The unrestricted freedom in building block design for the "metals as building block connectors" dendrimers has also led to higher dendritic generations than in the "metals as branching units" type, where a third generation metallodendrimer is yet to be reported (probably due to the compact and rigid structure of these dendrimers).

In our group, we have been concerned with metallodendrimer assembly of the "metals as building block connectors" type since 1994. First, however, the elegant studies of the groups of Puddephatt, Newkome, Constable, Takahashi and others will be discussed in some depth [10].

4.2 Organometallic Pt(II)/Pt(IV)-Containing Dendrimers

Puddephatt and Achar were the first to report the synthesis of organometallic dendrimers [62]. Their synthetic methodology is based on two well-known reactions in organometallic chemistry, namely the displacement of SMe_2 ligands from $[Pt_2Me_4(\mu\text{-}SMe_2)_2]$ by 2,2′-bipyridines and the oxidative addition of benzylic bromides to $[PtMe_2(bipy)]$-type complexes. As depicted in Fig. 14, a selective *trans* oxidative addition of the C-Br bonds of 4,4′-bis(bromomethyl)-2,2′-bipyridine to $[PtMe_2\{$4,4′-di-*tert*-butyl-2,2′-bipyridine] **10** gave the binuclear Pt(IV) complex **11**, which was subsequently reacted with $[Pt_2Me_4(\mu\text{-}SMe_2)_2]$ to the trinuclear complex

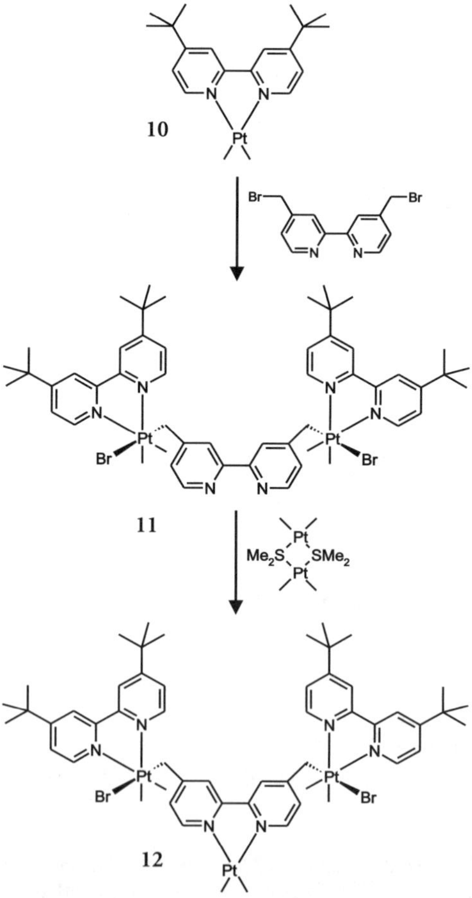

Fig. 14. Convergent dendritic growth strategy developed by Puddephatt and Achar

12. Further convergent dendritic growth occurs from the reactive Pt(II) center in **12** by a repetition of oxidative addition of 4,4′-bis(bromomethyl)-2,2′-bipyridine and conversion of the resulting bipyridine to the dimethylplatinum(II) complex.

After the third growth cycle (which yielded a Pt_{14} dendron), reaction at the focal bipyridine did not proceed well, and this was attributed to steric hindrance at the focal point, a common problem in dendritic chemistry. The dendrons were characterized by UV-VIS spectroscopy (useful for distinguishing between Pt(II) and Pt(IV)), ^{1}H and ^{195}Pt NMR spectroscopy, and gel permeation chromatography (GPC), which provided a rough estimate for the molecular weight.

Fourfold coupling of first and second generation Pt(II)/Pt(IV) complexes such as **12** (a first generation dendron) to 1,2,4,5-tetrakis(bromomethyl)benzene provided large (up to Pt_{28}) dendrimeric organoplatinum complexes, as reported in subsequent papers from the same group [63]. Moreover, by following the same methodology but employing the monofunctional 4-(bromomethyl)-4′-methyl-2,2′-bipyridine instead of 4,4′-bis(bromomethyl)-2,2′-bipyridine as the reagent in oxidative additions to Pt(II), oligomeric linear chains were obtained instead of dendrons, demonstrating the versatility of the synthetic approach [64]. Starting from bipyridines carrying redox-active ferrocene functions and using the same synthetic methodology, heterometallic dendrimers were synthesized (see, e.g., the example in Fig. 15) and electrochemically characterized by Achar and Catalano [65].

Finally, a divergent route to first generation organoplatinum and platinum-palladium dendrimers based on slightly modified building blocks has been re-

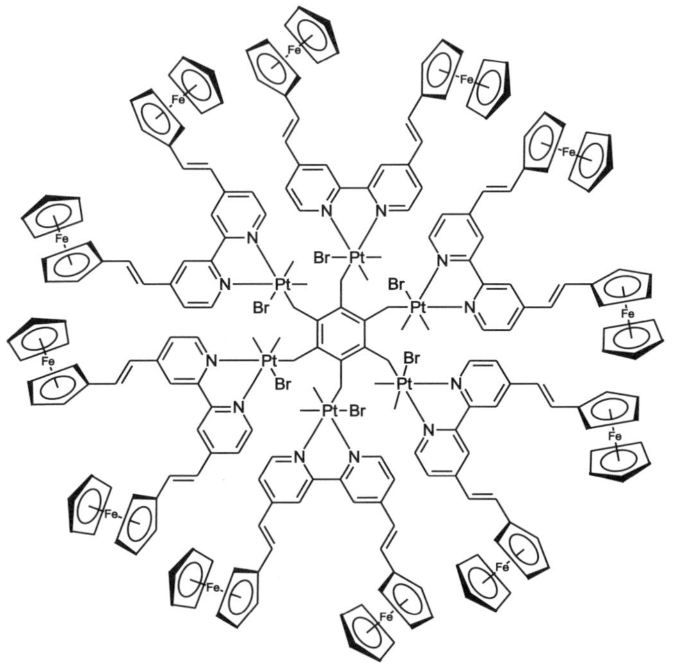

Fig. 15. Organometallic dendrimers containing peripheral ferrocene units

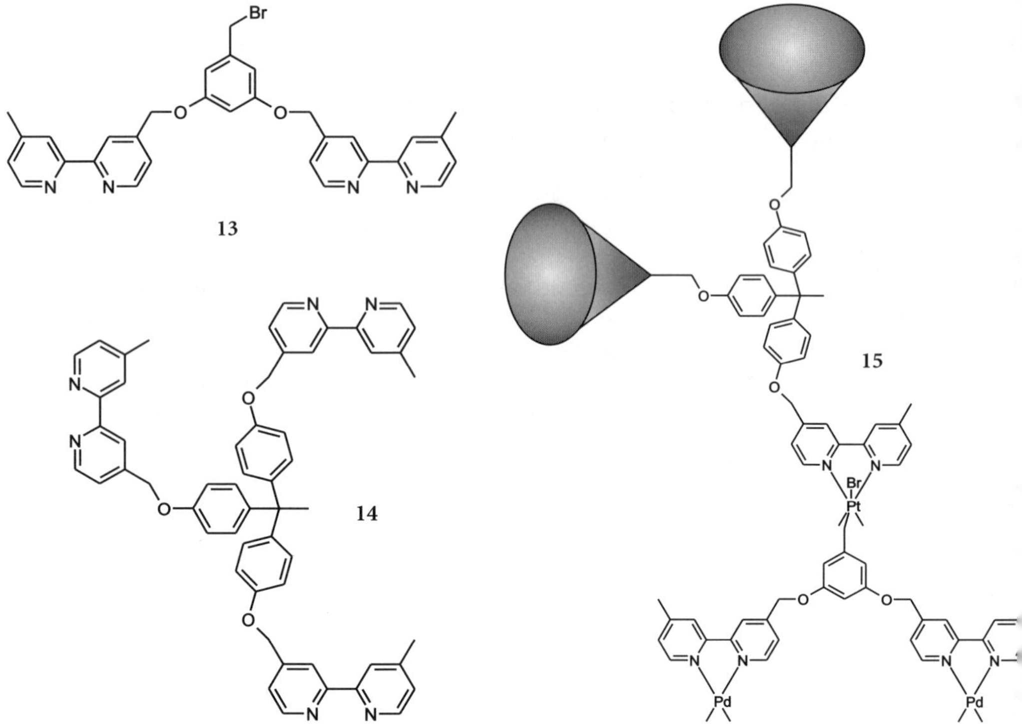

Fig. 16. Building blocks (**13** and **14**) in the synthesis of a first generation organometallic platinum-palladium metallodendrimer (**15**)

ported [66]. With the bis- and tris(2,2′-bipyridyl) reagents **13** and **14** as the dendritic building block and core, respectively, first generation complexes such as **15** were synthesized (Fig. 16). Unfortunately, solubility problems prevented further dendritic growth. The low stability of the Pd(IV) centers also restricted their position to the periphery of the dendrimer.

4.3 Dendrimers Based on Metal-Terpyridine Complexes

Transition metal-terpyridine moieties are not readily suitable for construction of dendrimers of the "metals as branching units" type since only two building blocks can be linked via terpyridines. Particularly the groups of Newkome [10a] and Constable [67] have investigated dendrimers containing Ru(II)-terpyridine units, both with the aim of exploiting the unusual and interesting (electrochemical, photochemical and photophysical) properties [68] that these new multinuclear species might exhibit. In one of the first reports [69], Newkome et al. described the synthesis of a dodecaruthenium(II) metallodendrimer by complexing small Ru(II)-terpyridine dendrons to a divergently synthesized dendrimer containing 12 terpyridines at the periphery (Fig. 17).

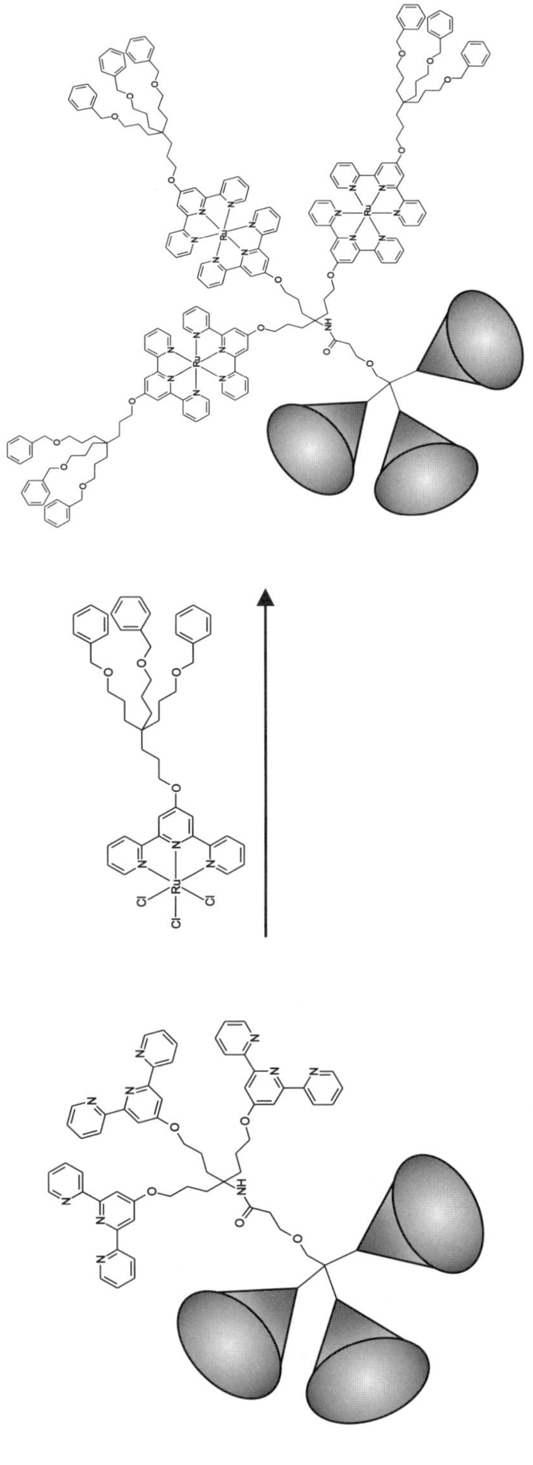

Fig. 17. Dodecaruthenium(II) metallodendrimer reported by Newkome, Constable and coworkers

This strategy of linking dendritic fragments via Ru(II)-terpyridines was later also applied in order to construct the Ru(II) bis-dendrimers discussed in Sect. 2. Expanding the stepwise assembly has resulted in double terpyridine-Ru(II) connectivity in each arm of a tetrahedral metallodendrimer [70]. The versatility of this synthetic methodology was demonstrated by the synthesis of isomeric metallodendrimers in which different dendrons (G1 and G2), attached to the core and peripheral terpyridine building blocks, were interchanged with respect to each other [71]. The two constitutional isomers depicted in Fig. 18 possess very different internal densities and void regions.

The cyclic voltammograms show a single, quasi-reversible pattern for the Ru(II) centers of isomer **16**, indicating that all Ru atoms are electrochemically equivalent. However, the two waves for the metal centers of isomer **17** led the authors to suggest that the interior of this dendrimer might be rigid enough to allow electrochemical communication between the metals [72]. Recently, Newkome and coworkers also reported dendrimers containing four internal bipyridine units which complex Ru(II) centers by reaction of the dendritic bipyridines with $[Ru(bpy)_2Cl_2]$ [73]. However, here the metals are not essential for the dendritic structure.

Constable and coworkers have predominantly focused on grafting linear oligonuclear complexes onto cores of different constitution and multiplicity [67]. Di- and trinuclear linear Ru(II)-terpyridine arms, incorporating nucleophilic HOtpy ligands on one side as the terminal terpyridine, were coupled to 1,4-bis(bromomethyl)benzene and 1,3,5-tris(bromomethyl)-2,4,6-trimethylbenzene to give the corresponding tetra-, hexa- and nonanuclear complexes, respectively [74]. The nonanuclear complex is depicted in Fig. 19. All polynuclear complexes could be purified by silica chromatography and were characterized by MALDI-TOF mass spectrometry. Their solubility is dictated by the counterion, the chloride salts are soluble in methanol and even water, and the corresponding PF_6^- salts are soluble in acetonitrile and acetone.

The trinuclear linear complex was also reacted with hexakis(bromomethyl)benzene to provide an octadecanuclear species [75]. These architectures are of the "star" type rather than the dendritic type, since branching only originates from the core. Using optical waveguide lightmode spectroscopy, it was shown that the adsorption of the nonanuclear star to silica-titania surfaces heavily depended on the bulk concentration [76].

Branched Ru(II)-terpyridine dendrimers (instead of stars) were synthesized by Constable et al. by twofold coupling of a Ru(II)-bisterpyridine complex monofunctionalized with a benzyl bromide to 4,4′-dihydroxy-2,2′-bipyridine, followed by coordination of three of these dendrons to either Fe(II) or Co(II) [77]. Recently Constable and coworkers have described pentaerythritol-based Ru(II)-terpyridine metallodendrimers [78]. The reaction of pentaerythritol with 4′-chloro-2,2′:6′,2″-terpyridine could be controlled to provide either a dendritic core containing four terpyridine metal-binding moieties or a building block containing one terpyridine and three pendant CH_2OH groups. These two structures were linked together by Ru(II) to give the tetranuclear **18** (Fig. 20), which remarkably could be reacted completely with $[(tpy)Ru(Cltpy)]^{2+}$ (**19**) to provide the hexadecanuclear complex **20**. The coordination of 4-halopyridines

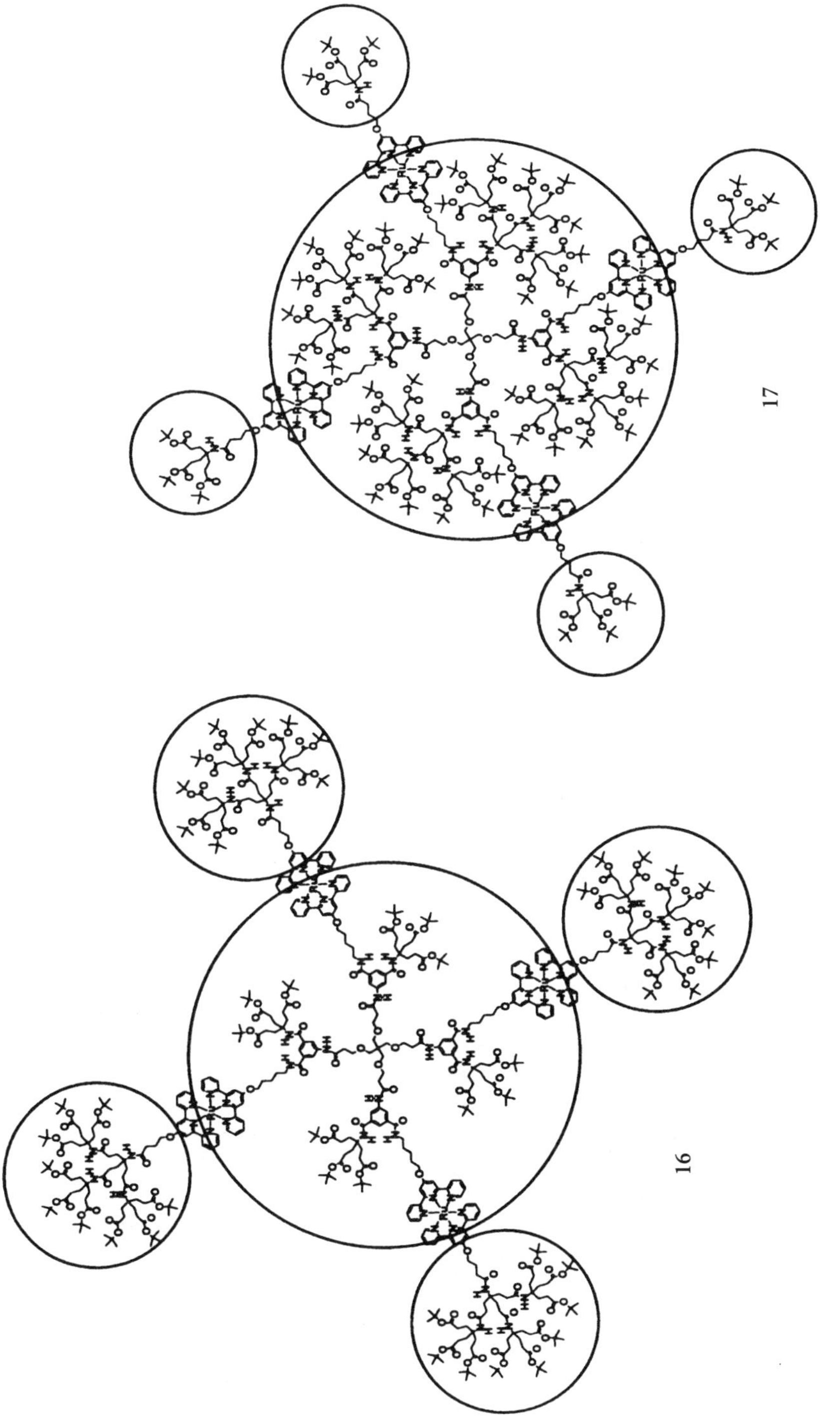

16

17

Fig. 18. Constitutionally isomeric metallodendrimer having different internal densities and void regions

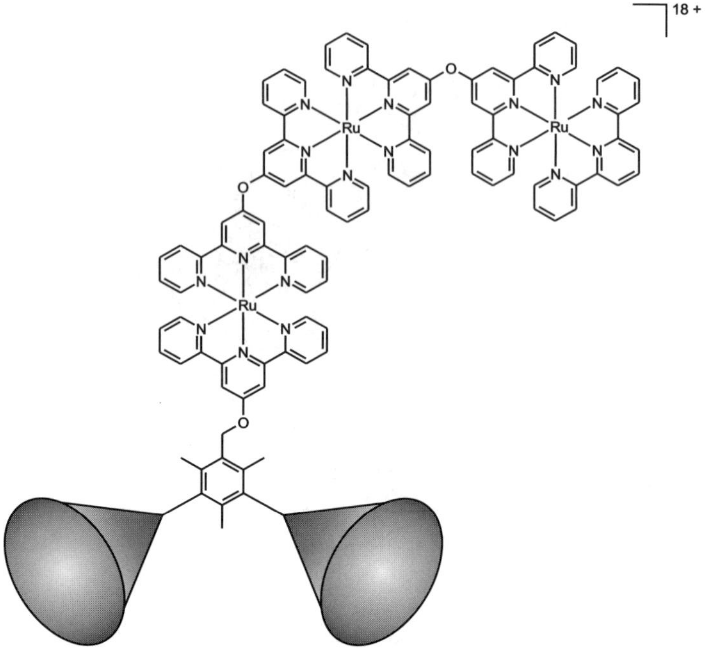

Fig. 19. Nonanuclear Ru(II)-terpyridine assembly reported by Constable and coworkers

to transition metals activates the 4-position for nucleophilic attack [79], and this is probably the reason why the fully functionalized complex **20** is formed.

The fourfold terpyridine-functionalized pentaerythritol unit was earlier described by Constable et al. in the synthesis of tetranuclear complexes bearing pendant carborane moieties [80]. Water-soluble boron-rich systems are of interest in the field of boron neutron capture therapy [81].

4.4
Dendrimers Based on Metal-Acetylide Complexes

Novel dendritic architectures based on Pt(II)-acetylide units have been reported by Takahashi and coworkers. The authors extended their previously developed strategy for transition metal poly-yne polymers [82] to the synthesis of metallodendrimers containing triynes as bridging ligands between metal centers. In the first example [83], 1,3,5-triethynyl-2,4,6-trimethylbenzene was treated with dichlorobis(tri-*n*-butylphosphine)platinum under CuCl catalysis to give the trinuclear complex **21** (Fig. 21), the molecular structure of which was determined by X-ray crystallography.

An excess of this complex could be reacted with 1,3,5-triethynyl-2,4,6-trimethylbenzene (under similar conditions, i. e., CuCl in Et_2NH) in order to prepare a metallodendrimer of higher generation, e. g., the nonanuclear **22**. The terminal chlorides were subsequently substituted by monofunctional phenylacetylene terminal groups. The resulting neutral second generation dendrimer was purified by column chromatography. Stang and coworkers also reported the divergent syn-

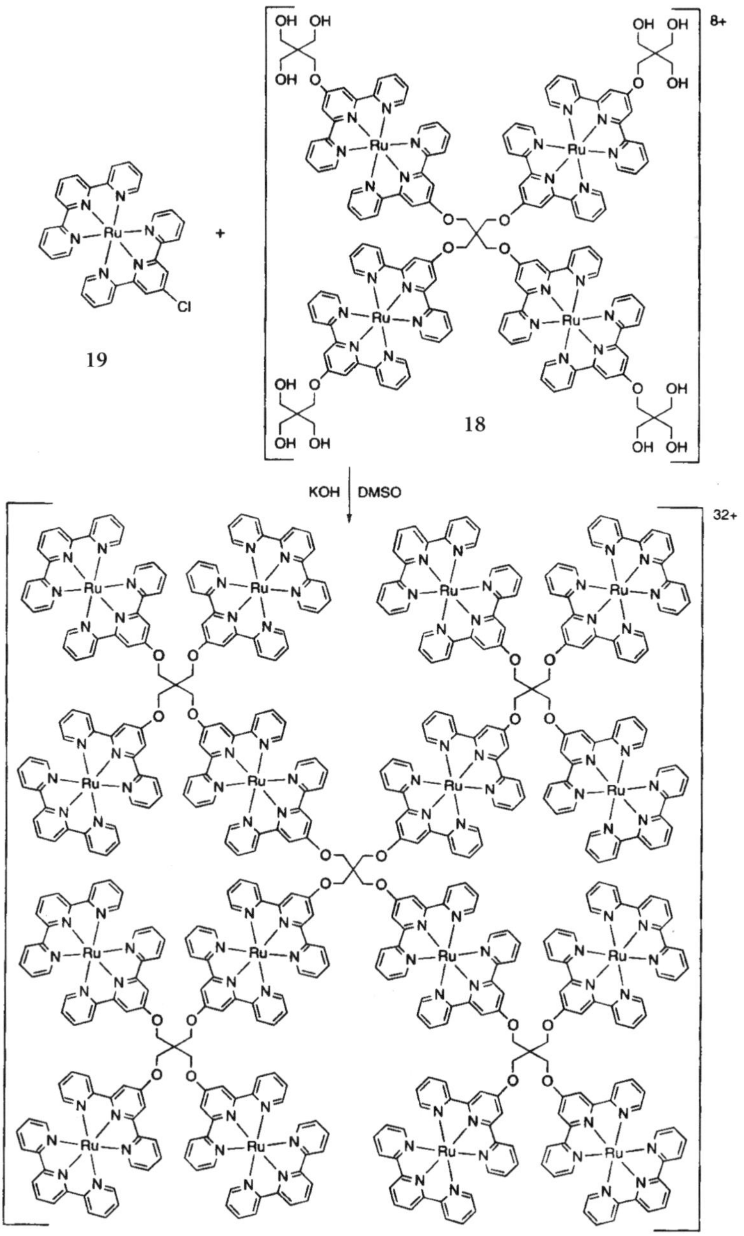

20

Fig. 20. Pentaerythritol-based Ru(II)-terpyridine metallodendrimers

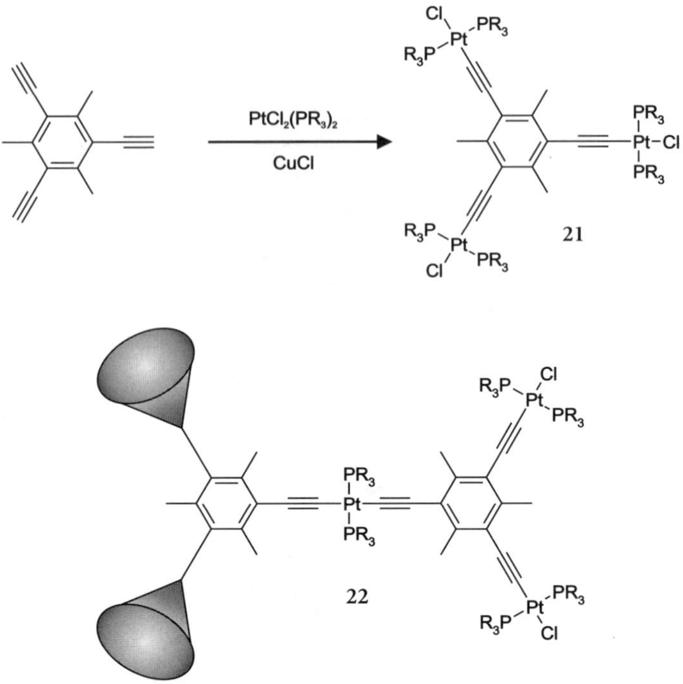

Fig. 21. Divergently synthesized tri- and nonanuclear Pt(II)-acetylide metallodendrimers

thesis of second generation organoplatinum dendrimers based on 1,3,5-triethynylbenzene building blocks [84]. Platinum iodide complexes were used instead of chloride complexes, because of the more labile halogen-metal bond in iodide complexes, facilitating their condensation with acetylenic compounds.

A convergent route to third generation Pt(II)-acetylide dendrimers was recently developed by Takahashi and coworkers [85]. Their convergent methodology relies on the use of two different trialkylsilyl protecting groups for the acetylene units of triethynylbenzene. Selective desilylation of the trimethylsilyl moiety in **23** (Fig. 22), followed by conversion of the terminal acetylenes to the corresponding *P*-methoxyphenylethynylplatinum(II) complexes, and final deprotection of the tri(isopropyl)silyl group gave the first generation dendritic wedge **24**. Second and third generation dendrons were similarly synthesized. Coupling of these wedges to the trifunctional complex **21** (Fig. 21, R = Et) gave the corresponding metallodendrimers, bearing up to 45 platinum centers (M = ± 26 kDa), in good yields (51 % for the third generation). Due to symmetry, the dendrimers were characterized by very simple ^{1}H and ^{31}P NMR spectra.

An alkynylruthenium dendrimer containing a large π-delocalized system was recently synthesized by Humphrey et al. as a new organometallic complex for nonlinear optics [86]. Bisacetylide Ru(II) units were incorporated into a dendritic wedge bearing a focal acetylene, which was subsequently coupled to a previously reported [87] trifunctional core to provide a rigid nonanuclear complex (Fig. 23).

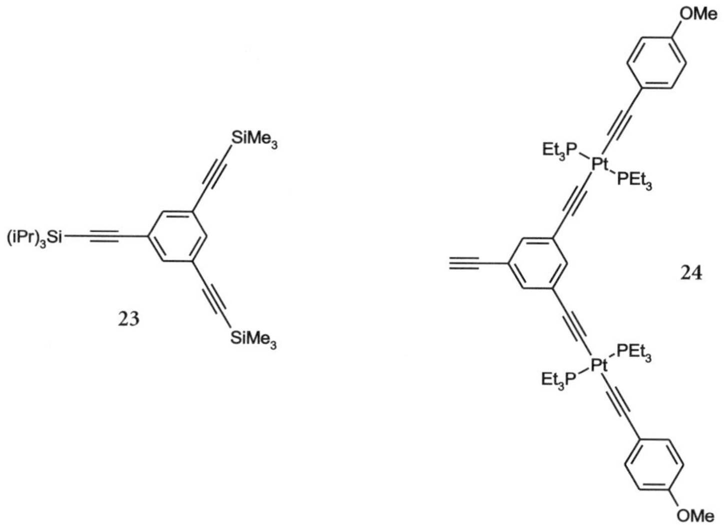

Fig. 22. Building block (**23**) and dendritic wedge (**24**) used in the convergent construction of Pt(II)-acetylide metallodendrimers

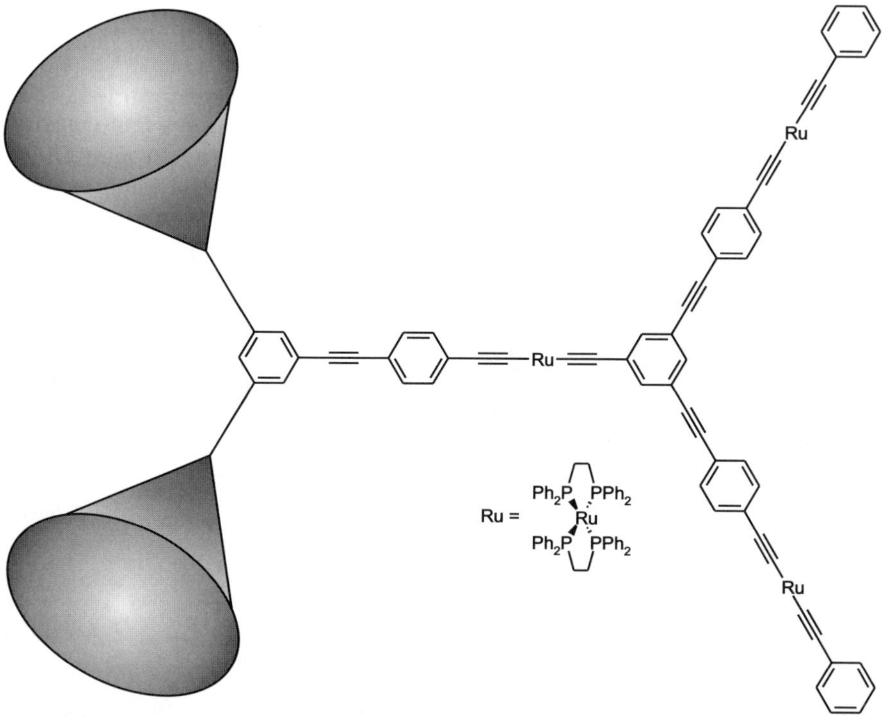

Fig. 23. Nonanuclear alkynylruthenium metallodendrimer used in nonlinear optics

Third-order NLO measurements for this organometallic dendrimer showed a significant enhancement of two-photon absorption (TPA) upon proceeding from the constituent building blocks to the nonanuclear complex. TPA materials are of interest for applications such as optical data storage and optical limiting [88].

Rigid dendrimers were synthesized by Osawa et al. [89] by grafting Ru(II)-terpyridine metallodendrons to a functionalized core Ni(II)-tris(bipyridine) unit. Metal-terpyridine and -bipyridine connectivities were combined before in the metallodendrimers reported by Constable and coworkers (see above) [77], with coordination of the bipyridine dendrons to Ru(II) as the final step. The final construction step used by Osawa and Wakatsuki is a covalent palladium-mediated coupling between aryl iodides connected to the core Ni(II)-tris(bipyridine) moiety and the focal acetylene of the Ru(II)-terpyridine dendrons. The largest (ca. 90 Å) dendrimer, containing 18 Ru(II) centers and 1 Ni(II) center, was characterized by transmission electron microscopy (TEM), which revealed individual molecules of nanosize dimensions.

4.5 Metallodendrimers Based on SCS Pd(II) Pincer Moieties

Pincer ligands are *meta*-xylene derivatives in which the two methylene moieties carry suitable donor atoms (e.g., N, P, or S) [90] for the complexation of transition metals. Cyclometallation (Fig. 24), in which a σ bond is formed between the metal and the aryl carbon, gives rise to a variety of systems displaying rich coordination chemistry [91].

The SCS Pd(II) pincer moiety, known since 1980 [92], has recently been exploited in molecular recognition [93], catalysis [94], and self-assembly [95]. Our group has been involved in the non-covalent construction of metallodendrimers based on this versatile motif since 1994. We first studied genuine self-assembly of molecules that have two pincers and a suitable ligand (e.g., RCN) in one building block. Compound **25a** (Fig. 25) was synthesized and its self-assembly upon removal of the labile acetonitrile ligands was investigated [96].

Intermolecular coordination of the nitrile groups to Pd(II) was evidenced by IR spectroscopy, in which a diagnostic shift of the C≡N stretch vibration from 2250 cm^{-1} (monomeric building block) to 2290 cm^{-1} was observed upon coordination. Further proof of the self-assembled structure was obtained from quasi-

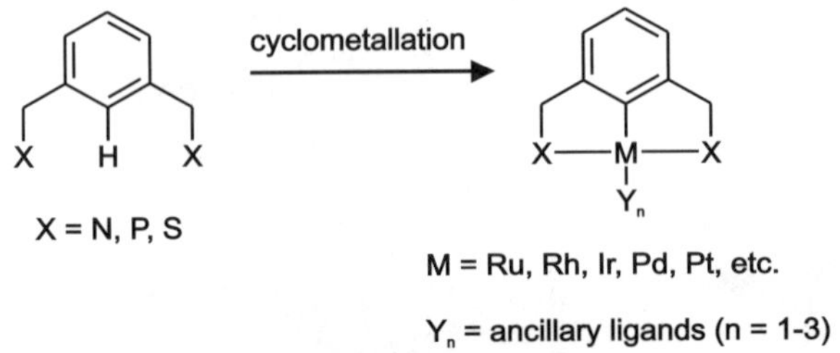

Fig. 24. Cyclometallation of pincer ligands

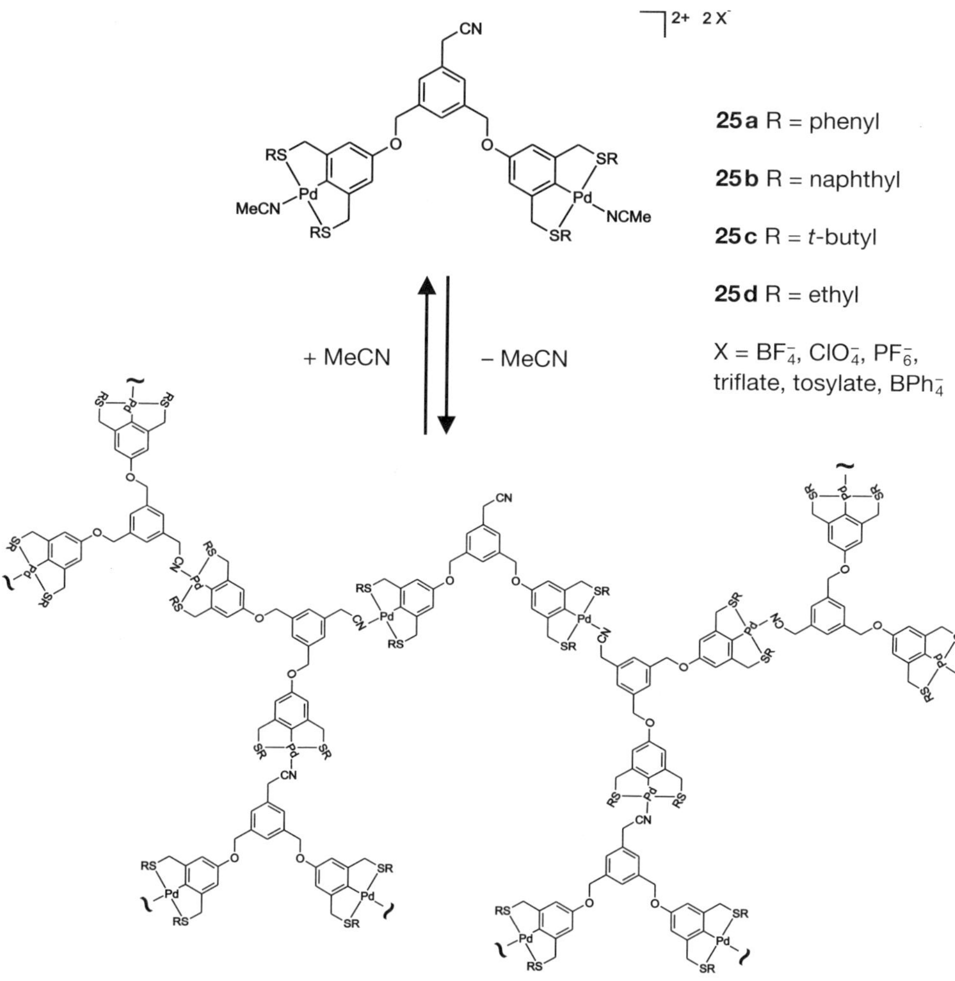

Fig. 25. Uncontrolled growth of self-assembled hyperbranched spheres from building blocks **25a–d**

elastic light scattering (QELS), which showed particles having a hydrodynamic diameter of 200 nm, and from both atomic force microscopy (AFM) and TEM, which revealed large globular-shaped assemblies having diameters in the 150–200 nm range. The reversible nature of the self-assembling process was proven by the addition of small amounts of acetonitrile to a nitromethane solution of the assembly, upon which sharp signals appeared in the ^{1}H NMR spectrum, indicating the presence of monomeric species.

Structural variations in both the building blocks (**25b–d** in Fig. 25) and the non-coordinating anions (X^- in Fig. 25) allowed the size of the self-assembled spheres to be controlled [97]. In general, larger anions and/or thioether groups gave smaller assemblies, as shown by QELS, AFM, and TEM. Presumably, in the dendritic model the anions will occupy the voids created by the branching of the monomers. When these cavities become too small the anions are forced out of the growing sphere and block further assembly by occupying the surface. The size at which this occurs is apparently also dependent on the bulkiness of the building blocks. The dense shell of the spheres was indicated by the slow disassembly upon the addition of benzonitrile, which even at higher temperature (70 °C) caused little degradation of the aggregates.

In contrast to the uncontrolled growth of the self-assembled spheres, a *controlled* assembly strategy would allow the synthesis of smaller metallodendrimers. Such a growth scheme requires a suitable dendritic core and a protection/deprotection strategy in order to control precisely the coordination of building blocks at the desired positions. We developed the building blocks depicted in Fig. 26.

In all the building blocks, the Pd(II) centers are protected by strongly coordinating chloride ligands. These chlorides can be removed, however, by precipitation with Ag(I) salts, thereby allowing the resulting vacant coordination site to be occupied by other ligands. Therefore, starting from the nucleus $\mathbf{G_0}$, the assembly of metallodendrimers can be controlled by the repetitive addition of Ag(I) salts and building blocks. At first, we investigated the divergent dendritic growth using $\mathbf{G_0}$ and $\mathbf{BB_{CN}}$**-Cl** [98]. As shown in Fig. 27, deprotection of $\mathbf{G_0}$ with 3 eq. of $AgBF_4$ followed by the addition of 3 eq. of $\mathbf{BB_{CN}}$**-Cl** resulted in the first-generation metallodendrimer $\mathbf{G_1}$, containing 6 terminal sites for further growth.

Repetition of the deprotection and building block addition procedure yielded the corresponding higher generation dendrimers (G2–G5). This synthesis is actually a one-pot procedure since intermediate generations need not be isolated. All metallodendrimers were characterized by ^{1}H NMR and FT-IR spectroscopy,

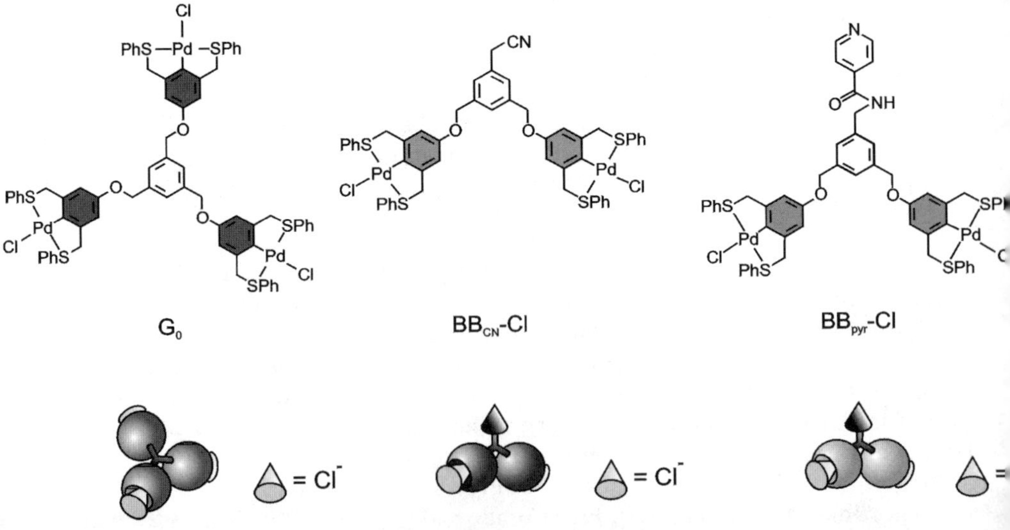

Fig. 26. Building blocks used in the controlled assembly of Pd(II)-containing metallodendrimers

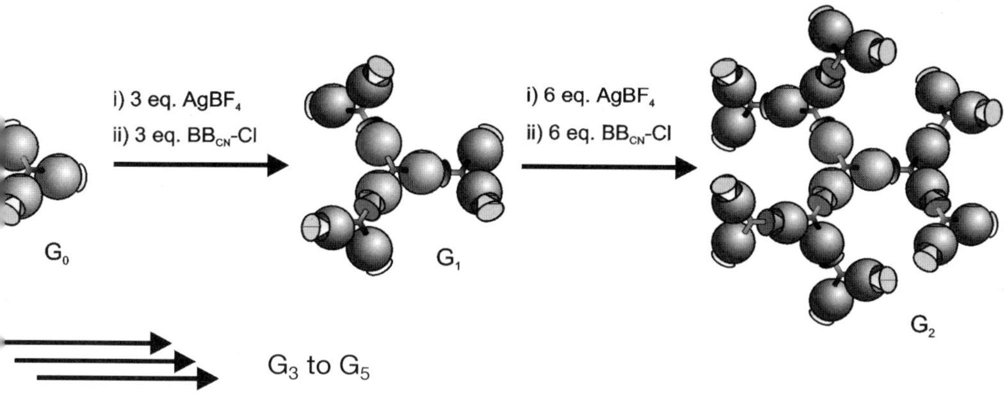

Fig. 27. Metallodendrimers of generation one to five assembled via a controlled divergent growth strategy

elemental analysis, electrospray mass spectrometry (except G5), and tapping mode AFM (TM-AFM) on graphite and mica (G5 only). A representative deconvoluted ES-MS spectrum of G4 is shown in Fig. 28 [99]. The higher generation dendrimers lack a large numbers of anions in the ES-MS spectra, which has also been observed for other metallodendrimers (Sect. 3.2) and for non-dendritic assemblies held together by coordination bonds [100].

Molecularly thin films of spincasted G5 were studied by TM-AFM measurements [99]. At low concentrations separate spheres were identified, having dimensions of approximately 15 nm in diameter, most likely corresponding to individual metallodendrimers.

In order to be able to construct metallodendrimers in a *convergent* fashion based on our strategy outlined above, additional requirements must be met. For example, if scrambling of building blocks is to be avoided in the convergent strategy, the coordination strength of the ligands bound to the Pd(II) pincers should decrease from the periphery towards the core. This necessitates the use of building blocks containing ligands other than the cyano moieties discussed earlier. Therefore we developed building blocks containing pyridine [101]

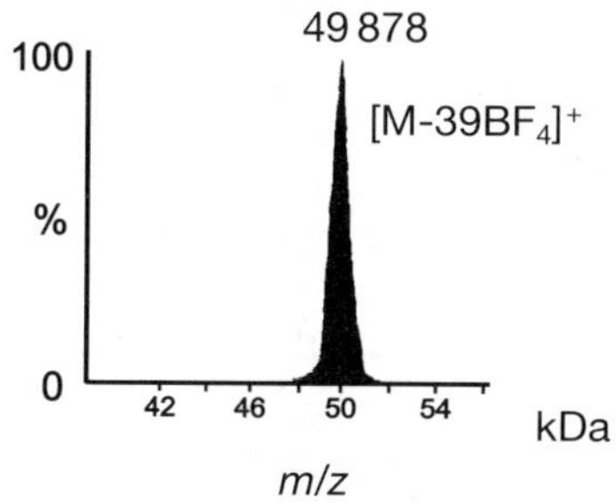

Fig. 28. Deconvoluted electrospray mass spectrum of a G4 metallodendrimer

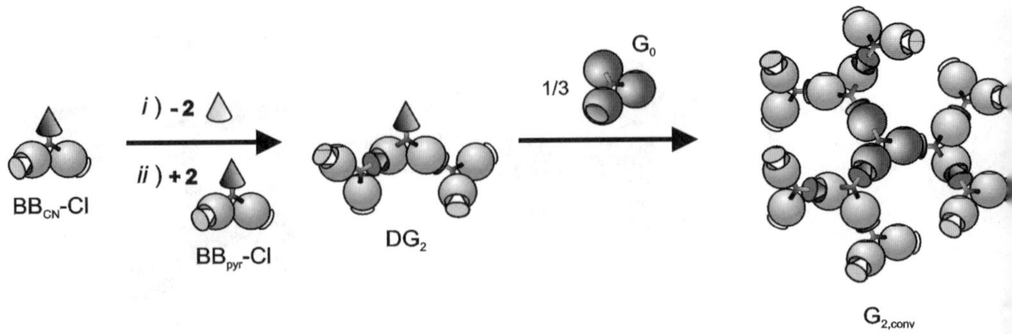

Fig. 29. Convergent growth of metallodendrimers using building blocks that differ in their ligand coordination strength

groups (**BB$_{pyr}$-Cl** in Fig. 26) and, recently, phosphines. Divergent growth using only the pyridine building blocks produced more stable but less soluble dendrimers than those based on the cyano building blocks. Convergent growth was carried out by coordinating two molecules of **BB$_{pyr}$-Cl** to deprotected **BB$_{CN}$-Cl**, producing a dendritic wedge containing a focal cyano ligand for further coordination, as schematically shown in Fig. 29. In this step the stronger pyridine ligand excludes the cyano group from coordination, as evidenced by ^{1}H NMR spectroscopy. In the final step three of the wedges were coordinated to activated **G$_0$** deprived of its chlorides. Both the dendritic wedge and the convergently assembled metallodendrimer were characterized by MALDI-TOF mass spectrometry, which revealed a signal corresponding to $[M-BF_4]^+$ in both cases.

Our convergent growth, based on coordination chemistry, allows the introduction of core and/or peripheral functionalities such as redox- or photoactive groups, provided that such moieties can be functionalized with either Pd(II) pincers or suitable ligands. For example, we have convergently constructed dendritic assemblies containing up to 12 peripheral porphyrins starting from 5-pyridyl-10,15,20-triphenylporphyrin [41g]. Also, metallodendrimers tailored with core *and* peripheral porphyrins have been synthesized, which might provide interesting energy- and electron-transfer behavior [41a]. Furthermore, we have recently begun to explore the photophysical properties of *individual* dendritic molecules containing a fluorescent rhodamine B core [102]. The rhodamine B core (1 eq.), functionalized with two Pd(II) pincer moieties, was added to 14 eq. of deprotected **BB$_{CN}$-Cl** (**25a** in Fig. 25) and the acetonitrile was evaporated (Fig. 30). In this way an uncontrolled self-assembly leads to a statistical mixture in which both metallodendrimers with and without rhodamine B core are present (the rhodamine B bis pincer, having no ligands, can only serve as a dendritic core).

The fluorescence of individual dendritic molecules was investigated by near-field scanning optical microscopy (NSOM). This technique has recently proven to be very useful in the visualization of single molecules [103]. Height and optical images of individual metallodendrimer molecules were obtained simultaneously, allowing co-localization of single-molecule fluorescence with dendritic

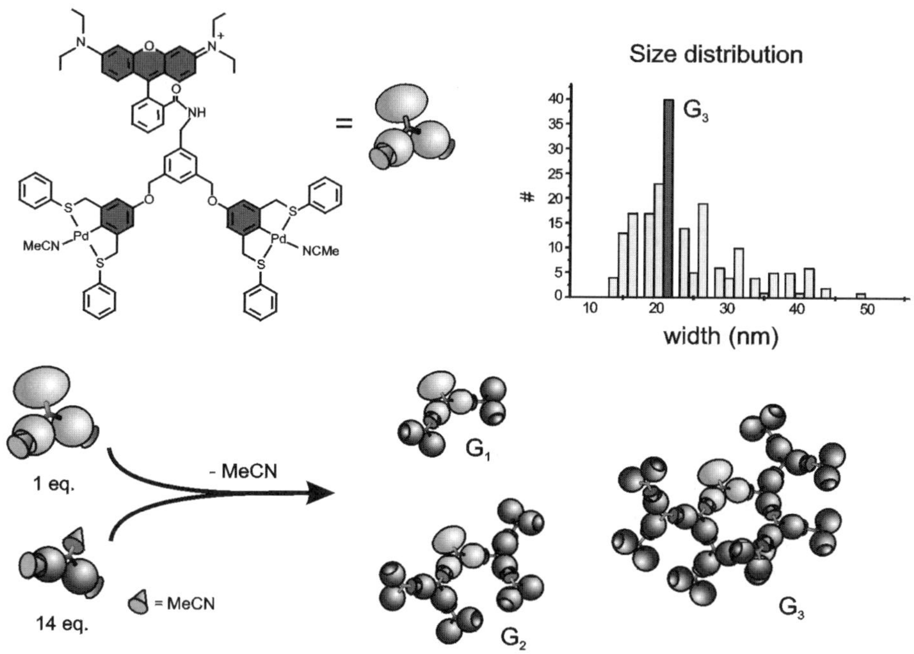

Fig. 30. Uncontrolled self-assembly of Rhodamine B-containing metallodendrimers

molecules. No differences were found in the height distribution of the dendrimers obtained with dynamic mode AFM and shear-force NSOM, emphasizing the non-destructiveness of both techniques. Moreover, the three-dimensional orientation of the individual fluorescent cores within the dendritic matrix could be determined, and these measurements indicated orientational mobility of the core. In combination with scanning probe techniques such as AFM, STM, and NSOM, the study of self-assembled monolayers [104] (SAMs) on surfaces (usually gold) also offers the possibility of investigating and, ultimately, manipulating the properties of single molecules. On gold, for example, functional molecules can be "isolated" by embedding them into a preformed self-assembled monolayer of (ω-functionalized) alkanethiols [105]. This can be achieved by an exchange mechanism in which some thiol adsorbates desorb from the surface upon exposure to solvent [106], and the created "defects" are subsequently filled by the adsorbates to be studied. To this end, the adsorbates must be functionalized with moieties having a significant affinity for gold, such as thiols, sulfides, and disulfides. Because of their large size, dendrimers are exquisite candidates for single molecule studies using scanning probe techniques [107]. In our group, we have been utilizing (metallo)dendrimers functionalized with long chain sulfides in order to visualize them on a gold surface by AFM [108]. Metallodendrimers (G1-G2) based on pyridine ligands were exposed to SAMs of decanethiol. Depending on the quality of the starting monolayer, the concentration of the dendrimer solution and the time of exposure of this solution to the gold surface, varying quantities of isolated features of nanosize dimensions could be distinguished by AFM (Fig. 31 bottom). In contrast, control experiments in which either the sulfide chain on the dendrimer was absent or the sulfur atom in the chain was replaced by a methylene group did not reveal such features (Fig. 31 top). This clearly demonstrates the necessity for an anchoring point for the gold surface.

As a final example, we have combined the coordination chemistry used to construct metallodendrimers with hydrogen bonding in the formation of dendritic *rosettes* [109]. The self-assembly of *covalent* dendrons containing focal bis(isophthalic acid) units into hydrogen-bonded hexameric cyclic arrays was reported earlier by Zimmerman and coworkers [110]. In our strategy [111] three metallodendrons, divergently synthesized using **BB$_{CN}$-Cl** and a barbituric acid-functionalized Pd(II) pincer, were bound via their focal barbituric acid groups to *N*-octadecanyl-*N′*-(2-*N*-*tert*-Boc-amino)phenylmelamine through 18 hydrogen bonds in a cyclic [3 + 3] fashion to form the hexameric rosette (Fig. 32), which was characterized using a variety of NMR techniques (low temperature ^{1}H, 2D-NOESY and TOCSY). NOE build-up curves were used to determine the rotation correlation times τ_c, and as expected these values increased with increasing size of the rosette dendrimers.

4.6 Conclusions

The class of metallodendrimers belonging to the "metals as building block connectors" type is the most diverse one of the three types discussed in this review. This is because coordination chemistry has provided this field with a large array of transition metals and corresponding ligand coordination geometries, thereby hardly posing any restrictions on the design of dendritic building blocks. This allows the resulting metallodendrimers to be tailored with specific functions, arising from the metals themselves (redox or catalytic activity, luminescence, etc.), the organic spacers bridging the metals (e.g., conducting acetylenes), or from their interplay.

5 Concluding Remarks

As evident from this review, metallodendrimers have grown from curiosities one decade ago towards common macromolecules nowadays. The exploitation of coordination chemistry has proven to be very useful in the non-covalent synthesis of metallodendrimers. At present, dendritic construction via metal-ligand coordination is straightforward in most cases. Moreover, and crucially, the number of techniques being used for (metallo)dendrimer characterization is ever increasing. NMR (from simple ^{1}H spectra to pulsed field gradient spin-echo measurements), IR, UV-VIS, and luminescence spectroscopy, mass spectrometry (in particular ES- and MALDI-TOF MS), size exclusion chromatography, cyclic voltammetry, and transmission electron, atomic force, and scanning tunneling microscopy are among the most useful characterization methods for these fas-

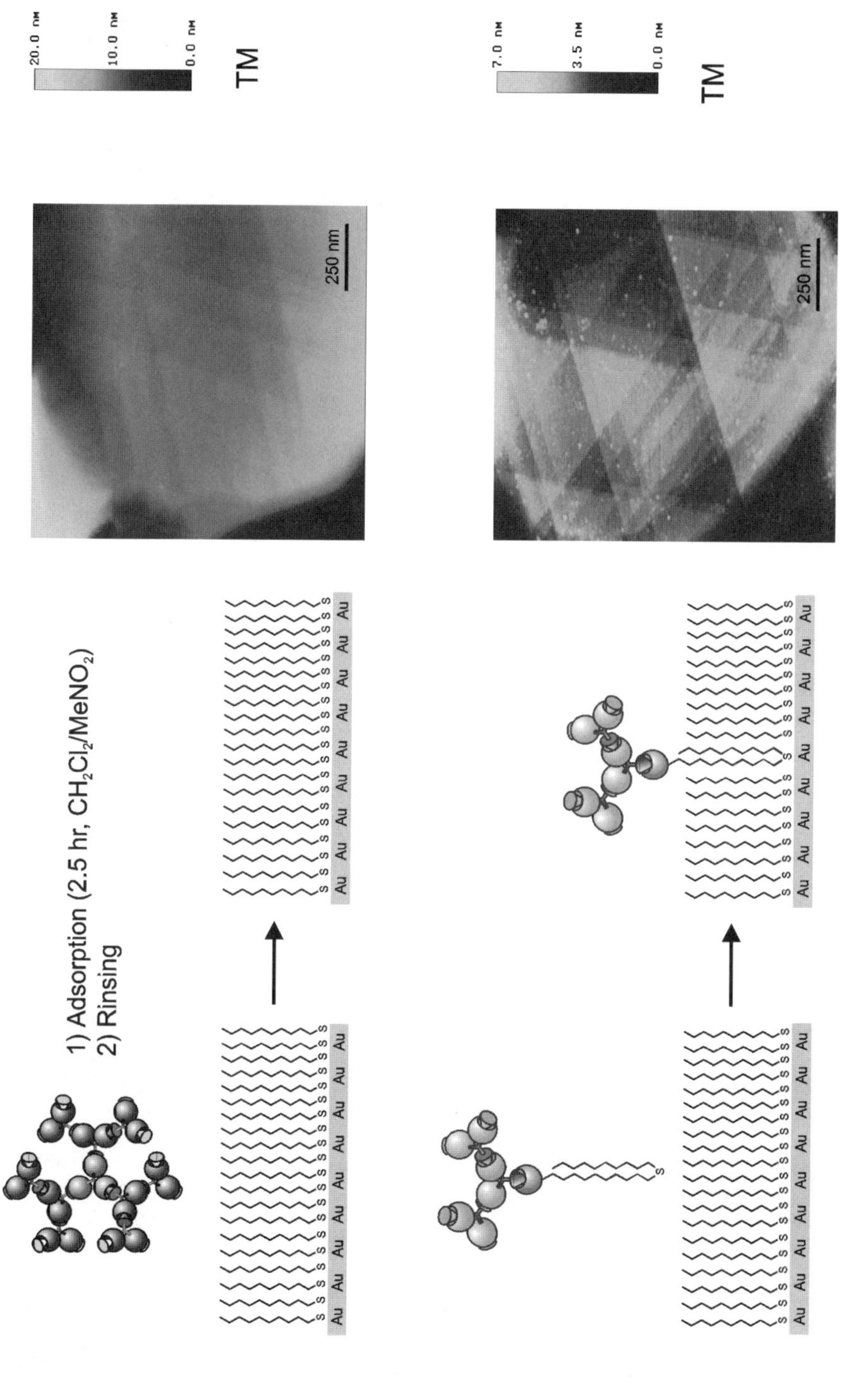

Fig. 31. Incorporation of metallodendrimers into self-assembled monolayers of decanethiol on gold. Tapping mode AFM pictures are displayed on the *right*, in which triangular gold terraces and nanosize features (*bottom*) can be clearly distinguished

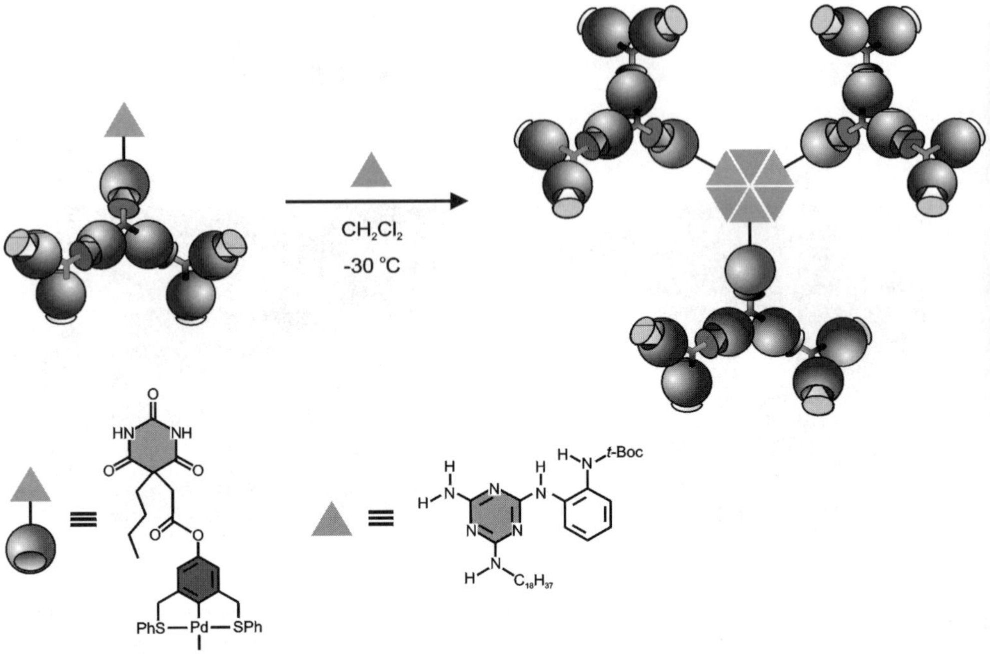

Fig. 32. Metallodendrimers held together by a combination of coordination and hydrogen bonds. The core contains the hydrogen-bonded rosette motif

cinating architectures. With respect to applications, their enormous foreseen potential has yet to be fully explored. However, promising utilitarian aspects have already been reported for metallodendrimers in the fields of catalysis (e.g., recyclable homogeneous catalysts) [7] and medical diagnostics (e.g., MRI contrast reagents) [15]. No doubt these initial successes will stimulate metallodendrimer research to be intensified for years to come.

6 References and Notes

1. (a) Flory PJ (1941) J Am Chem Soc 63:3083–3090; (b) (1941) J Am Chem Soc 63:3091–3096; (c) (1941) J Am Chem Soc 63:3096–3100
2. Buhleier E, Wehner W, Vögtle F (1978) Synthesis 155–158
3. Tomalia DA, Baker H, Dewald JR, Hall M, Kallos G, Martin S, Roeck J, Ryder J, Smith P (1985) Polym J 17:117–132
4. Newkome GR, Yao Z-Q, Baker GR, Gupta K (1985) J Org Chem 50:2003–2004
5. For reviews including discussions on dendritic growth strategies, see: (a) Matthews OA, Shipway AN, Stoddart JF (1998) Prog Polym Sci 23:1–56; (b) Chow H-F, Mong TK-K, Nongrum MF, Wan C-W (1998) Tetrahedron 54:8543–8660
6. For an overview, see: Cuadrado I, Morán M, Casado CM, Alonso B, Losada J (1999) Coord Chem Rev 193–195:395–445
7. For examples, see: (a) Knapen JWJ, van der Made AW, de Wilde JC, van Leeuwen PWNM, Wijkens P, Grove DM, van Koten G (1994) Nature 372:659–663; (b) Reetz MT, Lohmer G,

Schwickardi R (1997) Angew Chem Int Ed Engl 36:1526–1529; (c) Bourque SC, Maltais F, Xiao W-J, Tardif O, Alper H, Arya P, Manzer LE (1999) J Am Chem Soc 121:3035–3038
8. (a) Newkome GR, Moorefield CN (1993) Polym Prepr 34:75–76; (b) Newkome GR, Gross J, Moorefield CN, Woosley BD (1997) Chem Commun 515–516
9. (a) Ottaviani MF, Bossmann S, Turro NJ, Tomalia DA (1994) J Am Chem Soc 116:661–671; (b) Zhao M, Sun L, Crooks RM (1998) J Am Chem Soc 120:4877–4878; (c) Zhao M, Crooks RM (1999) Angew Chem Int Ed 38:364–366; (d) Balogh L, Tomalia DA (1998) J Am Chem Soc 120:7355–7356
10. (a) Newkome GR, He E, Moorefield CN (1999) Chem Rev 99:1689–1746; (b) Bosman AW, Janssen HM, Meijer EW (1999) Chem Rev 99:1665–1688; (c) Fischer M, Vögtle F (1999) Angew Chem Int Ed 38:884–905; (d) Hearshaw MA, Moss JR (1999) Chem Commun 1–8; (e) Gorman CB (1998) Adv Mater 10:295–309; (f) Zeng F, Zimmerman SC (1997) Chem Rev 97:1681–1712; (g) Newkome GR, Moorefield CN, Vögtle F (1996) Dendritic macromolecules: concepts, syntheses, perspectives. VCH, Weinheim, Germany
11. Majoral J-P, Caminade A-M (1999) Chem Rev 99:845–880
12. (a) Peerlings HWI, Meijer EW (1997) Chem Eur J 3:1563–1570; (b) Seebach D, Reiner PB, Greiveldinger G, Butz T, Sellner H (1998) Top Curr Chem 197:125–164
13. Jayaraman N, Nepogodiev SA, Stoddart JF (1997) Chem Eur 3:1193–1199
14. Schlüter AD, Rabe JP (2000) Angew Chem Int Ed 39:864–883
15. Krause W, Hackmann-Schlichter N, Maier FK, Müller R (2000) Top Curr Chem 210:261–308
16. Hawker CJ, Wooley KL, Fréchet JMJ (1993) J Am Chem Soc 115:4375–4376
17. (a) Bertini I, Gray HB, Lippard SJ, Valentine JS (1994) Bioinorganic chemistry. University Science Books, Mill Valley, USA; (b) Stryer L (1995) Biochemistry, 4th edn. Freeman, New York, USA
18. (a) Mahler G, May V, Schreiber M (eds) (1996) Molecular electronics: properties, dynamics and applications. Marcel Dekker, New York; (b) Lazarev PI (ed) (1991) Molecular electronics: materials and methods. Kluwer Academic, Dordrecht, The Netherlands
19. (a) Dandliker PJ, Diederich F, Gross M, Knobler CB, Louati A, Sanford EM (1994) Angew Chem Int Ed Engl 33:1739–1742; (b) Dandliker PJ, Diederich F, Gisselbrecht J-P, Louati A, Gross M (1995) Angew Chem Int Ed Engl 34:2725–2728; (c) Weyermann P, Gisselbrecht J-P, Boudon C, Diederich F, Gross M (1999) Angew Chem Int Ed 38:3215–3219
20. Gorman CB, Parkhurst BL, Su WY, Chen K-Y (1997) J Am Chem Soc 119:1141–1142
21. Gorman CB, Hager MW, Parkhurst BL, Smith JC (1998) Macromolecules 31:815–822
22. Gorman CB, Smith JC, Hager MW, Parkhurst BL, Sierzputowska-Gracz H, Haney CA (1999) J Am Chem Soc 121:9958–9966
23. Moore JS (1997) Acc Chem Res 30:402–413
24. Gorman CB, Su WY, Jiang H, Watson CM, Boyle P (1999) Chem Commun 877–878
25. Wang R, Zheng Z (1999) J Am Chem Soc 121:3549–3550
26. Juris A, Balzani V, Barigelletti F, Campagna S, Belser P, von Zelewsky A (1988) Coord Chem Rev 84:85–277
27. Issberner J, Vögtle F, De Cola L, Balzani V (1997) Chem Eur J 3:706–712
28. (a) Newkome GR, Nayak A, Behera RK, Moorefield CN, Baker GR (1992) J Org Chem 57:358–362; (b) Newkome GR, Moorefield CN, Baker GR, Behera RK, Escamilia GH, Saunders MJ (1992) Angew Chem Int Ed Engl 31:917–919
29. Mattai S, Seiler P, Diederich F, Gramlich V (1995) Helv Chim Acta 78:1904–1912
30. Plevoets M, Vögtle F, De Cola L, Balzani V (1999) New J Chem 63–69
31. Other dendritic antenna systems have been reported. See the work of Balzani and coworkers in the "metals as branching centers" section and also: (a) Steward GM, Fox MA (1996) J Am Chem Soc 118:4354–4360; (b) Shorttreed MR, Swallen SF, Shi ZY, Tan W, Xu Z, Devadoss C, Moore JS, Kopelman R (1997) J Phys Chem B 101:6318–6322
32. Vögtle F, Plevoets M, Nieger M, Azzellini GC, Credi A, De Cola L, De Marchis V, Venturi M, Balzani V (1999) J Am Chem Soc 121:6290–6298
33. Hoffman MZ, Bolletta F, Moggi L, Hug GL (1989) J Phys Chem Ref Data 18:219–544
34. Chow H-F, Chan IY-K, Chan DTW, Kwok RWM (1996) Chem Eur J 2:1085–1091

35. Armstrong FA, Hill HAO, Walton NJ (1988) Acc Chem Res 21:407–413
36. Gray HB, Winkler JR (1996) Ann Rev Biochem 65:537–561
37. Newkome GR, Güther R, Moorefield CN, Cardullo F, Echegoyen L, Pérez-Cordero E, Luftmann H (1995) Angew Chem Int Ed Engl 34:2023–2026
38. Tomoyoso Y, Jiang D-L, Jin R-H, Aida T, Yamashita T, Horie K, Yashima E, Okamoto Y (1996) Macromolecules 29:5236–5238
39. Darling SL, Mak CC, Bampos N, Feeder N, Teat SJ, Sanders JKM (1999) New J Chem 23:359–364
40. There is a vast amount of literature on multiporphyrin systems. For some representative covalent multiporphyrin arrays, see: (a) Wagner RW, Johnson TE, Lindsey JS (1996) J Am Chem Soc 118:11,166–11,180; (b) Officer DL, Burrell AK, Reid DC (1996) Chem Commun 1657–1658; (c) Mak CC, Bampos N, Sanders JKM (1998) Angew Chem Int Ed Engl 37:3020–3023; (d) Yeow EKL, Ghiggino KP, Reek JNH, Crossley MJ, Bosman AW, Schenning APHJ, Meijer EW (2000) J Phys Chem B 104:2596–2606; (e) Aratani N, Osuka A, Kim YH, Heong DH, Kim D (2000) Angew Chem Int Ed 39:1458–1462
41. Non-covalently constructed multiporphyrin arrays: (a) Sessler JL, Wang B, Springs SL, Brown CT (1996) Electron- and energy-transfer reactions in noncovalently linked supramolecular model systems. In: Lehn J-M, Atwood JL, Davis JED, MacNicol DD, Vögtle F (eds) Comprehensive supramolecular chemistry. Pergamon, Oxford, UK, 1987–1996, vol 4, chap 9; (b) Harriman A, Sauvage J-P (1996) Chem Soc Rev 25:41–48; (c) Ward MD (1997) Chem Soc Rev 26:365–375; (d) Drain CM, Nifiatis F, Vasenko A, Batteas JD (1998) Angew Chem Int Ed 37:2344–2347; (e) Drain CM, Russell KC, Lehn J-M (1996) Chem Commun 337–338; (f) Reek JNH, Schenning APHJ, Bosman AW, Meijer EW, Crossley MJ (1998) Chem Commun 11–12; (g) Huck WTS, Rohrer A, Anilkumar AT, Fokkens RH, Nibbering NMM, van Veggel FCJM, Reinhoudt DN (1998) New J Chem 22:165–168
42. Enomoto M, Aida T (1999) J Am Chem Soc 121:874–875
43. Catalano VJ, Parodi N (1997) Inorg Chem 36:537–541
44. Balch AL, Catalano VJ, Lee JW, Olmstead MM, Parkin SR (1991) J Am Chem Soc 113:8953–8955
45. Kawa M, Fréchet JMJ (1998) Chem Mater 10:286–296
46. Venturi M, Serroni S, Juris A, Campagna S, Balzani V (1998) Top Curr Chem 197:193–228
47. Campagna S, Denti G, Sabatino L, Serroni S, Ciano M, Balzani V (1989) J Chem Soc Chem Commun 1500–1501
48. Denti G, Campagna S, Serroni S, Ciano M, Balzani V (1992) J Am Chem Soc 114:2944–2950
49. Kim M-J, MacDonnell FM, Gimon-Kinsel ME, du Bois T, Asgharian N, Griener JC (2000) Angew Chem Int Ed 39:615–619
50. Juris A, Balzani V, Campagna S, Denti G, Serroni S, Frei G, Güdel HU (1994) Inorg Chem 33:1491–1496
51. Campagna S, Giannetto A, Serroni S, Denti G, Trusso S, Mallamace F, Micali N (1995) J Am Chem Soc 117:1754–1758
52. Serroni S, Denti G, Campagna S, Juris A, Ciano M, Balzani V (1992) Angew Chem Int Ed Engl 31:1493–1495
53. The protected decanuclear dendrimer and its deprotected analog were reported later in a paper describing generalized versions of the synthetic strategies: Campagna S, Denti G, Serroni S, Juris A, Venturi M, Ricevuto V, Balzani V (1995) Chem Eur J 1:211–221
54. Serroni S, Juris A, Venturi M, Campagna S, Resino IR, Denti G, Credi A, Balzani V (1997) J Mater Chem 7:1227–1236
55. For an overview of luminescent and redox-active polynuclear transition metal complexes, see: Balzani V, Juris A, Venturi M, Campagna S, Serroni S (1996) Chem Rev 96:759–833
56. Moucheron C, Kirsch-De Mesmaeker A, Dupont-Gervais A, Leize E, van Dorsselaer A (1996) J Am Chem Soc 118:12,834–12,835
57. Latterini L, Schweitzer G, De Schrijver FC, Moucheron C, Kirsch-De Mesmaeker A (1997) Chem Phys Lett 281:267–271

58. Latterini L, Pourtois G, Moucheron C, Lazzaroni R, Brédas J-L, Kirsch-De Mesmaeker A, De Schrijver FC (2000) Chem Eur J 6:1331–1336
59. MacDonnell FM, Bodige S (1996) Inorg Chem 35:5758–5759
60. Bodige S, Torres AS, Maloney DJ, Tate D, Kinsel GR, Walker AK, MacDonnell FM (1997) J Am Chem Soc 119:10,364–10,369
61. Kim M-J, MacDonnell FM, Gimon-Kinsel ME, Du Bois T, Asgharian N, Griener JC (2000) Angew Chem Int Ed 39:615–619
62. Achar S, Puddephatt RJ (1994) Angew Chem Int Ed Engl 33:847–849
63. (a) Achar S, Puddephatt RJ (1994) J Chem Soc Chem Commun 1895–1896; (b) Achar S, Vittal JJ, Puddephatt RJ (1996) Organometallics 15:43–50
64. Achar S, Puddephatt RJ (1995) Organometallics 14:1681–1687
65. Achar S, Immoos CE, Hill MG, Catalano V (1997) J Inorg Chem 36:2314–2320
66. Liu G-X, Puddephatt RJ (1996) Organometallics 15:5257–5259
67. For an overview, see: Constable EC (1997) Chem Commun 1073–1080
68. Sauvage J-P, Collin J-P, Chambron JC, Guillerez S, Coudret C, Balzani V, Barigelletti F, De Cola L, Flamigni L (1994) Chem Rev 94:993–1019
69. Newkome GR, Cardullo F, Constable EC, Moorefield CN, Cargill Thompson AMW (1993) J Chem Soc Chem Commun 925–927
70. Newkome GR, He E (1997) J Mater Chem 7:1237–1244
71. Newkome GR, He E (1998) Macromolecules 31:4382–4386
72. For another example of electronic communication between transition metals in dendrimers, see: Cuadrado I, Casado CM, Alonso B, Morán M, Losada J, Belsky V (1997) J Am Chem Soc 119:7613–7614
73. Newkome GR, Patri AK, Godínez LA (1999) Chem Eur J 5:1445–1451
74. Constable EC, Harverson P (1996) Chem Commun 33–34
75. Constable EC, Harverson P (1996) Inorg Chim Acta 252:281–291
76. Constable EC, Harverson P, Ramsden JJ (1997) Chem Commun 1683–1684
77. Constable EC, Harverson P, Oberholzer M (1996) Chem Commun 1821–1822
78. Constable EC, Housecroft CE, Cattalini M, Phillips D (1998) New J Chem 193–200
79. (a) Constable EC, Leese TA (1988) Inorg Chim Acta 146:55–58; (b) Constable EC, Cargill Thompson AMW, Harverson P, Macko L, Zehnder M (1995) Chem Eur J 1:360–367
80. Armspach D, Cattalini M, Constable EC, Housecroft CE, Philips D (1996) Chem Commun 1823–1824
81. (a) Hawthorne MF (1993) Angew Chem Int Ed Engl 32:950–984; (b) Soloway AH, Tjarks W, Barnum BA, Rong F-G, Barth RF, Codogni IM, Wilson J G (1998) Chem Rev 98:1515–1562
82. Hagihara N, Sonogashira K, Takahashi S (1981) Adv Polym Sci 41:149–179
83. Ohshiro N, Takei F, Onitsuka K, Takahashi S (1996) Chem Lett 871–872
84. Leininger S, Stang PJ, Huang S (1998) Organometallics 17:3981–3987
85. Onitsuka K, Fujimoto M, Ohshiro N, Takahashi S (1999) Angew Chem Int Ed 38:689–692
86. McDonagh AM, Humphrey MG, Samoc M, Luther-Davies B (1999) Organometallics 18:5195–5197
87. McDonagh AM, Humphrey MG, Samoc M, Luther-Davies B, Houbrechts S, Wada T, Sasabe H, Persoons A (1999) J Am Chem Soc 121:1405–1406
88. Bhawalker JD, He GS, Prasad PN (1996) Rep Prog Phys 59:1041–1070
89. Osawa M, Hoshino M, Horiuchi S, Wakatsuki Y (1999) Organometallics 18:112–114
90. These tridentate ligands are usually abbreviated as NCN, PCP and SCS pincer ligands
91. (a) Steenwinkel P, Gossage RA, van Koten G (19998) Chem Eur J 4:759–762, and references cited therein; (b) Weisman A, Gozin M, Kraatz H-B, Milstein D (1996) Inorg Chem 35:1792–1797, and references cited therein
92. Errington J, McDonald WS, Shaw BJ (1980) J Chem Soc Dalton Trans 2312–2314
93. (a) Kickham JE, Loeb SJ, Murphy SL (1997) Chem Eur J 3:1203–1213; (b) Cameron BR, Loeb SJ, Yap GPA (1997) Inorg Chem 36:5498–5504
94. Bergbreiter DE, Osburn PL, Liu Y-S (1999) J Am Chem Soc 121:9531–9538
95. Hall J, Loeb SJ, Shimizu GKH, Yap GPA (1998) Angew Chem Int Ed 37:121–123

96. Huck WTS, van Veggel FCJM, Kropman BL, Blank DHA, Keim EG, Smithers MMA, Reinhoudt DN (1995) J Am Chem Soc 117:8293–8294
97. (a) Huck WTS, Snellink-Ruël BHM, Lichtenbelt JWT, van Veggel FCJM, Reinhoudt DN (1997) Chem Commun 9–10; (b) Huck WTS, van Veggel FCJM, Reinhoudt DN (1997) J Mater Chem 7:1213–1219
98. Huck WTS, van Veggel FCJM, Reinhoudt DN (1996) Angew Chem Int Ed Engl 35: 1213–1215
99. Huck WTS, van Veggel FCJM, Sheiko SS, Möller M, Reinhoudt DN (1998) J Phys Org Chem 12:540–545
100. For a few representative examples, see: (a) Leize E, van Dorsselaer A, Krämer R, Lehn J-M (1993) J Chem Soc Chem Commun 990–993; (b) Hasenknopf B, Lehn J-M, Boumediene N, Dupont-Gervais A, van Dorsselaer A, Kneisel B, Fenske D (1997) J Am Chem Soc 119:10,956–10,962; (c) Fujita M, Fujita N, Ogura K, Yamaguchi K (1999) Nature 400:52–55; (d) Olenyuk B, Whiteford JA, Fechtenkötter A, Stang PJ (1999) Nature 398:796–799
101. Huck WTS, Prins LJ, Fokkens RH, Nibbering NMM, van Veggel FCJM, Reinhoudt DN (1998) J Am Chem Soc 120:6240–6246
102. Veerman JA, Levi SA, van Veggel FCJM, Reinhoudt DN, van Hulst NF (1999) J Phys Chem A 103:11,264–11,270
103. Dunn RC (1999) Chem Rev 99:2891–2927
104. Ulman A (1991) An introduction to ultrathin organic films: from Langmuir-Blodgett to self-assembly. Academic Press, San Diego
105. See, for example: (a) Bumm LA, Arnold JJ, Cygan MT, Dunbar TD, Burgin TP, Jones L II, Allara DL, Tour JM, Weiss PS (1996) Science 271:1705–1707; (b) Offord DA, Sachs SB, Ennis MS, Eberspacher TA, Griffin JH, Chidsey CED, Collman JP (1998) J Am Chem Soc 120:4478–4487; (c) Weck M, Jackiw JJ, Rossi RR, Weiss PS, Grubbs RH (1999) J Am Chem Soc 121:4088–4089
106. Schlenoff JB, Li M, Ly H (1995) J Am Chem Soc 117:12,528–12,536
107. (a) Hierlemann A, Campbell JK, Baker LA, Crooks RM, Ricco AJ (1998) J Am Chem Soc 120:5323–5324, and references cited therein; (b) Li J, Piehler LT, Qin D, Baker JR, Tomalia DA, Meier DJ (2000) Langmuir 16:5613–5616
108. (a) Huisman B-H, Schönherr H, Huck WTS, Friggeri A, van Manen H-J, Menozzi E, Vancso GJ, van Veggel FCJM, Reinhoudt DN (1999) Angew Chem Int Ed 38:2248–2251; (b) Friggeri A, Schönherr H, van Manen H-J, Huisman B-H, Vancso GJ, Huskens J, van Veggel FCJM, Reinhoudt DN (2000) Langmuir 16:7757–7763
109. More information about hydrogen-bonded rosettes can be found in: (a) Whitesides GM, Simanek EE, Mathias JP, Seto CT, Chin DN, Mammen M, Gordon DM (1995) Acc Chem Res 28:37–44; (b) Simanek EE, Li X, Choi I, Whitesides GM () 1996 Cyanuric acid and melamine: a platform for the construction of soluble aggregates and crystalline materials. In: Lehn J-M, Atwood JL, Davis JED, MacNicol DD, Vögtle F (eds) Comprehensive supramolecular chemistry. Pergamon, Oxford, UK, 1987–1996, vol 9, chap 17; (c) Timmerman P, Vreekamp RH, Hulst R, Verboom W, Reinhoudt DN, Rissanen K, Udachin KA, Ripmeester J (1997) Chem Eur J 3:1823–1832
110. Zimmerman SC, Zeng F, Reichert DEC, Kolotuchin SV (1996) Science 271:1095–1099
111. Huck WTS, Hulst R, Timmerman P, van Veggel FCJM, Reinhoudt DN (1997) Angew Chem Int Ed Engl 36:1006–1008

Dendritic Catalysts

Robert Kreiter · Arjan W. Kleij · Robertus J. M. Klein Gebbink · Gerard van Koten

Utrecht University, Debye Institute, Department of Metal-Mediated Synthesis, Padualaan 8, 3584 CH Utrecht, The Netherlands
E-mail: g.vankoten@chem.uu.nl

Abstract. After the publication of the first papers on dendritic catalysts in 1994, many different examples in this class of (macro)molecular catalysts have been reported in recent years. This chapter provides an overview of (recent) highlights and developments in the field of dendrimer catalysis, with an emphasis on homogeneous catalysis. The distinctive features of periphery-functionalized, chiral and non-chiral metallo-dendrimers are discussed and are compared to those of core-functionalized metallo-dendrimers and metallo-dendrimers containing metal complexes throughout their structure. Furthermore, the class of non-metal-containing dendritic catalysts is described. Special attention is focused on the different types of selectivity encountered in dendrimer catalysis and the concept of dendritic catalyst recycling. A summary of the various reactions catalyzed by dendritic catalysts is provided at the end of this chapter.

Keywords: Dendrimers, (Homogeneous) catalysis, Metals

1 Introduction

Dendrimers have a particular position within the broad spectrum of macromolecules. One of the most striking features of dendrimers is surely their well-defined structure, in contrast to many other types of macromolecules. The elegance often expressed in the fractal-like dendrimer structure has inspired many research groups over the years [1]. Many of the dendrimer properties are introduced by the applied iterative synthesis, and the use of either convergent or divergent strategies allows for the fine-tuning thereof. Among these are monodispersity, often pseudo spherical structure, and amplification of functional groups. The wide range of possibilities offered by dendritic molecular systems has led to the description of many applications in several fields of science [2]. Potential (bio)chemical applications include host-guest chemistry, drug delivery, self-assembly, and usage as sensor materials. One of the most promising applications of dendrimers is found in homogeneous catalysis, in which the usage of a wide variety of dendritic catalysts and catalyst supports is currently being pursued. Some of these systems are based mainly on the amplification of functional groups at the periphery of the structure. This amplification could lead to dendritic catalysts that are large enough to be recovered from a reaction mixture by ultrafiltration or size-exclusion techniques, thereby solving one of the classical separation problems in homogeneous catalysis. It is also possible that the amplification of functional groups enables cooperative effects between peripheral catalytic sites. Other systems make use of a (single) catalytic group at the interior of a dendrimer. In this way, interactions of the catalyst with the reaction medium or with other catalytic sites can be diminished, possibly resulting in substrate selective catalysis. Dendritic catalysts can then become selective and/or tailor-made catalysts, with properties reminiscent of those often encountered for enzymes. Whereas the high degree of perfection of enzymes might be an unreachable goal, the idea of designing catalytic systems with tunable properties, is a true challenge. A last class of dendritic systems combines the properties of a larger structure with the amplification of functional groups within the structure. In these systems the dendrimer backbone functions not only as a "support", but also holds ligating groups in a highly repetitive and uniform manner. This can result in a high catalyst-to-dendrimer ratio, thereby preventing extensive dilution of active material.

Here, we present an overview of the more recent and important earlier achievements in the field of dendritic catalysis, with an emphasis on homogeneous (organo)metallic catalysis. Dendrimers that are functionalized with metal complexes at their periphery (Sect. 2) as well as a their core (Sect. 3) are discussed. Subsequently, dendrimers that contain metal complexes throughout their structure (Sect. 4) and dendritic catalysts that operate without metals (Sect. 5) will be discussed. At the end, a graphical summary of the catalyzed reactions involved is provided (Sect. 6).

2 Metal Functionalities at the Periphery of a Dendrimer

Starting from the concept of attaching metal complexes to the periphery of a dendrimer, many new dendritic catalysts have been developed. Numerous examples involve the attachment of earlier-developed complexes to a dendrimer "support". These dendritic species exhibit new properties including catalyst deactivation due to proximity effects, catalyst stabilization, and even site co-operativity. Dendrimers with non-chiral and chiral peripheral catalytic sites are described below.

2.1 Non-Chiral Metal Complexes at the Periphery of a Dendrimer

The first example of a catalytically active metallo-dendrimer, having catalytic groups at the periphery, was prepared in the group of Van Koten [3], in a collaboration with the group of Van Leeuwen. The synthesis of this metallo-dendrimer started from carbosilane molecules [4] containing silicon-chlorine bonds at their periphery to which (NCN)-type terdendate ligands $\{NCN=[C_6H_3(CH_2NMe_2)_2\text{-}2,6]\}$ were connected. To prevent possible interactions between the different metal sites 1,4-butanediol linkers were placed between the carbosilane backbone and the ligating site. Nickel was introduced in the activated position of the ligands by oxidative addition to a zero-valent nickel source [e.g., $Ni(PPh_3)_4$]. The resulting dendritic aryl-nickel(II)-species **1** (Fig. 1) was applied as a homogeneous catalyst in the *a*tom *t*ransfer *r*adical *a*ddition reaction (ATRA or Kharasch addition reaction) of CCl_4 to methyl methacrylate (MMA). Compared to the mononuclear catalyst, having a similar *para*-functional group, this polynuclear system shows similar behavior, indicating that each NCN-NiX-site (X=Cl, Br) acts independently. The activity per Ni-site is only slightly lower than that of the mononuclear system, which has been ascribed to the non-perfect composition of the metallo-dendrimer. Furthermore, the reaction catalyzed by the polynuclear system involves a clean, regioselective 1:1 addition without telomerization/polymerization or the formation of side products. Due to the dimensions of this metallo-dendrimer (2.5 nm) it was the first example of a metallo-dendritic catalyst that was, in principle, suitable for recovery by membrane filtration techniques.

Stimulated by these results, other periphery-functionalized metallo-dendrimer catalysts based on similar carbosilane backbones were prepared, having NCN-metal units connected directly to the carbosilane backbone [5]. Metal introduction in these systems was possible via lithiation followed by a transmetallation of the polylithiated species using an appropriate d^8 metal salt. This new procedure yielded different generations of polynuclear nickelated carbosilane dendrimers (**2**, Fig. 2) [6], which were again tested as homogeneous catalysts in the Kharasch addition reaction [7]. For this series of dendrimers an interesting dependency of the activity on the generation number was found. The G_0-$(NCN\text{-}NiX)_4$ dendrimer showed an activity comparable to that of the mononuclear catalyst. However, for the G_1-$(NCN\text{-}NiX)_{12}$ and G_2-$(NCN\text{-}NiX)_{36}$ dendrimers a dramatic decrease in activity was observed. A nearly complete loss of activity was found for these

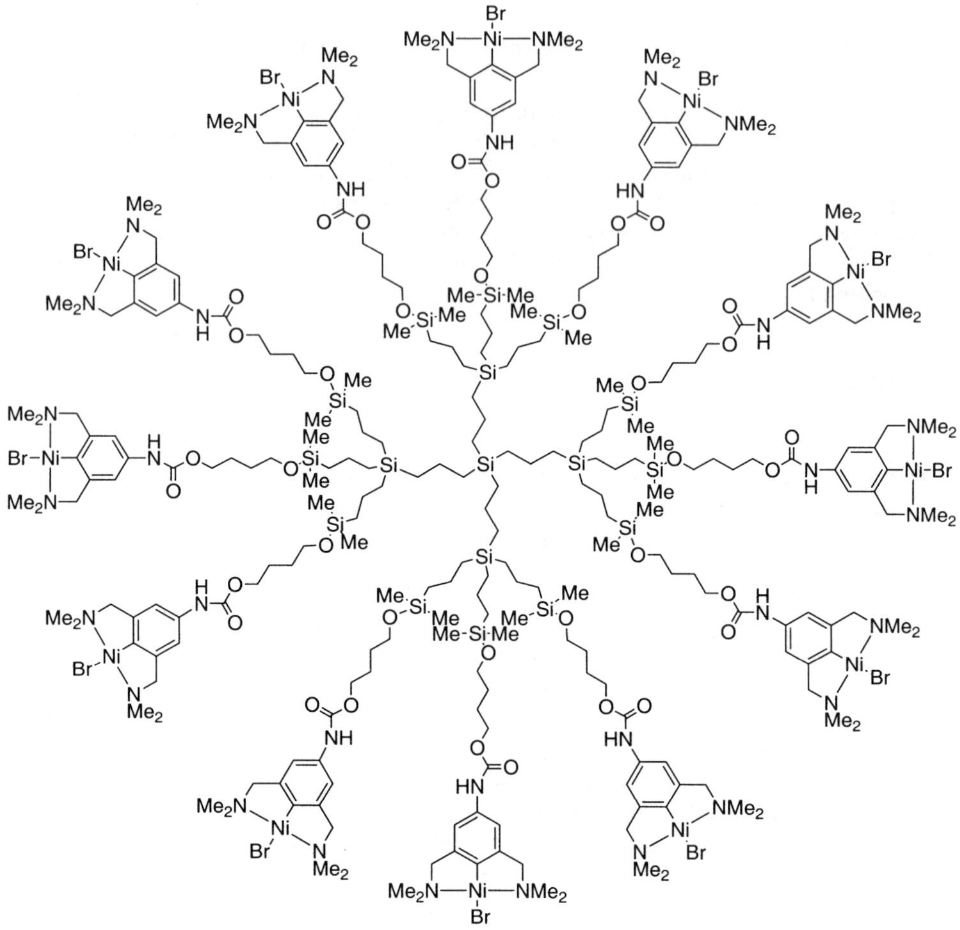

1, G_1-$O(CH_2)_4OC(O)N(H)NCN.NiBr$

Fig. 1. Van Koten's dendritic polynickel complex

higher generation dendrimers with the conversions of MMA being only 18 and 1.5%, respectively. The loss of activity was ascribed to a proximity effect between different Ni(II)-sites, i.e., a (negative) dendritic effect. During the catalytic process a Ni(III) center can interact with a neighboring site (forming a mixed valence complex) rather than with the transient radicals in solution. This effect is very pronounced in the ATRA catalytic process, which involves a Ni(II)/Ni(III) redox couple. In order to test this hypothesis, modifications of the carbosilane backbone were carried out to yield modified dendrimers **3** and **4** (Fig. 2), which have less congested dendrimer peripheries. These species were successfully applied as homogeneous catalysts in the ATRA reaction and indeed showed activities that were again comparable to the mononuclear catalyst. The dendritic G_0-$(NCN\text{-}NiX)_4$ and G_1-$(NCN\text{-}NiX)_{12}$ (**2**) complexes were also tested in

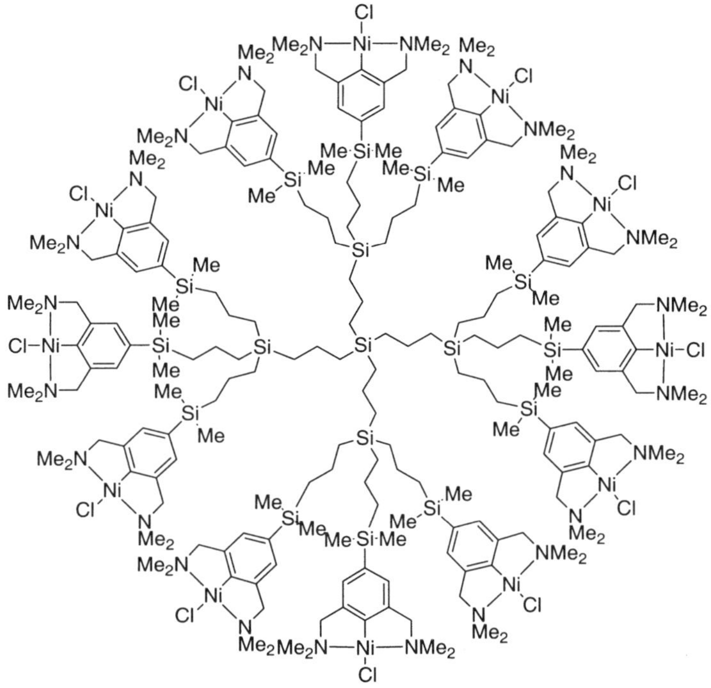

2, G_1-(NCN.NiCl)$_{12}$

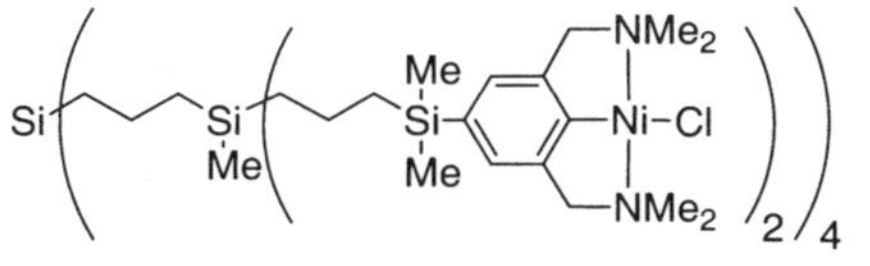

3, G_1'-(NCN.NiCl)$_8$

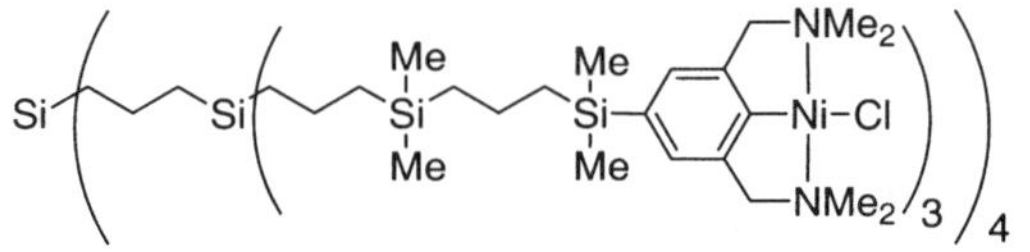

4, G_1-((CH$_2$)$_3$Si(Me)$_2$NCN.NiCl)$_{12}$

Fig. 2. Modified dendritic polynickel complexes prepared by Van Koten et al.

a continuous membrane reactor equipped with a SelRO MPF-50 nanofiltration membrane [8]. These species showed retentions of 97.4% and 99.8%, respectively, which should be sufficient for many applications. In conclusion, it was shown that the dendritic NCN-NiX complex can very well be applied as a recyclable homogeneous catalyst in the ATRA reaction if proximity effects are taken into account.

Organometallic NCN-NiX catalysts were also connected to the periphery of a dendritic framework built up from amino acids (5, Fig. 3), in order to investigate systems that are suitable supports for other functionalized materials. A series of these highly polar compounds was prepared and tested as catalysts in the

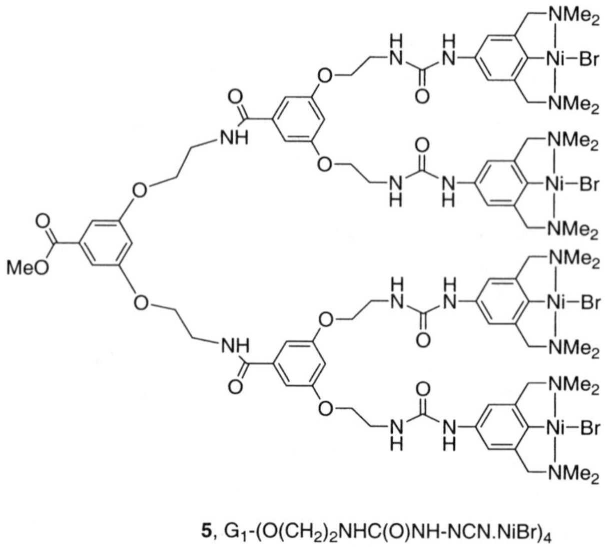

Fig. 3. Amino acid-based dendrimers containing NCN-NiBr catalytic groups

ATRA reaction of CCl_4 with MMA [9]. The catalytic activities of these compounds were in the same order of magnitude as the activity of the mononuclear complex. This indicates that the catalytic reaction is not influenced by the presence of polar groups in the dendrimer backbone.

Recently, a general procedure towards periphery-functionalized carbosilane dendrimers (Scheme 1) was reported [10]. Polylithiated carbosilane dendrimers of different generations were prepared as precursors for various ligand systems including phosphines. One of these dendritic ligands, the 2-pyridyl alcohol-functionalized dendrimer, was reacted with a suitable ruthenium source to form complexes (**6**, Fig. 4) that were a suitable catalysts for the ring closing metathesis (RCM) reaction of bifunctional olefins. In this reaction, the described dendrimer catalysts showed activities that were comparable to that of a unimolecular

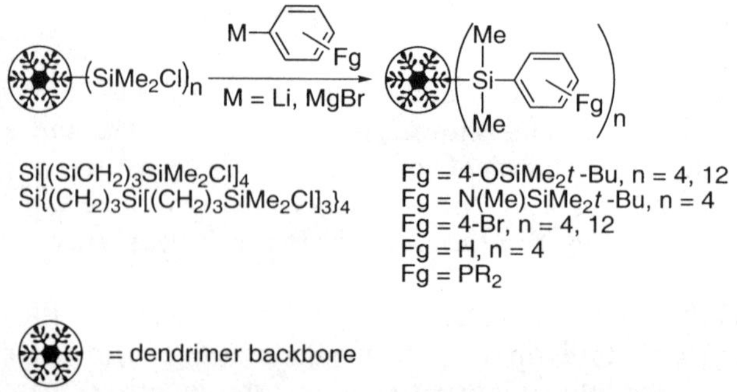

Scheme 1. General route towards periphery functionalized carbosilane dendrimers

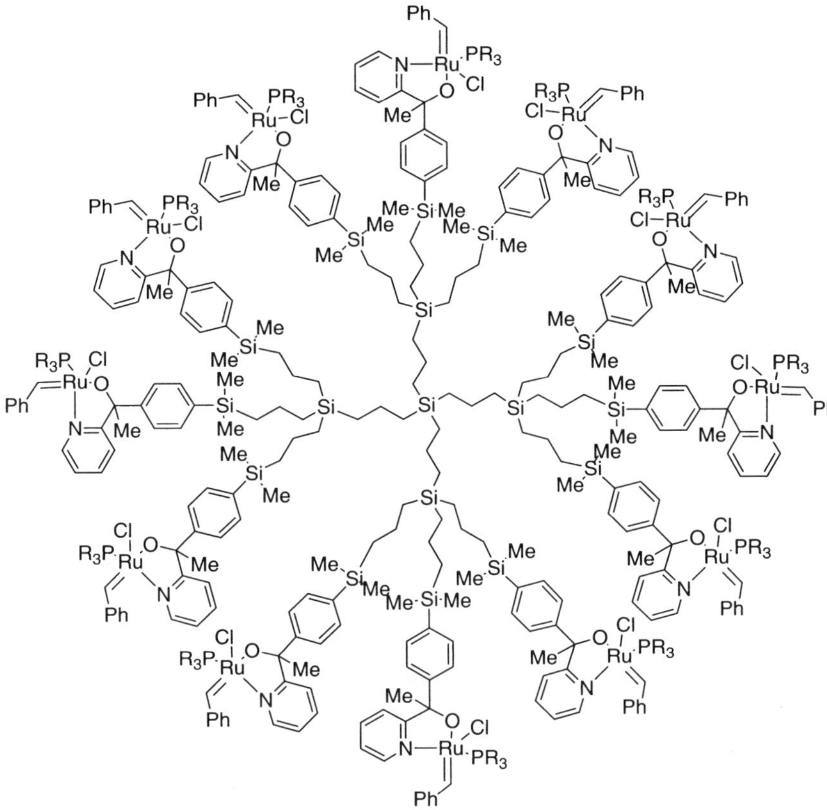

6, G_1-($(CH_2)_3Si(Me)_2C_6H_4C(Me)(py)ORuCl(CHPh)(PPh_3))_{12}$

Fig. 4. Dendritic carbosilane ruthenium complex suitable as catalyst for the RCM reaction of bifunctional olefins reported by Van Koten et al.

catalyst (based on the amount of ruthenium). Furthermore, these catalysts were tested in a commercially available nanofiltration membrane (SelRO MPS-60) in order to separate the catalyst from products and reactants. Although it was shown that leaching of the catalyst through the membrane did not occur under these conditions, the conversion stopped at 20%. Extensive decomposition of the catalyst was observed, which was ascribed to a reaction on the membrane surface.

Similar monomeric and dendrimer-bound ruthenium complexes based on styrenyl ether and 1,3,5-dimesityl-4,5-dihydroimidazol-2-ylidene ligands were reported by Hoveyda et al. [11]. These complexes were applied as metathesis catalysts and the dendritic species could be recovered by silica gel chromatography.

An elegant demonstration of the use of membrane technology for the effective recovery of metallo-dendritic catalysts and for selective product formation was presented by the Van Koten group in collaboration with the group of Vogt

[12]. In this work, carbosilane dendrimers were functionalized at the periphery with various ω-diphenylphosphinocarboxylic acid ester end groups, which can act as hemi-labile bidentate ligands to d^8 metal fragments. The Pd-complexes of these systems were prepared in situ by addition of $[(\eta^3\text{-}C_4H_7)Pd(cod)]BF_4$ (7, Fig. 5) and were subsequently tested in the Pd(II)-catalyzed hydrovinylation of styrene. One of the major problems generally encountered in this reaction was solved using this approach. Since at higher conversions subsequent isomerization of the product (i.e., 3-phenyl-1-butene) to internal olefins (both *E*- and *Z*-isomers) occurs, this reaction has to be run at low conversion with continuous removal of the 3-phenyl-1-butene, or otherwise carried out at high styrene concentrations. A strategy was developed to selectively produce the desired 3-phenyl-1-butene at low conversions under membrane reactor conditions. Under these specific conditions using the G_0-Pd_4 catalyst, a highly selective conversion of styrene to 3-phenyl-1-butene was achieved with no significant isomerization or generation of side products, albeit in very low yield per time unit. A modest retention of this small dendritic species in a nanofiltration membrane system (MPF-60 NF) ($\geq 85\%$) was found, which is far from ideal for continuous operations. Although palladium black was formed inside the reactor,

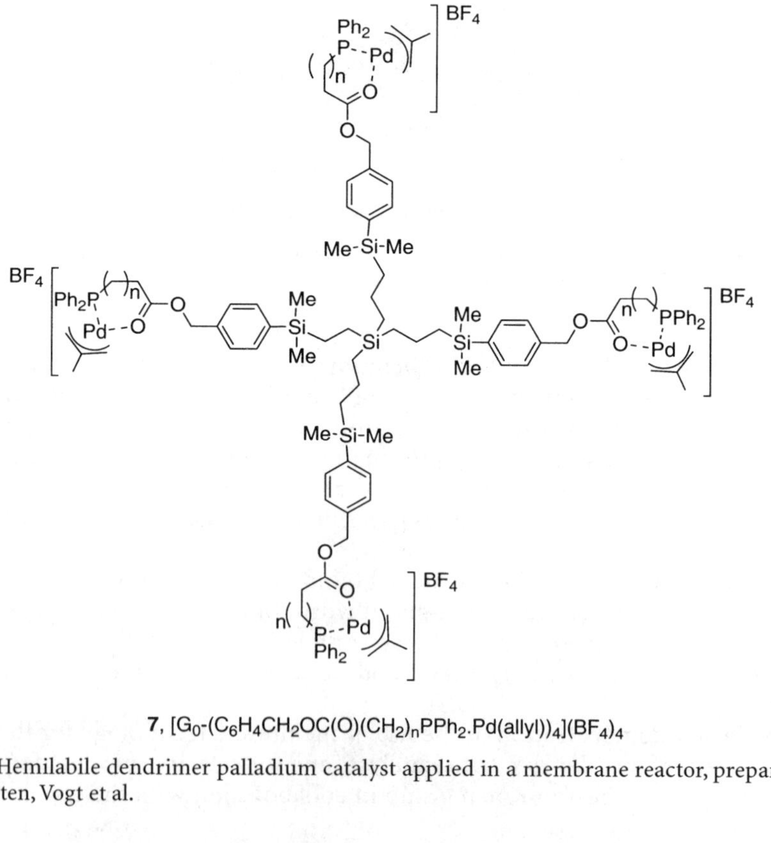

7, $[G_0\text{-}(C_6H_4CH_2OC(O)(CH_2)_nPPh_2.Pd(allyl))_4](BF_4)_4$

Fig. 5. Hemilabile dendrimer palladium catalyst applied in a membrane reactor, prepared by Van Koten, Vogt et al.

the G_0-Pd_4 catalyst did produce 3-phenyl-1-butene during a period of 80 h. The authors expect that a G_1-Pd_{12} catalyst derived from the next generation of dendritic ligands will show sufficient retention in a nanomembrane reactor to give effective catalyst immobilization. The decomposition of the Pd-catalyst is ascribed to the intrinsic properties of this type of palladium catalysis and has also been observed in experiments carried out by Reetz et al., as described below [13]. It should be noted that, in the latter palladium catalytic species, the metal is exclusively bonded via heteroatom donor coordination. This leads to a higher degree of leaching compared to the NCN-metal containing dendrimers in which the metal is bonded via a covalent M–C σ-bond.

An example of phosphine-containing dendrimers (see also Scheme 1) was reported by Reetz et al. [13]. These authors described a DAB-based poly(propylene imine) dendrimer (DAB = 1,4-diaminobutane) which is functionalized at the periphery with diphenylphosphine groups (**8a**, Fig. 6). The phosphine groups together with the nitrogen branching point form a potentially terdentate P,N,P-ligand. A [$PdMe_2$] complex of such a dendrimer was tested as a catalyst in the Heck reaction of bromobenzene and styrene with formation of stilbene. The activity of the dendrimer catalyst is, surprisingly, higher than that of the corresponding monomeric catalyst. Furthermore, unlike their monomeric analogues the dendritic catalysts do not decompose to elemental Pd. This is an interesting example of a positive dendritic effect on catalyst stability in homogeneous catalysis. These dendrimers also showed good activities as precatalysts in the allylic substitution of methyl (3-phenyl-2-propenyl)acetate with morpholine. More recently, the use of these metallo-dendrimers in a continuously operated membrane reactor was demonstrated [14]. The authors also prepared Rh and Ir complexes of these dendrimers. Preliminary results indicate that the Rh-complexes are effective hydroformylation catalysts.

Recently, Reetz and co-workers have shown that sulfonylated DAB-based poly(propylene imine) dendrimers can be cross-linked using scandium triflate [15]. This yields a material that can serve as a heterogeneous catalyst in several reactions, such as the reaction of benzaldehyde, aniline, and an enolsilane to

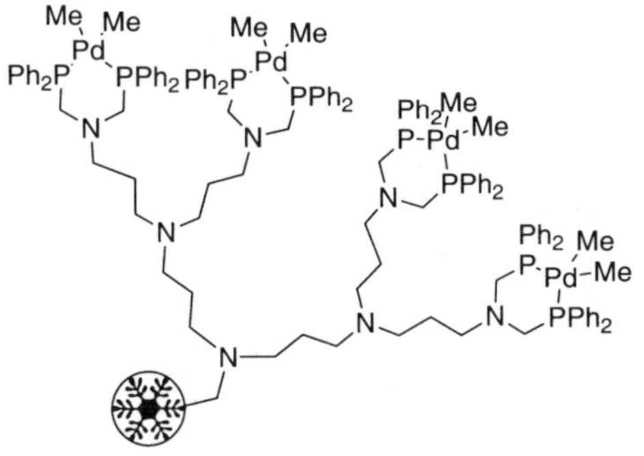

8a, G_2-($PNP.PdMe_2$)$_8$
8b, G_3-($PNP.PdCl_2$)$_{16}$

Fig. 6. Phosphine dendrimer catalyst prepared by Reetz et al.

yield β-amino ketones, Mukaiyama aldol additions, and the Diels-Alder reaction of methyl vinyl ketone with cyclopentadiene. The authors showed that the cross-linked dendrimer material could be recycled without loss of activity.

Kaneda and co-workers applied a ligand system comparable to that of Reetz et al. [16]. These ligands were used to introduce $[PdCl_2]$ units to form dendritic Pd(II) complexes (**8b**, Fig. 6) that were applied in the hydrogenation of conjugated and non-conjugated olefins. In the case of the conjugated olefins the dendrimer complex proved to be a highly effective hydrogenation catalyst. Remarkably as observed for **8a**, the activity of this polynuclear complex was higher compared to that of the corresponding mononuclear complex. The authors also performed the same hydrogenation reaction under heterogeneous conditions and recovered the dendritic catalyst. They showed that the activity as well as the XPS and IR spectra of the spent catalyst were comparable to those of the fresh catalyst.

The research group of Van Leeuwen reported on carbosilane dendrimers appended with peripheral diphenylphosphino end groups [17]. After in situ complexation with allylpalladium chloride, the resultant metallo-dendrimer (**9**, Fig. 7) was used as catalyst in the allylic alkylation of sodium diethyl malonate with allyl trifluoroacetate in a continuous flow reactor. Unlike in the batch reaction, in which a very high activity of the dendrimer catalyst and quantitative conversion of the substrate was observed, a rapid decrease in space-time yield of the product was noted inside the membrane reactor. The authors concluded that this can most probably be ascribed to catalyst decomposition. The product flow (i.e., outside the membrane reactor) was also investigated and it was shown that no active catalyst had leached through the membrane.

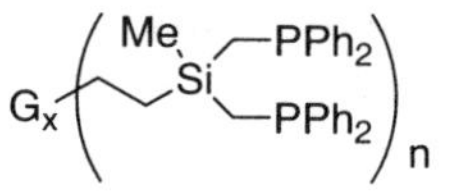

9

For **9**: x =1 and n = 12

Fig. 7. Carbosilane dendrimer-based phosphine ligand prepared by Van Leeuwen et al.

Recently, the same authors reported on rhodium complexes of these phosphine dendrimers that were applied as catalysts in the hydroformylation of 1-octene[18]. They describe monodentate and bidentate phosphine ligands attached to carbosilane dendrimers containing 2 and 3 carbon atoms between the branching points. The ratio of linear to branched product is about the same for all catalysts reported. However, the monodentate phophines showed higher activities than their bidentate counterparts. Furthermore, for the monodentate phosphines the C_3-spacer dendrimers showed higher activities than the more compact C_2-spacer dendrimers, in contrast to the bidentate phosphines where no effect of the spacer was observed. Higher generation dendrimers generally gave slower rates. The authors suggested that the change in activity for the monodentate phophines is due to the distance between the individual phosphines and thus the (dendrimer)-P-Rh-P ring size. Preliminary results on membrane ultrafiltration using a commercially available membrane (SelRO MPF-60) showed that this membrane was not compatible with the applied hydroformylation conditions, due to solvent and temperature restrictions.

Other phosphorus-based dendrimers have been described by Majoral, Chaudret and co-workers (**10**, Fig. 8). The authors describe the incorporation of Pd, Pt, and Rh in the diphosphine moieties of these dendrimers [19]. Furthermore, they describe organometallic experiments on the surfaces of these complexes, which indicate that such complexes can serve as homogeneous catalysts. More recently, the authors showed that palladium(II) and ruthenium(II) complexes of these phosphorus-based dendrimers can indeed be applied as homogeneous catalysts in organic transformations such as Stille couplings, Knoevenagel condensations, and Michael additions [20].

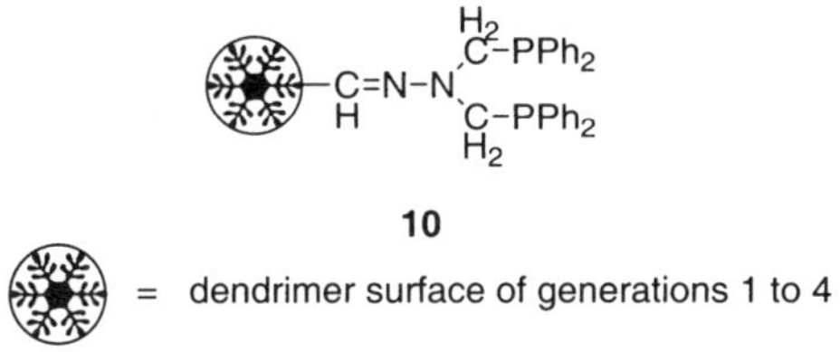

Fig. 8. Phosphine dendrimers prepared by Majoral, Chaudret and co-workers

2.2 Chiral Metal Complexes at the Periphery of a Dendrimer

Meijer et al. prepared different generations of DAB-based poly(propylene imine) dendrimers, which were substituted at the periphery with chiral amino alcohols (see, e.g., **11**, Fig. 9) [21]. The latter functionalities act as chiral ligand sites from which chiral alkylzinc aminoalcoholate catalyst sites can be generated in situ. The dendritic ligands were tested as catalyst precursors in the reductive allylation of benzaldehyde, a reaction that was successfully studied by Seebach et al. (vide infra). The 5th generation dendrimer showed almost no enantioselectivity in this particular reaction and almost no measurable optical rotation for these chiral dendrimers was observed. The decrease in conversion as well as product selectivity was explained in terms of multiple interactions between the terminal groupings at the periphery as a result of increased steric congestion.

Polyamidoamine (PAMAM) dendrimers were applied by Soai et al. as a support for chiral ephedrine groups (**12**, Fig. 10) [22]. These dendritic ligands were applied as catalysts in the chiral addition of diethylzinc to *N*-diphenylphosphinylimines. The dendritic species exhibited only a moderate effect on the product ees, whereas a high chiral induction was found for the non-dendritic model species. Furthermore, quite large amounts of dendrimer catalyst (up to 50 mol %) were needed to obtain these moderate ees. Recently, the same authors applied dendrimers based on rigid hydrocarbon backbones using the same

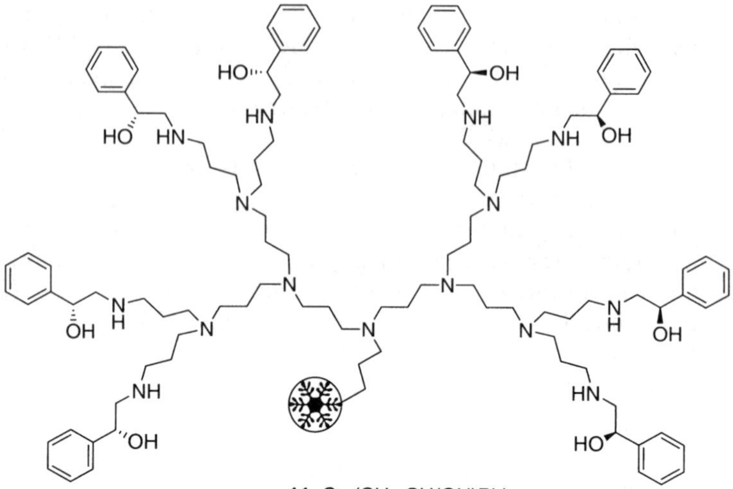

11, G_3-(CH$_2$-CH(OH)Ph)

Fig. 9. Chiral periphery-functionalized dendrimer reported by Meijer et al.

ephedrine groups (**13**, Fig. 10) [23]. With these systems ees up to 86% were obtained with a catalyst concentration of only 3.3 mol %. The authors explain the different results of the flexible and rigid systems by suggesting that due to the flexibility of the PAMAM dendrimer arms, different ephedrine groups can interact with each other. This may prevent the effective transfer of chiral information during the C–C coupling reaction. In contrast, the hydrocarbon backbone in **13** is much more rigid, thereby diminishing interaction between the chiral groups and resulting in a stereoselective process.

A further example of the application of chiral dendrimers is provided by the group of Togni that has developed dendrimers with asymmetric diphosphine ferrocenyl groups [so-called (*R*)-(*S*)-Josiphos] attached to the surface [24]. The dendrimer backbone is constructed from benzene-1,3,5-tricarboxylic acid (**14a**) or adamantane-1,3,5,7-tetracarboxylic acid (**14b**) as core. In a more recent paper Togni et al. describe the same type of dendrons, attached to a cyclophosphazene core (**14c**, Fig. 11) [25]. These dendrimer ligands were converted in situ into the Rh(I) complexes, by a complexation reaction of the dendritic ligand with [Rh(COD)]BF_4. The adamantane core-based dendrimer complexes were tested as hydrogenation catalysts in the asymmetric hydrogenation of dimethyl itaconate in methanol. In all cases, the measured enantioselectivity was high (98.0–98.7%), and only slightly lower than the ees found for the monomeric rhodium complex of Josiphos. Likewise, for the cyclophosphazene-core based dendrimers preliminary catalysis experiments indicated that comparable ees were obtained. The authors concluded that in both cases the catalytic units act as independent units (compare polynickel complexes **1**–**4**) and that the dendritic structure was not influencing the stereoselection process. Furthermore, preliminary experiments for the adamantane core-based dendrimer indicate that it is completely retained by a commercially available nanofiltration membrane.

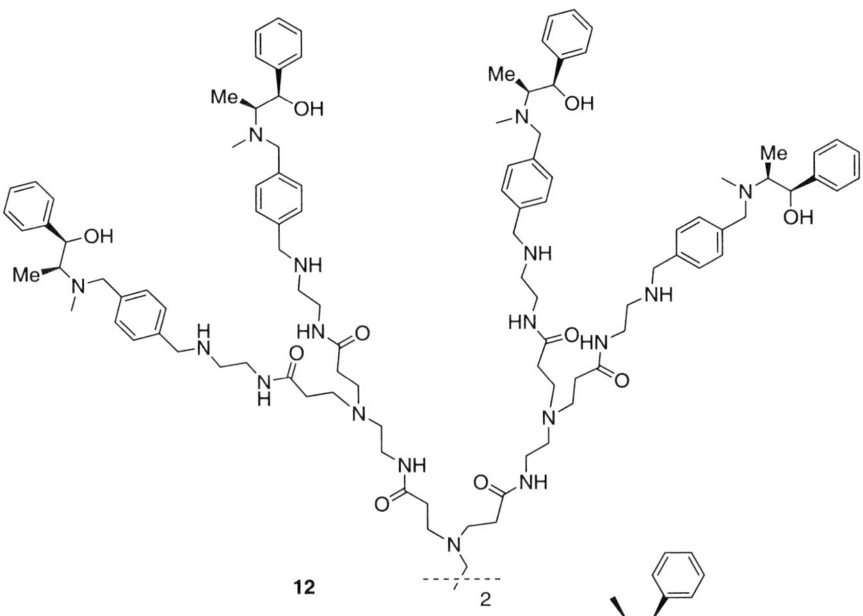

12

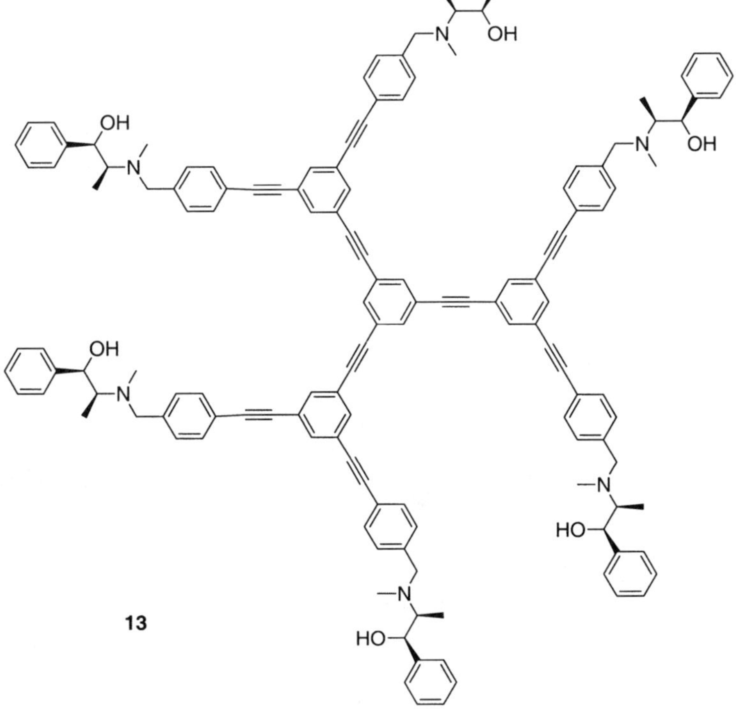

13

Fig. 10. Flexible and rigid chiral dendrimers prepared by Soai et al.

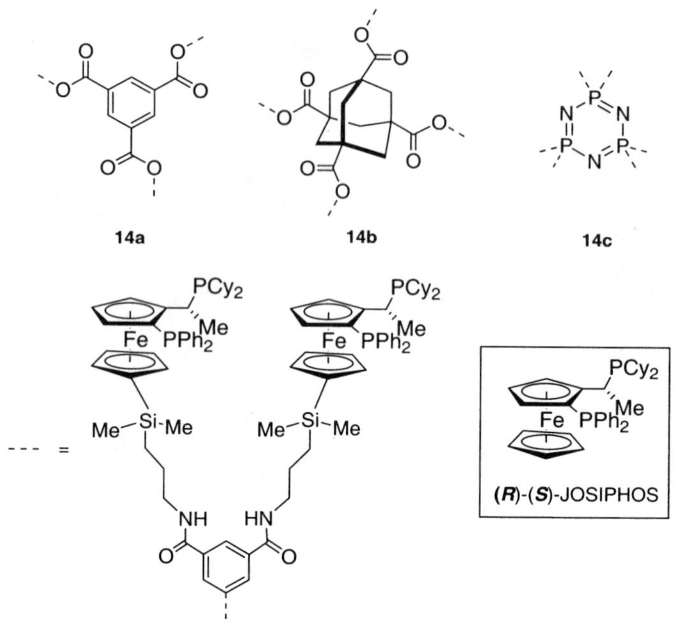

Fig. 11. Ferrocenyl-based dendrimers described by Togni et al.

Another approach was followed by Bolm et al., who prepared dendron ligands consisting of a chiral pyridyl alcohol connected to the focal point of Fréchet-type dendrons (**15**, Fig. 12) [26]. The dendritic chiral ligands were used for the in situ generation of ethylzinc dendritic complexes, to catalyze the addition of

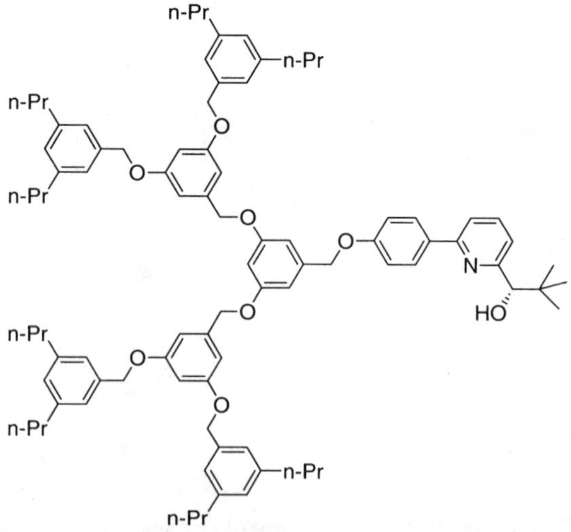

15, G_2-(C_6H_4-py-CH_2(OH)*t*-Bu)

Fig. 12. Monofunctional dendritic catalyst described by Bolm et al.

diethylzinc to benzaldehyde. Although reasonably high ees were obtained, the size of the dendron appeared to have practically no influence on the enantioselectivity of this reaction.

Van Koten and co-workers reported a stoichiometric approach towards chiral dendrimer catalysis. They used enantiopure carbosilane dendrimers (**16**, Fig. 13) as substrates in the ester enolate-imine condensation leading to β-lactams [27]. Different dendrimers were applied, which in all cases gave high *trans*-selectivity. Moreover, it was demonstrated that, before the condensation reaction, functionalization reactions can be carried out (e.g., Suzuki couplings). The level of stereo-induction that was obtained was similar to that found earlier for systems without dendritic supports. The use of enantiopure dendritic supports did not affect the enantioselectivity of C–C bond formation significantly. An interesting detail of this work is that the dendrimer backbone could be separated from the product by GPC techniques, which opened the way to recyclable soluble supports in organic synthesis.

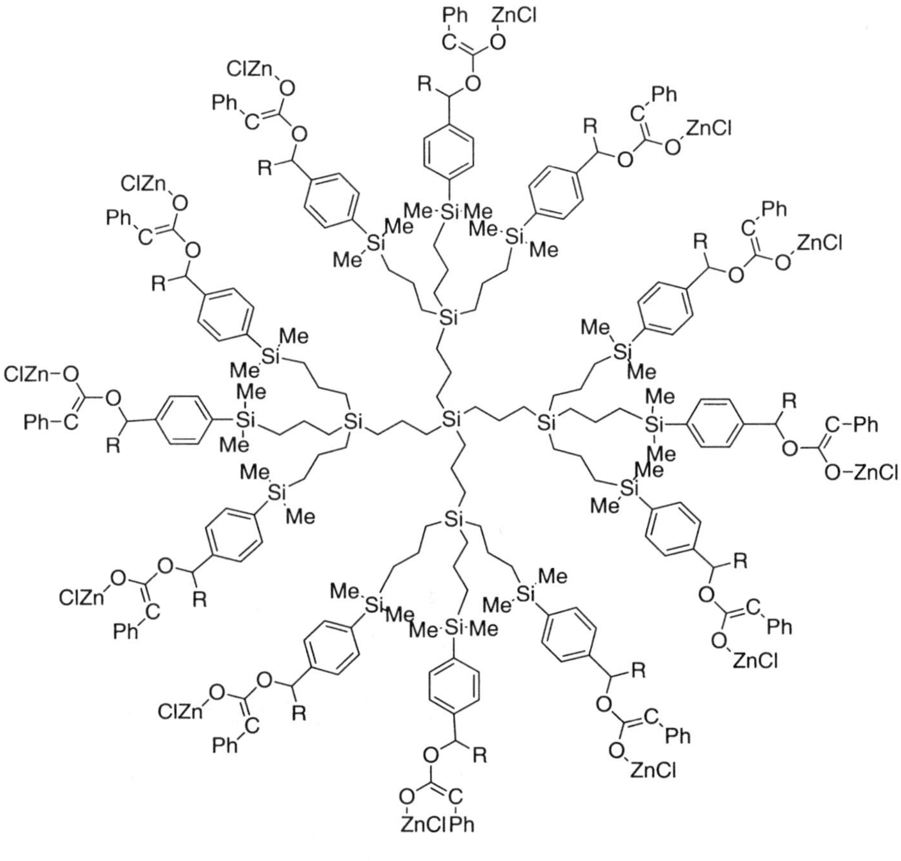

16, G_2-$CH_2(R)OC(CPh)OZnCl$

Fig. 13. Chiral dendrimers applied as stoichiometric reagents by Van Koten et al.

2.3
Miscellaneous Periphery-Functionalized Dendritic Metal Complexes

As a model for the O_2-transport copper protein hemocyanin, multicopper dendrimers were prepared by Nolte and co-workers [28]. The dendrimers employed consist of a DAB-based poly(propylene imine) backbone, which is functionalized with pyridylethyl moieties (**17**, Fig. 14). These terdentate ligands are able to form copper(II) and zinc(II) complexes upon addition of the metal ion perchlorates. The authors also succeeded in the preparation of the corresponding

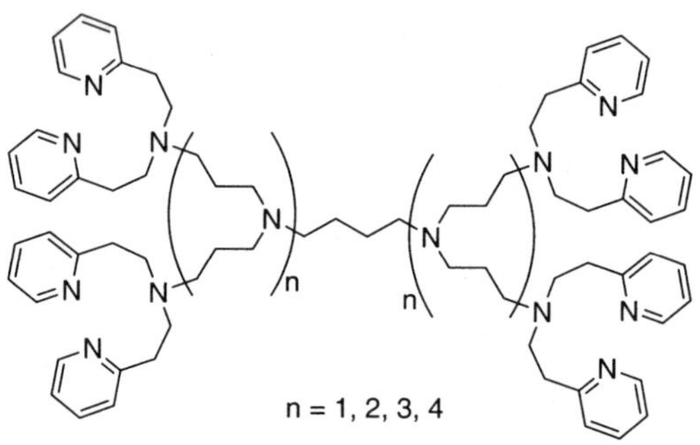

17

Fig. 14. Dendritic ligand for the synthesis of a potential oxidation catalyst described by Nolte et al.

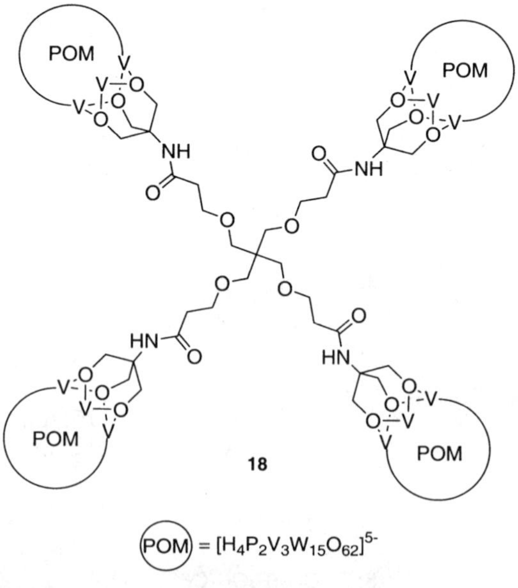

18

Fig. 15. Simple dendritic structures with polyoxometalate functionalities described by Newkome et al.

copper(I) complexes and showed for the $Cu(I)_{32}$ complex that, upon treatment of this complex with O_2, 65–70% of the copper centers are involved in dioxygen binding. These complexed dioxygen molecules may be regarded as highly activated, which would make this dendritic complex a good candidate for an oxidation catalyst.

The attachment of inorganic polyoxometalates (POM) to simple dendrimer surfaces was reported by Newkome and Hill [29]. The authors describe the attachment of four $[H_4P_2V_3W_{15}O_{62}]^{5-}$ units to simple dendrimer backbones by ester bonds (**18**, Fig. 15). These POM derivatives were applied as homogeneous catalysts in the oxidation of tetrahydrothiophene (THT) by both *t*-BuOOH and H_2O_2. Although this reaction is catalyzed by strong acids, catalysis by the (less acidic) POM derivatives is more efficient than catalysis by *p*-toluenesulfonic

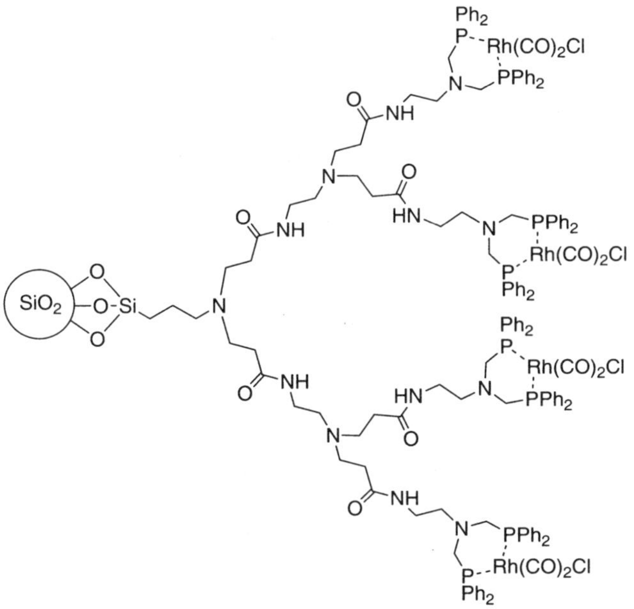

19

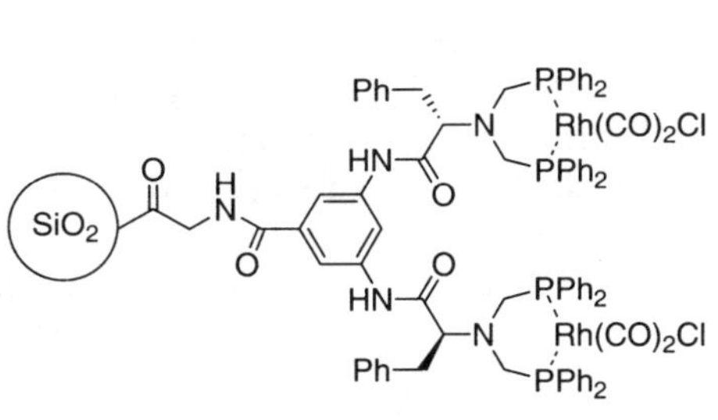

20

Fig. 16. Heterogeneous dendrimer catalysts prepared by Alper et al.

acid. Furthermore, these systems can be recovered by precipitation followed by filtration, and used again without loss of activity.

Heterogeneous periphery-functionalized dendrimer catalysts were described by the group of Alper, who made use of dendritic wedges grafted on silica beads [30]. For this purpose PAMAM and branched phenyl propionaldehyde dendritic wedges were used to which diphosphine ligands were connected. Rhodium(I) complexes of these dendrimers were applied as hydroformylation catalysts (**19**, **20**, Fig. 16). The rhodium content of these compounds was estimated using ICP analysis. In catalysis, both kinds of complexes are highly active and show excellent selectivities towards branched aldehydes using a variety of olefins. However, higher generation dendrimers of these systems showed lower activities. The authors ascribe this to steric congestion of the dendrimer surface and tested this hypothesis by introducing a spacer arm. Although slow leaching of rhodium was observed, the dendrimers containing a spacer showed higher activities, even after 4 cycles, than the ones without spacer [31]. Compared to polymer-supported catalysts these systems on silica beads show very high activities. The authors propose that this is caused by the well-exposed ligands on the outer-core of these systems. They suggest that cooperativity is another factor leading to high reactivity [32]. In conclusion, this is a successful example of the use of a dendrimer-based heterogeneous catalyst system with an activity comparable to that of homogeneous systems. At the same time catalyst recovery might be possible via size exclusion techniques.

3
Metal Complexes at the Core of a Dendrimer

Several examples of dendrimers consisting of a metal complex as core and a surrounding shell of dendritic wedges have been reported. These dendrimers were applied as shape-, size-, or enantio-selective catalysts. With this kind of dendritic complexes it is possible to isolate the catalytic site in order to create a chiral environment, or to isolate catalytic sites from each other or from the reaction medium.

3.1
Shape-Selective or Regioselective Catalysis in the Core of a Metallo-Dendrimer

A first example of shape-selective catalysis by metallo-dendrimers was reported by the groups of Moore and Suslick [33, 34]. Using manganese porphyrins to which phenylpolyester dendrons were attached (**21**, Fig. 17) and iodosylbenzene as the oxygen source, the authors investigated the effect of dendron size on the rate of epoxidation for different non-conjugated dienes and for 1:1 mixtures of linear and cyclic alkenes. When compared to the non-substituted manganese porphyrin complex, the dendritic species showed a clear increase in selectivity for conversion of the alkenes with external, i.e., less hindered double bonds and a higher affinity toward electron-rich olefins. Furthermore, an increased stability of the metalloporphyrin core towards oxidation was observed for the dendritic species.

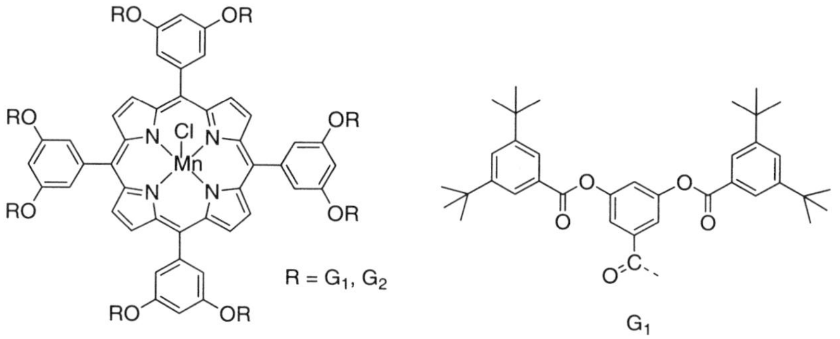

G1

21, Mn{T(3',5'-G$_n$Ph)P}(Cl)

Fig. 17. Shape-selective manganese porphyrins prepared by Moore and Suslick et al.

Another example of the use of dendritic catalysts in shape-selective catalysis was provided by the work of Chow and Mak [35, 36]. These authors described the synthesis of dendritic bisoxazoline ligands, which were used for the in situ preparation of Cu(II) complexes (**22**, Fig. 18). These complexes were employed as homogeneous catalysts in the Diels-Alder reaction of cyclopentadiene with a crotonylimide. The G_0, G_1, and G_2 dendrimer catalysts all catalyzed the cycloaddition reactions with similar rates, whereas a significant drop in activity was encountered for the G_3 catalyst. The authors rationalized these observations by assuming that the G_3 catalyst has structural features that differ from that of the G_0, G_1, and G_2 ones. They suggest that backfolding of the wedges, due to steric constraints, occurs in the G_3 dendritic species, which results in steric blocking of the metal center. Although less apparent than in the work of Moore and Suslick [33, 34], a slight increase in substrate selectivity for the smaller crotonylimides was found for this G_3 generation copper catalyst.

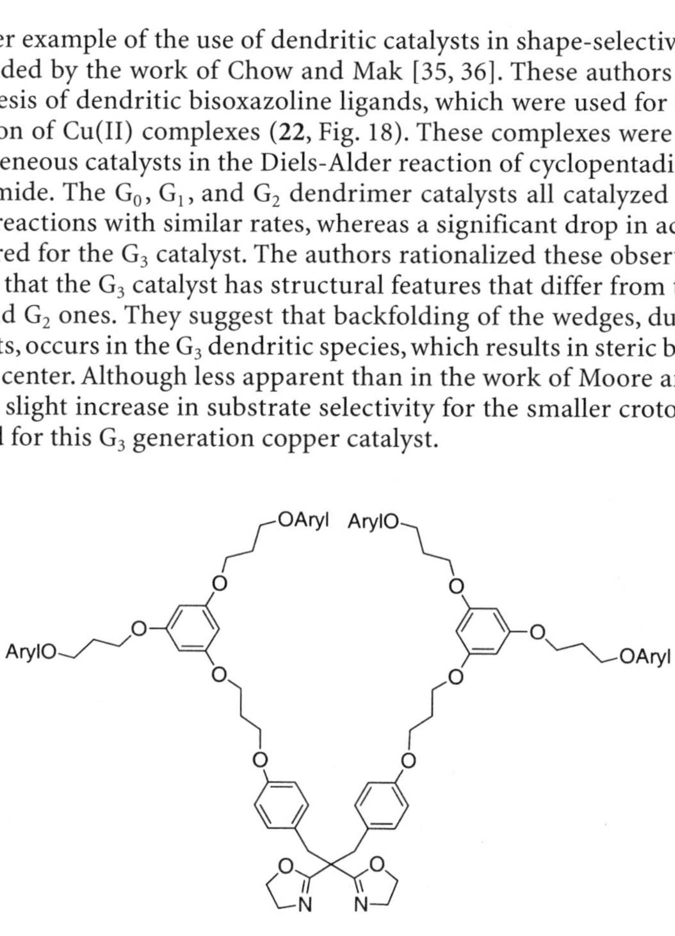

22

Fig. 18. Dendritic bisoxazoline ligand prepared by Chow and Mak

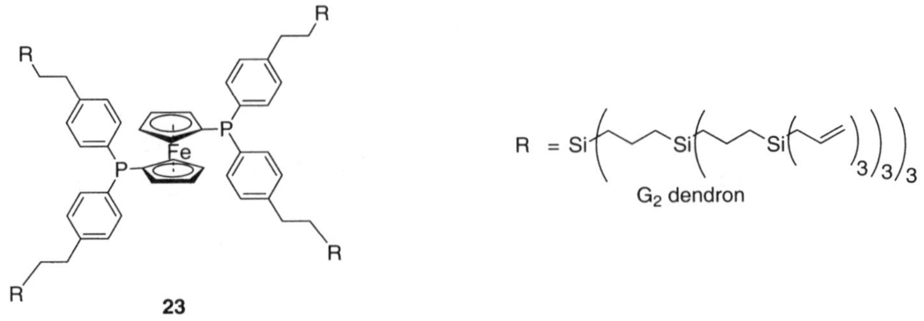

Fig. 19. Ferrocenyldiphosphine core dendrimer prepared by Van Leeuwen et al.

Stereo-selective catalysis at the core of a metallo-dendrimer was pursued by the group of Van Leeuwen, who prepared a ferrocenyldiphosphine (DPPF) core functionalized with different generations of carbosilane dendrons (**23**, Fig. 19) [37]. The resulting dendritic ligands were used for the in situ complexation of $PdCl_2$ from a suitable Pd(II) precursor to yield $PdCl_2$-dendrimer complexes. These complexes were used as catalysts for the allylic alkylation of 3-phenyl-acetate with diethyl 2-sodio-2-methylmalonate. For this reaction, a decrease in catalyst activity with increasing generation number of the carbosilane dendrons was found. In addition, the selectivity for the *trans*-product slightly decreased from 90% for a DPPF model compound to 76% for a G_3 carbosilane dendron ligand.

Dendritic phthalocyanines were prepared and studied by Kimura et al. [38]. Cobalt complexes of these phthalocyanines (**24**, Fig. 20) were applied as catalysts for the oxidation of 2-mercaptoethanol. The catalytic activity of the dendrimer catalyst was 20% less than that of a non-dendritic phthalocyanine. However, the stability of the dendritic catalyst was higher than that of the non-dendritic species, most likely due to encapsulation of the metallo-phthalocyanine in the dendritic structure. The dendritic structure also prevented molecular aggregation of the phthalocyanines in polar solvents and thin films. The authors suggest that these dendritic phthalocyanines can be used as shape-selective catalysts.

Recently the same authors reported a poly(propylene imine) dendrimer that was functionalized with poly(*N*-isopropylacrylamide) (PIPAAm) (**25**, Fig. 20) [39]. This dendrimer was applied as a dendritic host for modified, anionic cobalt(II) phthalocyanine complexes (**26**, **27**, Fig. 20), which bind via electrostatic and hydrophobic interactions. These dendritic complexes were applied as homogeneous catalysts in the thiol oxidation described above and showed an interesting temperature dependency; above 34 °C the reaction rate increases sharply. The authors propose that above the lower critical solubility temperature (LCST) the PIPAAm arms phase-separate and contract. At that point, the phthalocyanine complexes are better accessible for substrates and the reaction rate increases.

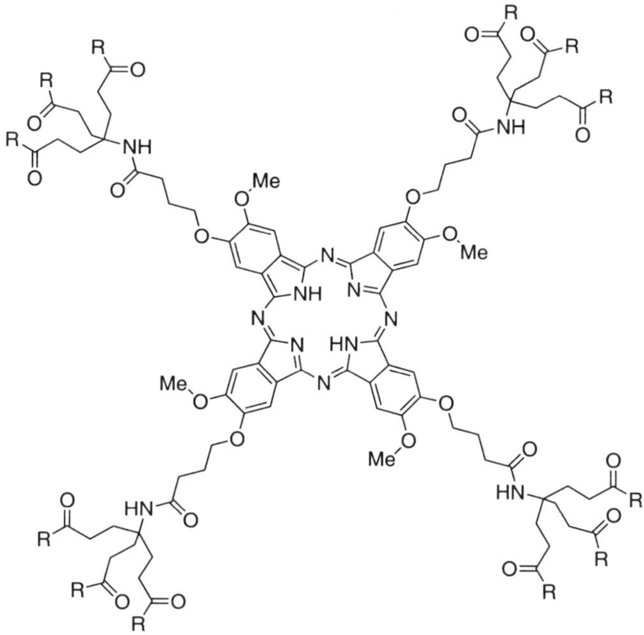

24

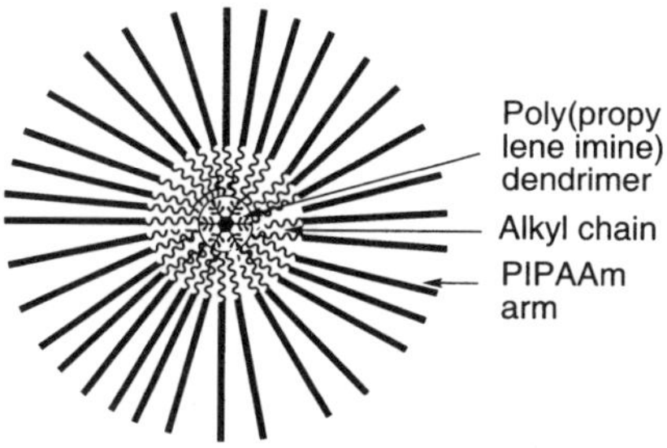

25

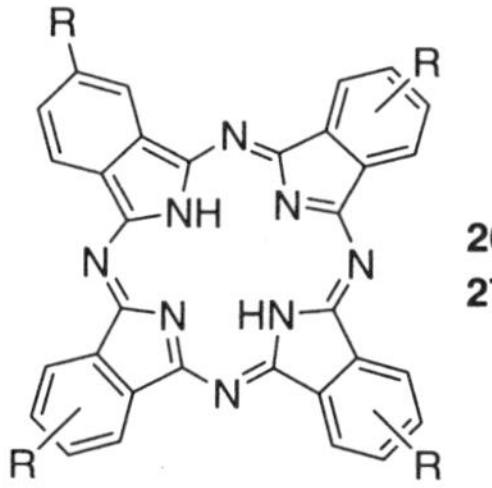

26, R = COO^-
27, R = $O(CH_2)_2N^+(CH_3)_3$

Fig. 20. Dendritic phthalocyanine and host-guest system reported by Kimura et al.

3.2
Enantioselective Catalysis in the Core of a Metallo-Dendrimer

Brunner et al. synthesized and applied metal complexes with a shell of dendrons, which he called *dendrizymes*, in enantioselective catalysis [40]. These catalysts are based on core phosphine donor atoms, which are complexed to transition metal salts. The phosphines carry dendrons that are functionalized with chiral groups like menthol or borneol ligands (see **28**, Fig. 21). The core phosphine donor-atoms can be complexed to (transition) metals salts, like rhodium, resulting in a dendron-enlarged 1,2-bis(diphosphino)ethane Rh(I) complex that was used as catalyst in the hydrogenation of acetamidocinnamic acid to yield

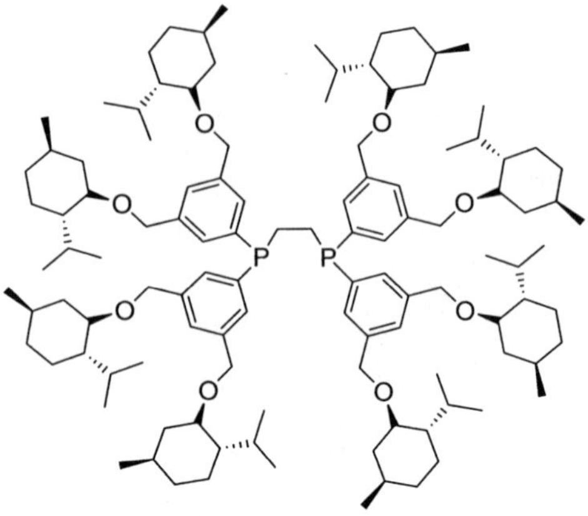

28

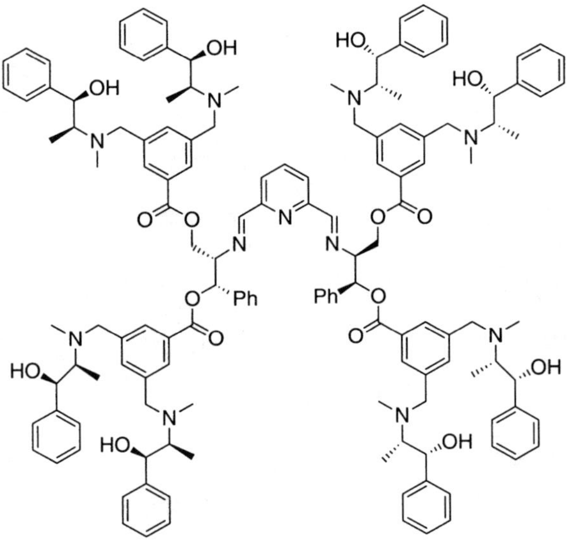

29

Fig. 21. Dendron-enlarged 1,2-bis(diphospino)ethane and 2,6-pyridinediimine derivatives reported by Brunner et al.

N-acetylphenylalanine. A small retardation of the hydrogenation of the substrate was encountered, which was ascribed to the *meta*-positioned dendron substituents. No products that were significantly enantiomerically enriched were isolated. However, using different dendritic species based on a 2,6-pyridinediimine CuOTf complex as the core (see, e.g., **29**, Fig. 21), a low enantioselectivity, up to 10–11% ee, was observed in the cyclopropanation of styrene with ethyl diazoacetate [41].

An elegant example of dendrimer catalysts having a chiral metal complex as core is found in the work of Seebach et al., who applied $\alpha,\alpha,\alpha',\alpha'$-tetraaryl-1,3-dioxolane-4,5-dimethanols (TADDOLS) as core for dendritic ligands [42,43].

Different dendrons were attached to these core complexes: classical Fréchet dendrons (e.g., **30**, Fig. 22), chiral dendrons (derived from 3-hydroxybutanoic acid), and Fréchet dendrons with peripheral octyl groups. TitaniumTADDOL dendritic complexes with either non-chiral or chiral dendrons were applied as homogeneous catalysts in the asymmetric reductive alkylation of benzaldehyde with diethylzinc, yielding the secondary alcohol with ees up to 96%. These results are almost identical to the results obtained for model TADDOL complexes in this type of catalysis. The dendritic TADDOLates show high catalytic activities that slightly decrease from G_3 upwards. The authors assume that this effect is due to the accessibility of the catalytic site. Furthermore, the enantioselectivity was not influenced by the presence or absence of chirality in the dendrons, due to the distance between the metal center and the additional chiral building blocks.

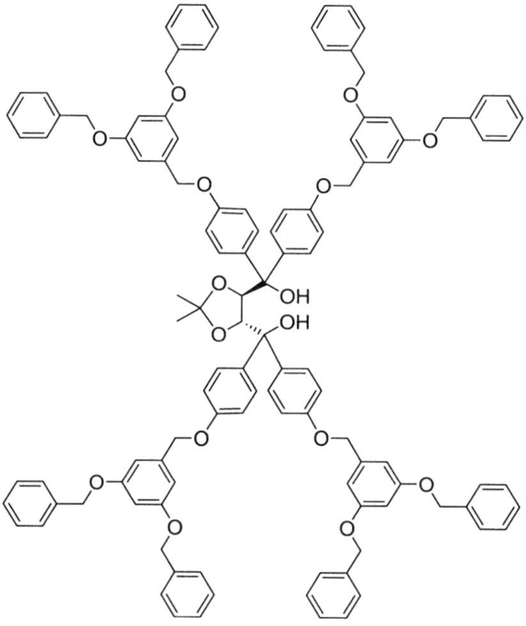

30

Fig. 22. Enantioselective dendritic TADDOL ligand prepared by Seebach et al.

Binaphthol (BINOL) ligands modified with Fréchet-type dendrons of different generations (**31**, Fig. 23) were applied as homogeneous catalysts in the allylation of benzaldehyde with allylstannane by Yoshida et al. [44]. Compared to a parent binaphthol ligand, these ligands show similar activities and ees, indicating that comparable titanium complexes are formed in situ.

Similar dendritic ligands, BINAP ligands equipped with Fréchet-type dendrons of different generations (**32**, Fig. 23), were applied by Chan et al. [45]. In situ prepared ruthenium(II) complexes were tested for their activity in the

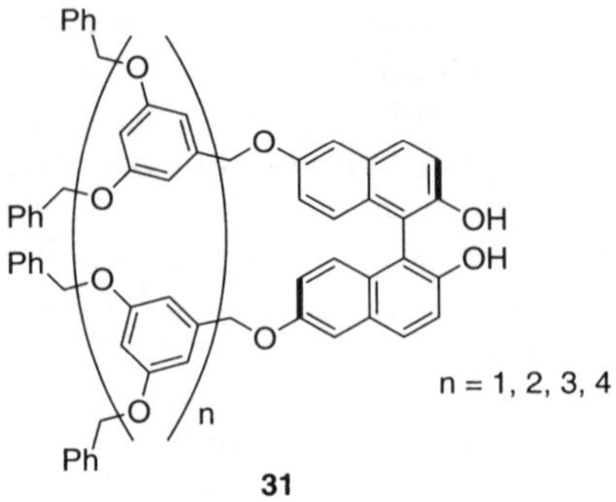

31

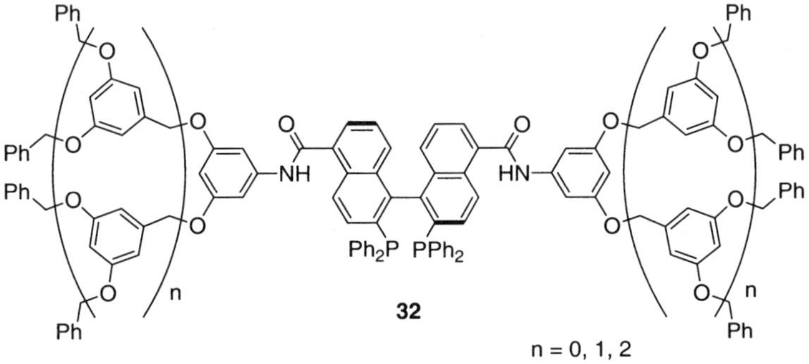

32

n = 0, 1, 2

Fig. 23. Dendritic binaphthol ligands described by Yoshida et al. and dendritic BINAP ligands reported by Chan et al.

asymmetric hydrogenation of 2-[*p*-(2-methylpropyl)phenyl]acrylic acid. The authors report slightly higher activities and ees (91.8–92.6%) for the G_0 and G_1 dendron complexes, than obtained with the parent BINAP-ruthenium complex. However, for the catalyst derived from the G_2 dendron ligand a much higher conversion was observed with a comparable ee (91.6%). This was ascribed to the presence of increased steric bulk on the BINAP ligand, which affects the ligand's bite-angle. The latter ligand could be recovered by precipitation and filtration, and reused three times with similar activity and enantioselectivity.

4
Metal Complexes Throughout the Dendritic Structure

Besides metallo-dendrimers having metal complexes at either the periphery or at the core, some examples exist of dendrimers with ligating sites in the backbone itself to which metal ions can be complexed. Apart from the striking fact that for this kind of dendrimers the backbone is at the same time the "ligating site", an advantage is found in the fact that the ratio of metal-to-support (i.e., the dendrimer) is much higher than for periphery- or core-functionalized dendrimers. Examples of this kind of metallo-dendrimers are described below: dendrimer-stabilized, zero-valent metal-nanoparticles and dendrimer backbone-complexed metal-ions.

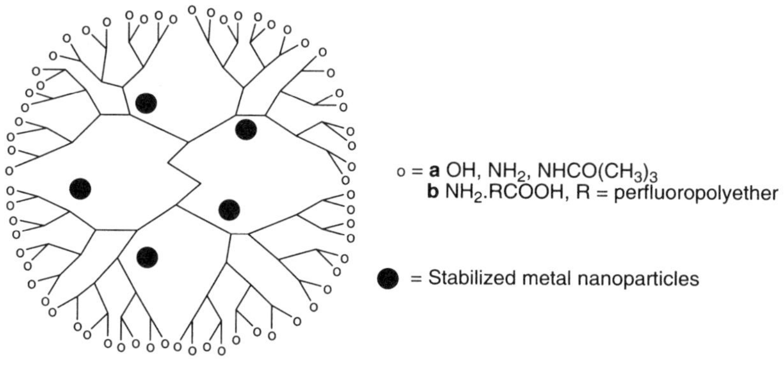

33

Fig. 24. PAMAM dendrimer-stabilized, zero-valent metal clusters described by Tomalia et al. and further applied by Crooks et al.

A very elegant route towards metallo-dendrimers was described by Tomalia, who used G_4 and G_5 PAMAM dendrimers as hosts for complexation of Cu(II) ion guests. The cavities of these dendrimers can act as a template for metal ion complexation and, interestingly, also as a stabilizer for zero-valent metal clusters obtained after reduction of complexed ions [46] (**33a**, Fig. 24). This concept was further elaborated by Crooks and co-workers who prepared G_4 and higher generation PAMAM dendrimer stabilized Pt(0) and Pd(0) nanoparticles and used these as catalysts for hydrogenation [47] and O_2 reduction [48]. They also reported the exchange of Cu(0) nanoparticles by other more noble metal ions to form Pt(0), Pd(0), Ag(0), and Au(0) nanoparticles [49]. Interestingly, these reactions go to completion and the resulting particles are stable. These noble metal dendrimer catalysts have proven to be suitable for the reduction of O_2. The Cu^{2+} ions resulting from the exchange reaction can be retained inside the dendrimer, depending on the pH at which the displacement is carried out. A very recent application of such dendrimer-encapsulated metal nanoparticles is their use in fluorous biphasic catalysis (**33b**, Fig. 24) [50]. Crooks et al. reported the non-covalent modification of a PAMAM-Pd nanoparticle complex with perfluoropolyethers by complexation of the carboxylic end groups of these polyethers with the terminal amine groups of the dendrimer. This modification afforded dendrimers that dissolved exclusively in the fluorous phase of a THF/FC-75 (fluorous solvent) mixture. A biphasic Pd(0)-dendrimer system was successfully applied in the reduction of alkenes and the catalyst could be recycled for 12 times without appreciable loss of activity.

Homogeneous organophosphine dendrimers containing metal sites throughout the structure were reported by Dubois et al. [51]. The authors prepared different small dendritic structures with phosphorus branching points, which can serve as binding sites for metals. The resulting terdentate (P,P,P)-ligating sites were palladated using a Pd(II) salt in the presence of PEt_3. The resultant cationic complexes (e.g., **34**, **35**, Fig. 25) were successfully applied as homogeneous catalysts for the electrochemical reduction of CO_2 to CO. The observed reaction rates

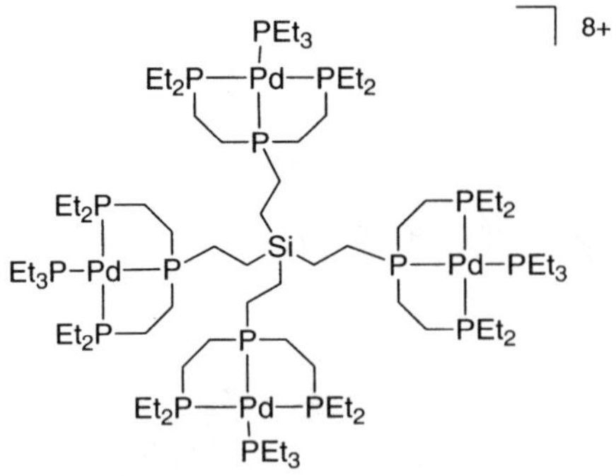

34

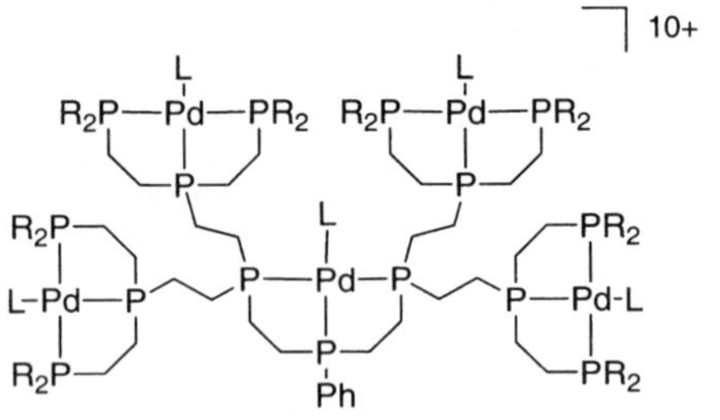

35

R = Et, Ph
L = CH_3CN

Fig. 25. Phosphorus-based dendrimer-Pd complexes having ligating sites throughout the structure reported by Dubois et al.

and selectivities were comparable to those found for an analogous monomeric palladium complex, indicating that no cooperative effects between the different metal sites are present.

Another phosphorus-based example of dendrimers containing ligating sites throughout the structure was prepared by Kakkar et al. [52]. They succeeded in the synthesis of various generations of metallophosphino dendrimers by acid-base hydrolysis of aminosilanes with molecules possessing terminal OH-groups. Each phosphorus branching point could serve as a ligating site for Rh(I) centers. These Rh-complexes (e.g., **36**, Fig. 26) were successfully applied as homoge-

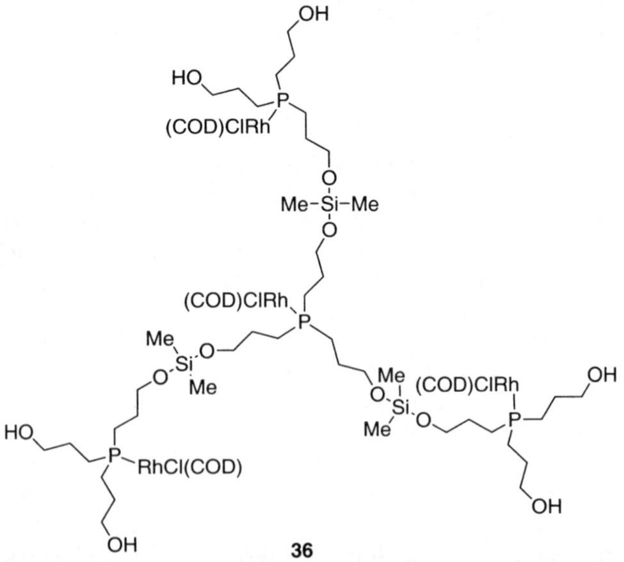

36

Fig. 26. Phosphorus-based dendrimers containing P-Rh sites throughout the structure as reported by Kakkar et al.

neous hydrogenation catalysts. 1-Decene was hydrogenated with activities similar to that of the monomeric Rh(I) model complex. Furthermore, the authors demonstrated that the catalyst could be recovered by a simple precipitation in a batch process and reused without significant loss of activity.

5 Dendrimer Catalysts Without Metals

Although the emphasis in the field of dendrimer catalysis lies on metallodendritic catalysts, an important class is formed by dendritic catalysts without metals. A few examples of this type are described, which can act as stabilizer for ions, as shape-selective catalysts, or as chiral auxiliaries.

The first example of the use of a non-metal containing dendrimer for catalysis was provided by the work of Ford et al. [53], who made use of polyether dendrimers that were functionalized at the periphery with quaternary ammonium ions (**37**, Fig. 27). These dendrimers were used in the unimolecular decarboxylation of 6-nitrobenzisoxazole-3-carboxylate and the bimolecular hydrolysis of *p*-nitrophenyl diphenyl phosphate catalyzed by *o*-iodobenzoate. In aqueous media, these reactions are accelerated by the polycationic dendrimer through stabilization of the organic anions, which presumably bind in high concentration to the polycationic periphery of the dendrimer. The pseudo micellar environment of the dendrimer promotes the formation of organic anions. In the same group poly(propylene imine) dendrimer complexes with Cu(II), Zn(II), and Co(III) ions were investigated as catalysts in the reaction described above [54]. Reaction rates were found to be 1.3–6.3 times faster than in the absence of metal ions.

Recently, the same authors reported DAB-based dendrimers fuctionalized with triethyleneoxy methyl ether (TEO) and octyl chains at each amine [55]. These dendrimers could be quaternized using methyl iodide (**38**, Fig. 28) after which chloride ions were introduced via ion exchange. The resulting quaternary

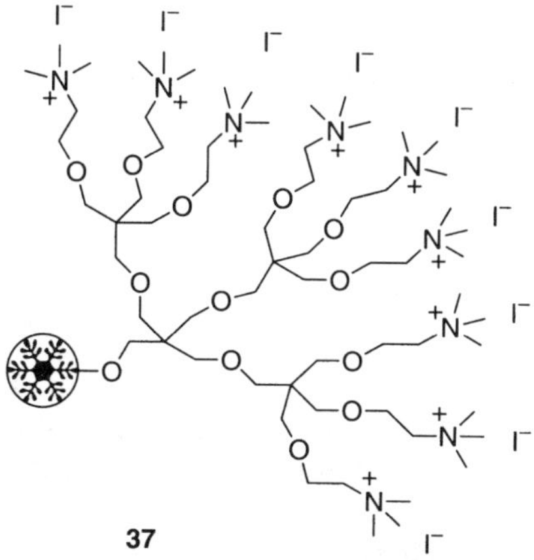

37

Fig. 27. Polyether dendrimer functionalized with peripheral quaternary ammonium groups reported by Ford et al.

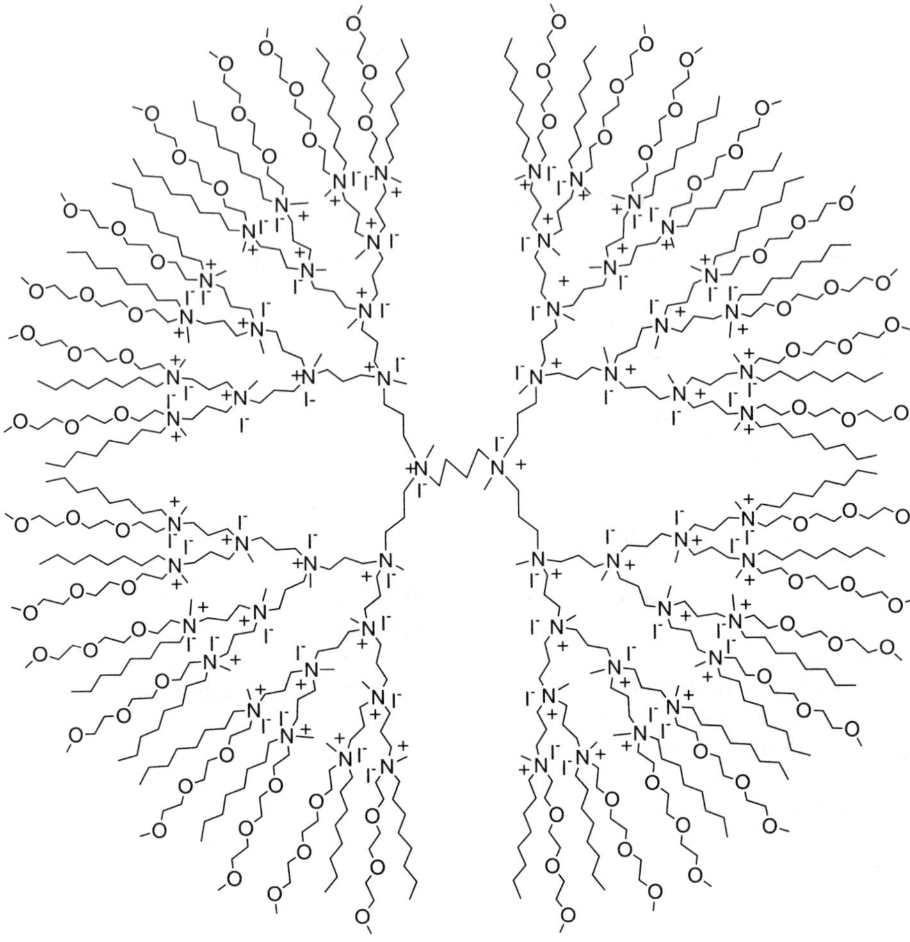

38

Fig. 28. DAB-based dendrimer containing internal quarternary ammonium groups reported by Ford et al.

ammonium chloride dendrimers were applied in the decarboxylation reaction described above. These dendrimers show much higher rates than the external ammonium ion dendrimers **37**. The authors explain this difference by stating that hydrogen bonding of the solvent (water) to the carboxylate anion reduces the rate, whereas dipolar aprotic solvents increase the rate of decarboxylation. Use of the modified dendrimers (**38**) would lead to a less hydrated substrate, which then would react faster. A fourth generation dendrimer resulted in a higher rate than a second generation dendrimer, indicating that the substrate is less hydrated in the higher generation dendrimer.

Another example of a non-metal containing dendrimer in catalysis is the use of a single amine core, functionalized with Fréchet-type polyether dendrons

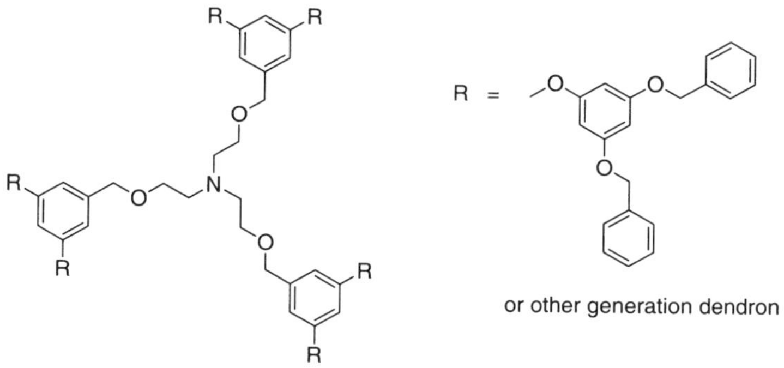

39

Fig. 29. Dendron-enlarged single amine core reported by Morao and Cossío

(**39**, Fig. 29) [56]. These amines can catalyze the nitroaldol or Henry reaction between aromatic aldehydes and nitroalkanes. Molecular modeling studies indicate that the cavities around the amine core should be able to accommodate the reactants, although in the higher generation dendrimers the core is more shielded by the dendrons. Catalyst activity is found to decrease with increasing generation number, suggesting that shielding of the center by the dendrons indeed takes place.

Chiral amphiphilic dendrimers were applied by Rico-Lattes and co-workers [57]. These water-soluble, but-THF insoluble, dendrimers consist of amine-based dendrimers functionalized with glucose derivatives (e.g., **40**, Fig. 30) and can be used in the homogeneously (water) and heterogeneously (THF) catalyzed reduction of prochiral aromatic ketones by sodium borohydride to

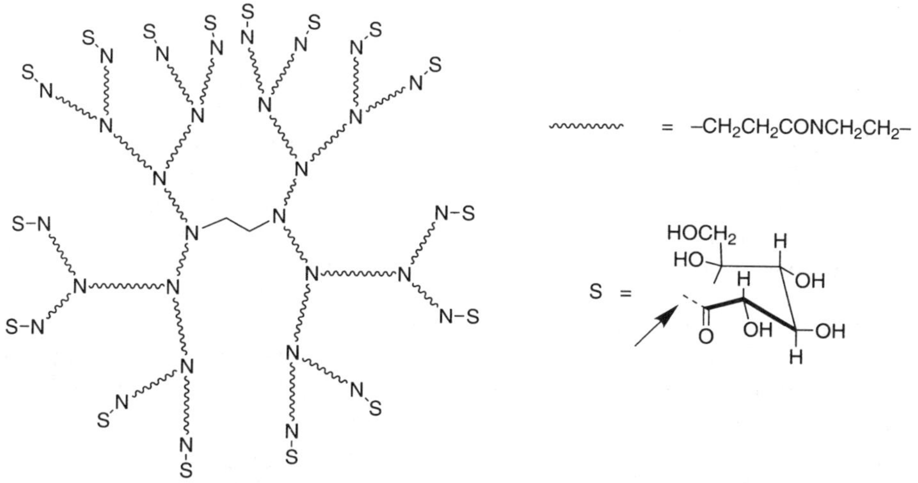

40

Fig. 30. Chiral water-soluble dendritic catalysts reported by Rico-Lattes et al.

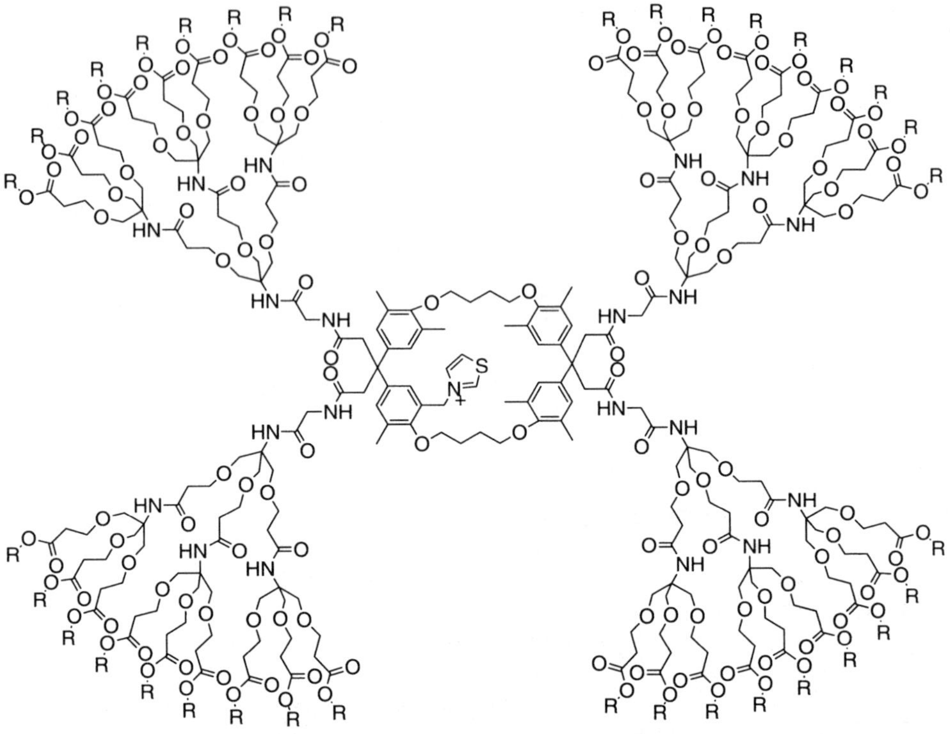

41

R = Me, ⌄⌃O⌃⌄O⌃⌄O⌃

Fig. 31. Dendrophane structure reported by Diederich et al.

yield the corresponding alcohol derivatives in high optical purity. Interestingly, the reactions take place at the pseudomicellar chiral interface in an aqueous medium.

As a functional mimic of the enzyme pyruvate oxidase, catalytic dendrophanes were prepared by Diederich et al. [58]. The authors prepared thiazolium cyclophanes modified with dendrons to yield the so-called dendrophanes (**41**, Fig. 31). These dendrophanes were applied as oxidation catalysts, but showed low activities compared to a model cyclophane. The authors propose that the unfavorable steric hindrance of the dendrons causes this low activity, and suggest that the interior of densely packed dendrimers might not be a good environment for catalysis.

Dendrimers containing potential cation stabilizer sites were reported by Fréchet and co-workers, who prepared dendritic inverse micelles (**42**, Fig. 32) [59]. These dendrimers contain internal hydrogen bonding sites and an external apolar shell, and can be applied as a stabilizer for cationic intermediates in analogy with enzymes. The authors tested these systems in E_1 elimination reactions in an apolar solvent (cyclohexane). Conversions were high for these systems. The authors claim that the polar interior of the dendrimer provides an ideal nanoenvironment for reactions that involve polar transition states.

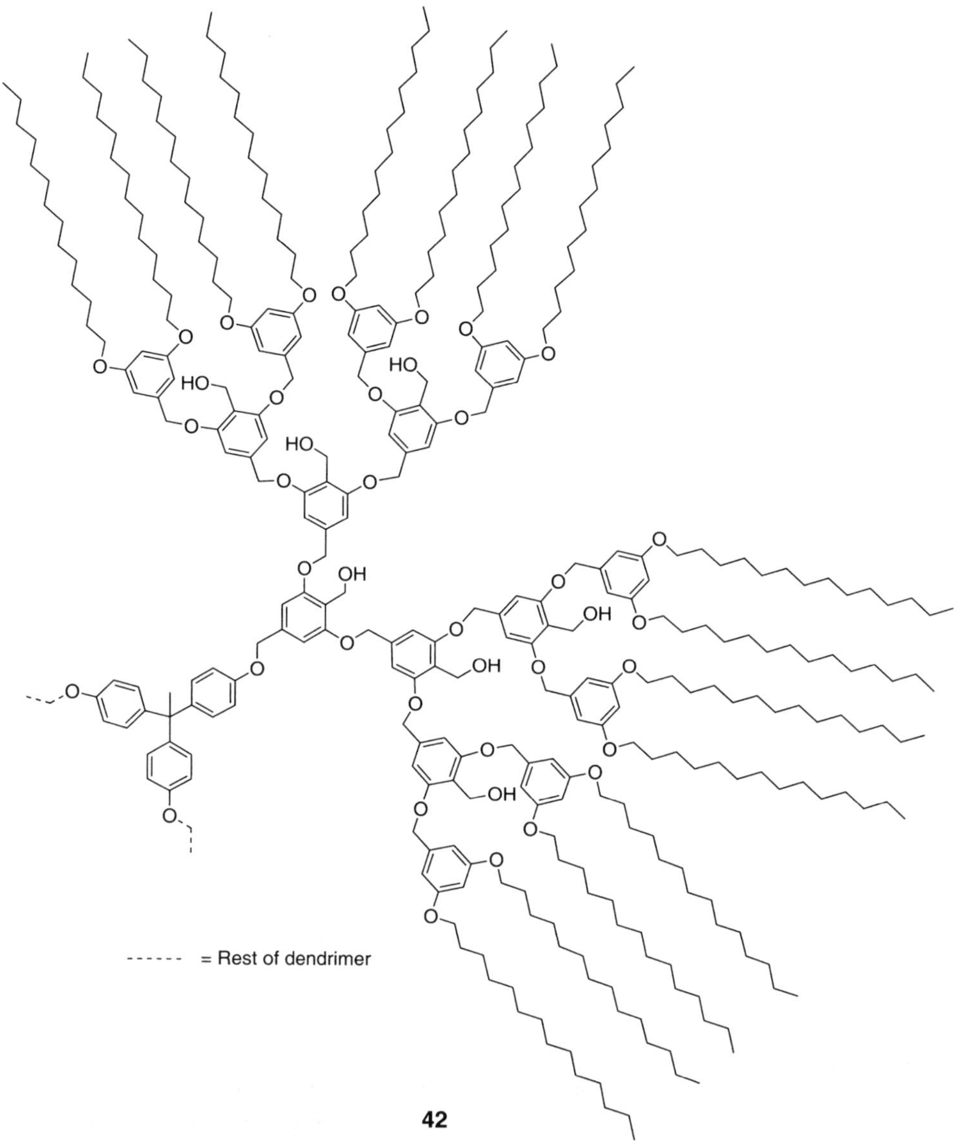

42

Fig. 32. Dendritic inverse micelles described by Fréchet et al.

Recently, polycationic dendrimers, that consist of tetra(NCN)-silane molecules quaternarized with G_1 and G_2 Fréchet-type dendrons (**43**, Fig. 33), were prepared by Van Koten and co-workers [60]. These dendrimers were tested as phase-transfer catalysts in the S_N2 reaction between (excess) potassium cyanide and benzyl bromide to afford benzyl cyanide in an H_2O/CH_2Cl_2 biphasic system. In the absence of catalyst only minor amounts of benzyl cyanide (= 6% after 20 h) were formed, whereas in the presence of **43** (0.01 mol % catalytic ammo-

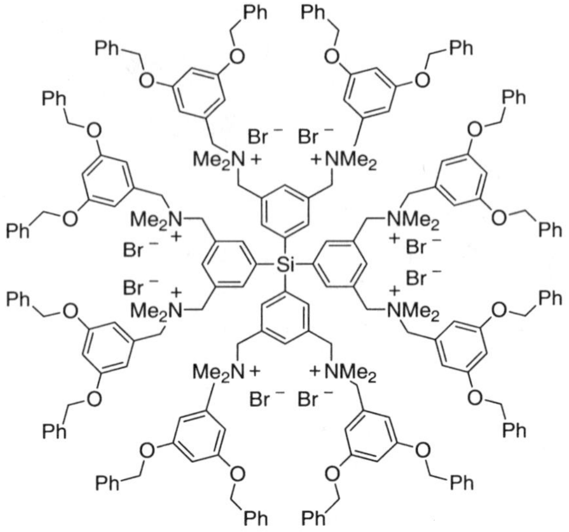

43, $Si(NCN)_4.G_1Br$

Fig. 33. Polycationic dendrimer reported by Van Koten et al.

nium groups) a clear increase in conversion was found (59% after 20 h). These results indicate that these polycationic dendrimers can act as microreactor systems in substitution reactions.

6 Summary of Reactions

Reactions catalyzed by dendritic catalysts are summarized in Table 1.

7 Summary and Perspectives

Although many different dendrimer catalysts have now been described in the literature, it is clear that the field of dendritic catalysis is still rapidly evolving. The amount of publications that appears each year follows an exponential curve upward [1b] and many new applications are still being described. One of the latest developments in dendrimer catalysis is the recovery of the catalyst by means of nanomembrane filtration techniques. After pioneering work by Kragl and co-workers [61, 62], using polymer-enlarged oxazaborolidines, some interesting dendritic examples were reported as described above [3, 10, 12, 13, 17]. However, we are still quite far from a stable dendrimer catalyst system that can be applied in a continuous membrane reactor system, without loss of activity. We believe that a great challenge lies in this field, in particular when more suitable membrane systems are developed. The stability of metal complexes depends very much on the way of metal binding. From the results presented

Table 1. Summary of reactions catalyzed by dendritic catalysts.

Cat.[a]	Catalyzed reaction	M-complex[b]	Recovery[c]
1–5	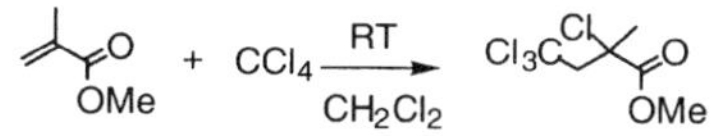	isol.	yes
6	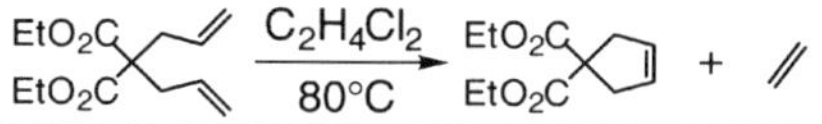	isol.	yes
7	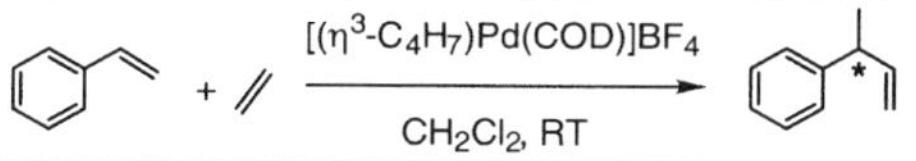	in situ	yes
8a	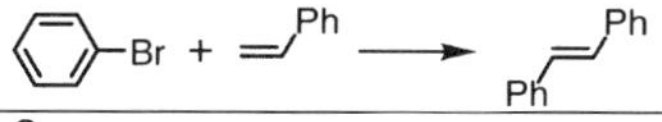	isol.	yes
8a	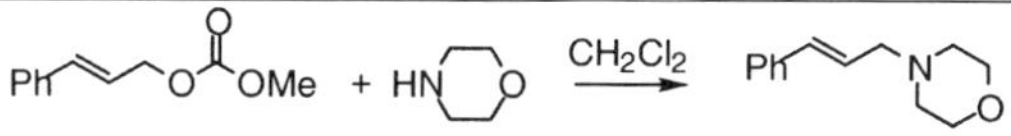	isol.	yes
8b 33b	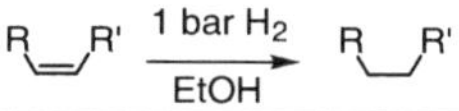	isol. isol.	yes yes
9 23	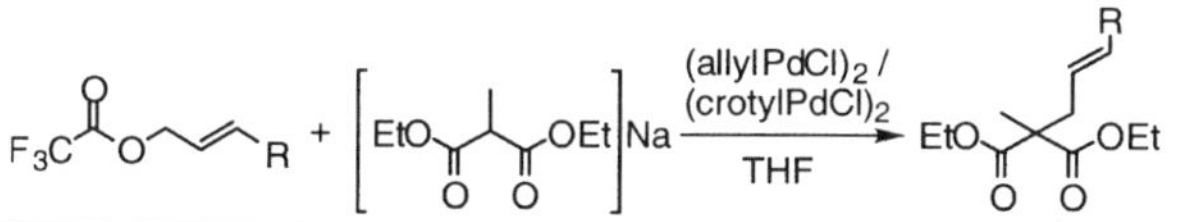	in situ in situ	yes no
10	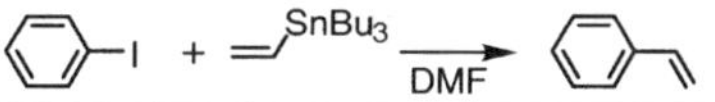	isol.	yes
10	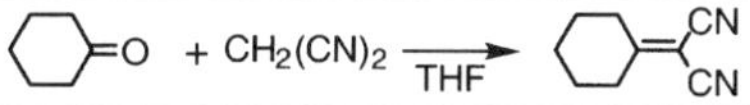	isol.	yes
10	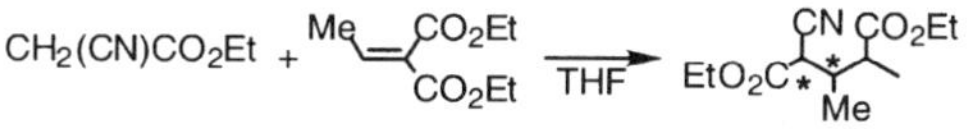	isol.	yes
11 15 30	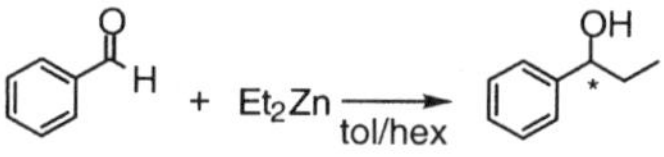	in situ in situ in situ	no no no
12 13	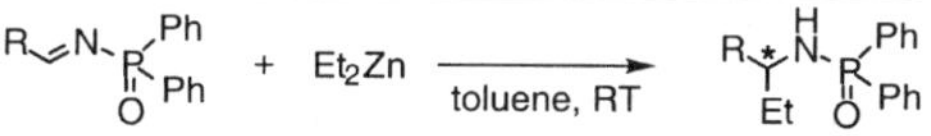	in situ in situ	no no
14	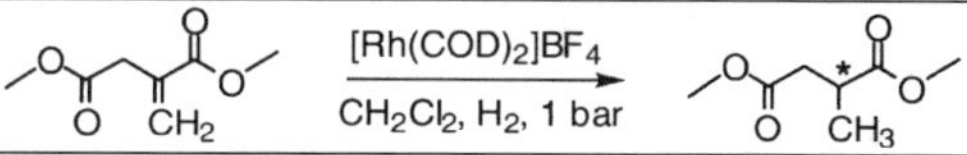	in situ	sugg.
16	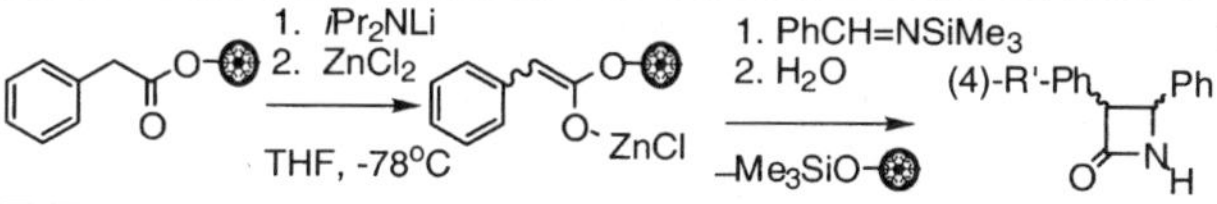	in situ	yes
18	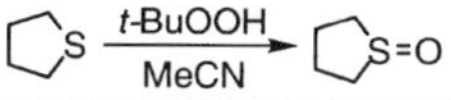	isol.	yes
9 19 20	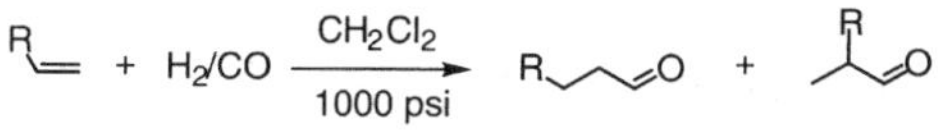	in situ isol. isol.	yes sugg. sugg.
21	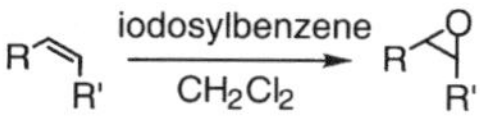	isol.	no

Table 1 *(Continued)*

Cat.[a]	Catalyzed reaction	M-complex[b]	Recovery[c]
22	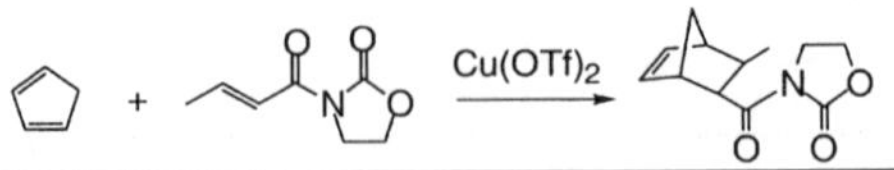	in situ	no
24 **25**	$4\ RSH + O_2 \longrightarrow 2\ RS{-}SR + 2\ H_2O$	isol. isol.	no no
28	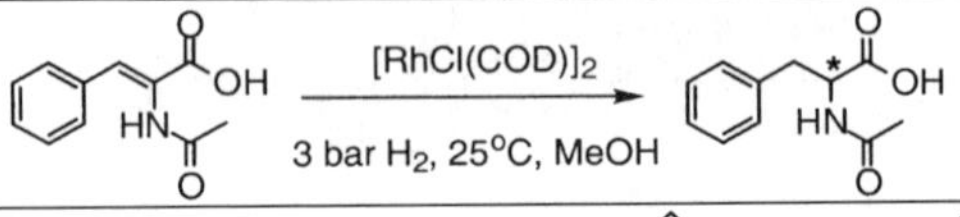	in situ	no
29	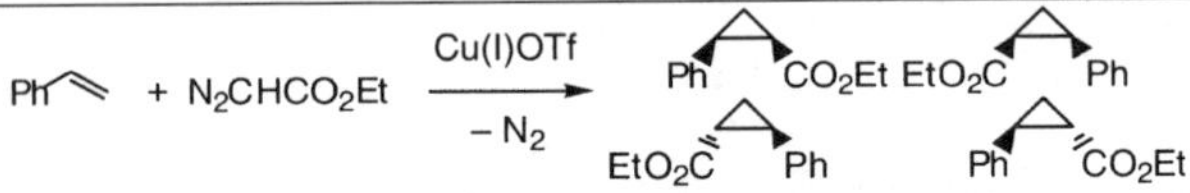	in situ	no
31	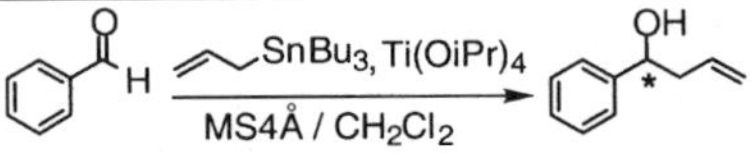	in situ	no
32	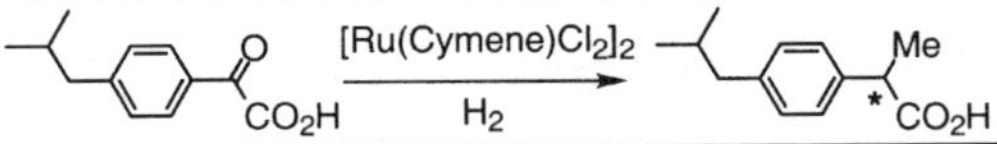	in situ	yes
33a	$O_2 \xrightarrow{H^+, e^-} 2\ H_2O$	isol.	no
34 **35**	$CO_2 + H^+ \xrightarrow[-H_2O]{e^-} CO$	isol. isol.	no no
36	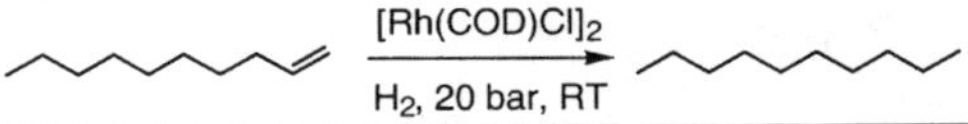	isol.	yes
37 **38**	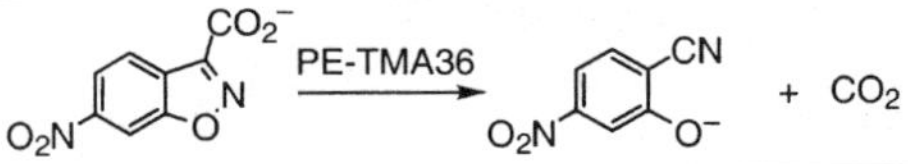	none none	no no
39	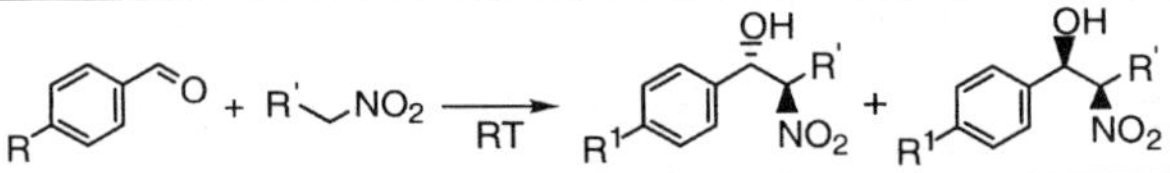	none	no
40	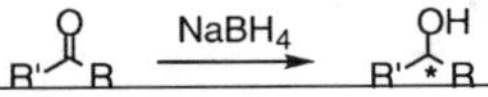	none	no
41	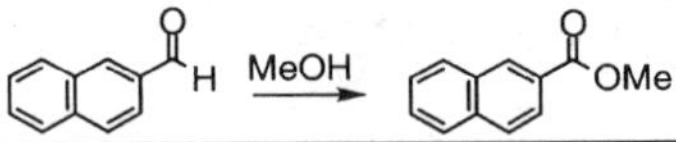	none	no
42	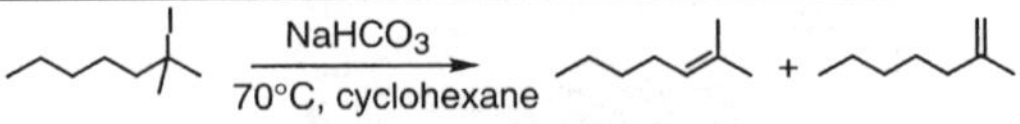	none	no
43	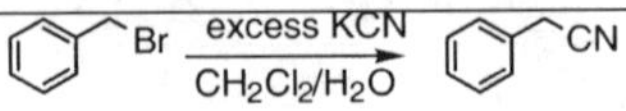	none	no

[a] Catalysts are numbered as they appear in the text.
[b] Preparation of a metal complex in situ, as an isolated compound (isol.), or the absence of metal (none).
[c] Recovery of the dendritic catalyst that was successful (yes), suggested as possible (sugg.), and not performed (no).

above it can also be concluded that binding of the metal via a metal-carbon σ-bond results in metallo-dendritic catalysts with more stable metal sites. Consequently, these catalyst systems do not suffer from the leaching problems encountered in the case of the metallo-dendritic catalysts with peripheral coordination complexes as active sites. Another unsolved problem in dendrimer catalysis is the non-random distribution of active sites, in contrast to the homogeneous distribution of sites in solution encountered for monomeric catalysts.

Even though very interesting and appealing from an academic point of view, perfect structures, such as dendrimers ideally are, are not a prerequisite for many applications. Alternatively, polymers with very narrow weight distributions, such as hyperbranched polymers [63], could be applied. These polymers can, in contrast to dendrimers, often be prepared in one step. One of the challenges in future research will be the transformation of principles designed for dendrimers to hyperbranched polymers. In our group an example of such a system was prepared using polyallylsilane molecules [64].

Furthermore, the preparation of dendritic systems that are more versatile than the current, often very specialized, catalysts could be of interest. One could think of a dendritic host as a support for catalytically active guest molecules, such as prepared by Kimura et al. [39]. Developments of this type of host systems would widen the scope of the dendritic systems developed so far, and allow more versatile applications in catalysis.

In the near future many research efforts will continue to be devoted to the field of dendrimer catalysis. The latest developments in this area show very attractive concepts that need further exploration. Although we are still far from using the level of molecular recognition and substrate selectivity exemplified by enzymes, usage of dendritic structures in catalyst design and biomimicry continues to be highly promising. As described above, one of the greatest challenges for the future would be the transformation of our present systems into more versatile and flexible, and therefore more economically attractive, and applicable systems.

Acknowledgements. The authors wish to thank all their co-workers, who have made contributions to dendrimer chemistry described in this review. We would also like to thank the different groups that were involved in collaborations, Prof. P.W.N.M. Van Leeuwen, Prof. U. Kragl, Prof. D. Vogt, Prof. H. Frey and their co-workers and Dr. J. Verweij (DSM-GB). Furthermore, STW, CW-NWO, EU, Shell, Dow, and DSM-GB are acknowledged for financial support.

8
References

1. For reviews on dendrimers see: (a) Newkome GR, Moorefield CN, Vögtle F (1996) Dendritic Molecules: Concepts, Synthesis and Perspectives, VHC: Weinheim, Germany; (b) Fischer M, Vögtle F (1999) Angew Chem Int Ed Engl 38:884; (c) Hearshaw MA, Moss JR (1999) Chem Commun 1
2. For a review on properties and applications of dendrimers see: Bosman AW, Janssen HM, Meijer EW (1999) Chem Rev 99:1665
3. Knapen JWJ, van der Made AW, De Wilde JC, van Leeuwen PWMN, Wijkens P, Grove DM, van Koten, G (1994) Nature 372:659
4. (a) van der Made AW, van Leeuwen PWMN, De Wilde JC, Brandes AC (1993) Adv Mater 5:466 (b); van der Made AW, van Leeuwen PWMN (1992) J Chem Soc Chem Commun 1400

5. Kleij AW, Kleijn H, Jastrzebski JTBH, Smeets WJJ, Spek AL, van Koten G (1999) Organometallics 18:268
6. Kleij AW, Kleijn H, Jastrzebski JTBH, Smeets WJJ, Spek AL, van Koten G (1999) Organometallics 18:277
7. Kleij AW, Gossage RA, Jastrzebski JTBH, Lutz M, Spek AL, van Koten G (2000) Angew Chem Int Ed Engl 39:176
8. Kleij AW, Gossage RA, Klein Gebbink RJM, Brinkman N, Kragl U, Reyerse EJ, Lutz M, Spek AL, van Koten G (2000) J Am Chem Soc 122:12112
9. Gossage RA, Jastrzebski JTBH, van Ameijde J, Mulders SJE, Brouwer AJ, Liskamp RMJ, van Koten G (1999) Tetrahedron Lett 40:1413
10. Wijkens P, Jastrzebski JTBH, van der Schaaf PA, Kolly R, Hafner A, van Koten G (2000) Org Lett 2:1621
11. Garber SB, Kingsbury JS, Gray BL, Hoveyda AH (2000) J Am Chem Soc 122:8168
12. Hovestad NJ, Eggeling EB, Heidbüchel HJ, Jastrzebski JTBH, Kragl U, Keim W, Vogt D, van Koten G (1999) Angew Chem Int Ed Engl 38:1655
13. Reetz MT, Lohmer G, Schwickardi R (1997) Angew Chem Int Ed Engl 36:1526
14. Brinkmann N, Giebel D, Lohmer G, Reetz MT, Kragl UJ (1999) J Cat 183:163
15. Reetz MT, Giebel D (2000) Angew Chem Int Ed 39:2498
16. Mizugaki T, Ooe M, Ebitani K, Kaneda K (1999) J Mol Cat A: Chem 145:329
17. de Groot D, Eggeling EB, de Wilde JC, Kooijman H, Haaren RJ, van der Made AW, Spek AL, Vogt D, Reek JNH, Kamer PCJ, van Leeuwen PWNM (1999) Chem Commun 1623
18. de Groot D, Emmerink PG, Coucke C, Reek JNH, Kamer PCJ, van Leeuwen PWNM (2000) Inorg Chem Commun 3:711
19. Bardaji M, Kustos M, Caminade A-M, Majoral J-P, Chaudret B (1997) Organometallics 16:403
20. Maraval V, Laurent R, Caminade A-M, Majoral J-P (2000) Organometallics 19:4025
21. Sanders-Hovens MSTH, Jansen JFGA, Vekemans JAJM, Meijer EW (1995) Polym Mater Sci Eng 73:338
22. Suzuki T, Hirokawa Y, Ohtake K, Shibata T, Soai K (1997) Tetrahedron Asymmetry 8:4033
23. Sato I, Shibata T, Ohtake K, Kodaka R, Hirokawa Y, Shirai N, Soai K (2000) Tetrahedron Lett 41:3123
24. Köllner C, Pugin B, Togni A, (1998) J Am Chem Soc 120:10274
25. Schneider R, Köllner C, Weber I, Togni A (1999) Chem Commun 2415
26. Bolm C, Derrien N, Seger A (1996) Synlett 387
27. Hovestad NJ, Jastrzebski JTBH, van Koten G (1999) Polym Mater Sci Eng 80:53
28. Klein Gebbink RJM, Bosman AW, Feiters MC, Meijer EW, Nolte RJM (1999) Chem Eur J 5:65
29. Zeng H, Newkome GR, Hill CL (2000) Angew Chem 112:1842
30. Bourque SC, Maltais F, Xiao WJ, Tardiff O, Alper H, Arya P, Manzer L (1999) J Am Chem Soc 121:3035
31. Bourque SC, Alper H, Manzer L, Arya P (2000) J Am Chem Soc 122:956
32. Arya P, Rao NV, Singkhonrat J, Alper H, Bourque SC, Manzer L (2000) J Org Chem 65:1881
33. Bhyrappa P, Young JK, Moore JS, Suslick KS (1996) J Mol Cat A Chem 113:109
34. Bhyrappa P, Young JK, Moore JS, Suslick KS (1996) J Am Chem Soc 118:5708
35. Chow HF, Mak CC (1997) J Org Chem 62:5116
36. Mak CC, Chow HF (1997) Macromolecules 30:1228
37. Oosterom GE, van Haaren RJ, Reek JNH, Kamer PCJ, van Leeuwen PWNM (1999) Chem Commun 1119
38. Kimura M, Sugihara Y, Muto T, Hanabusa K, Shirai H, Koboyashi N (1999) Chem Eur J 5:3495
39. Kimura M, Kato M, Muto T, Hanabusa K, Shirai H (2000) Macromolecules 33:1117
40. Brunner H (1995) J Organomet Chem 500:39
41. Brunner H, Altman S (1994) Chem Ber 127:2285
42. Seebach D, Marti RE, Hintermann T (1996) Helv Chim Acta 79:1710
43. Rheiner PB, Seebach D (1999) Chem Eur J 5:3221

44. Yamago S, Furukawa M, Azuma A, Yoshida J-I (1998) Tetrahedron Lett 39:3783
45. Fan Q-H, Chen Y-M, Chen X-M, Jiang D-Z, Xi F, Chan ASC (2000) Chem Commun 789
46. Balogh L, Tomalia DA (1998) J Am Chem Soc 120:7355
47. Zhao M, Crooks RM (1999) Angew Chem 38:364
48. Zhao M, Crooks RM (1999) Adv Mater 11:217
49. Zhao M, Crooks RM (1999) Chem Mater 11:3379
50. Chechik V, Crooks RM (2000) J Am Chem Soc 122:1243
51. Miedaner A, Curtis CJ, Barkley RM, Dubois DL (1994) Inorg Chem 33:5482
52. Petrucci-Samija M, Guillemette V, Dasgupta M, Kakkar AK (1999) J Am Chem Soc 121:1968
53. Lee JJ, Ford WT, Moore JA, Li Y (1994) Macromolecules 27:4632
54. Vassilev K, Ford WT (1999) J Polym Sci A: Polym Chem 37:2727
55. Pan Y, Ford WT (2000) Macromolecules 33:3731
56. Morao I, Cossío FP (1997) Tetrahedron Lett 38:6461
57. Schmitzer A, Perez E, Rico-Lattes I, Lattes A (1999) Tetrahedron Lett 40:2947
58. Habicher T, Diederich F, Gramlich V (1999) Helv Chim Acta 82:1066
59. Piotti ME, Rivera F Jr, Bond R, Hawker CJ, Frechet JMJ (1999) J Am Chem Soc 121:9471
60. Kleij AW, Klein Gebbink RJM, van de Coevering R, Noordman A-M, Spek AL, van Koten G (2001) Chem Eur J 7:181
61. Kragl U, Dreischbach C (1996) Angew Chem Int Ed Engl 35:642
62. Giffels G, Beliczey J, Felder M, Kragl U (1998) Tetrahedron Assymetry 9:691
63. For some recent examples of hyperbranched molecules see: (a) Chang HT, Fréchet JM (1999) J Am Chem Soc 121:2313; (b) Jikei M, Chon SH, Kakimoto M, Kawauchi S, Imase T, Watanebe J (1999) Macromolecules 32:2061: (c) Sunder A, Hanselmann R, Frey H, Mülhaupt R (1999) Macromolecules 32:4240
64. Schlenk C, Kleij AW, Frey H, van Koten G (2000) Angew Chem Int Ed 39:3445

Glycodendrimers

Niels Röckendorf · Thisbe K. Lindhorst

Institut für Organische Chemie, Christian-Albrechts-Universität zu Kiel, Otto-Hahn-Platz 4, 24098 Kiel, Germany
E-mail: tklind@oc.uni-kiel.de

Glycodendrimers have become valuable tools in glycobiology especially in the context of multivalency which is an important principle of carbohydrate-protein interactions. Multivalency effects observed in glycobiology are being currently discussed controversially and a conclusive understanding of this phenomenon has not yet been obtained. Rather than discuss the many biological data which have been collected using glycodendrimers as molecular tools in glycobiology, the principal design and molecular architectures are, therefore, discussed and highlighted with representative examples. The term "glycodendrimer" was interpreted as a designation for carbohydrate-containing molecules which can be grown generationwise following an iterative repetitive synthesis. For classification of the diverse examples, four groups have been distinguished, (i) glyco-coated non-carbohydrate dendrimers, (ii) glycodendrimers by convergent assembly of dendritic carbohydrate-containing wedges, (iii) glycodendrimers grown from carbohydrate-building blocks, and (iv) dendrimers containing a carbohydrate core molecule.

Keywords. Glycodendrimers, Glycoconjugates, Glycobiology, Multivalency, Ligand-receptor interactions, Oligosaccharides, Carbohydrates

1 Introduction

In molecules called "glycodendrimers", saccharide portions are conjugated according to the principles of dendritic growth or they are ligated to dendrimers, respectively. The idea of supplementing carbohydrate chemistry by the concepts, which have made dendrimer chemistry the intriguing field it is, has essentially been triggered by questions addressed in glycobiology [1–4]. During recent years, it has become clear that in carbohydrate-protein interactions, which are essential molecular recognition events in cell biology [5, 6], multivalency plays an important role [7, 8]. Therefore, chemists and biochemists have sought carbohydrates and glycoconjugates which can be used as molecular tools for the investigation and possibly manipulation of carbohydrate-protein interactions [9], especially with regard to the multivalency effect which has been observed [10].

To obtain the required carbohydrate molecules, one possibility is to isolate them from natural sources. This, however, is difficult, especially because homogeneous material can hardly be harvested in sufficient quantities. This is due to the fact that the carbohydrate moieties of a particular glycoconjugate, which is expressed by a cell, comprises significant structural diversity. This phenomenon can be understood from the biosynthesis of these molecules and is known as "microheterogeniety"[11]. An alternative approach includes the chemical [12–14], enzymatic [15–17] or chemoenzymatic synthesis [18, 19] of the required molecules according to the natural example structures. Even though enormous progress has been made in this area during recent years [20, 21], the synthesis of complex carbohydrates and glycoconjugates is still a major adventure and extremely time-consuming work for every single example. Another approach for getting access to the wide structural variety of branched and hyperbranched oligosaccharides found in nature includes combinatorial techniques, which are currently being evaluated in carbohydrate chemistry [22–25].

Quite different from everything mentioned so far, is the idea of providing the structural requirements needed in molecular recognition of carbohydrates by artificial design. This strategy would furnish saccharide-containing molecules, which are more or less dramatically different from their natural counterparts. The variety of structures thus obtained may be gathered under the term "glycomimetics" [26, 27]. This is not a strictly defined class of molecules, however, it is implied that its representatives have the capability to mimic the biological properties of their natural saccharide counterparts or even surpass their activities in a given system [28]. Concurrently, the synthesis of designed glycomimetics is easier compared to classical oligosaccharide synthesis, normally larger amounts are accessible and the variation of structural characteristics is also easier. Glycomimetic design may, e.g., include the substitution of glycosidic linkages by other chemistries such as peptide coupling [29] or thiourea-bridging [30, 31], e.g.

As the oligosaccharides found in glycoconjugates resemble fractal molecules, similar to the typical structures of dendrimers [32, 33], synthesis of "glyco-dendrimers" [34, 35] became an important approach in glycosciences to mimic the hyperbranched character of oligoantennary carbohydrates. Since the first example of a glycodendrimer less than ten years ago [36], many different examples have followed. Also a large variety of carbohydrate-containing molecules, which can be used as molecular wedges, have been synthesized with regard to the design of multivalent glycoconjugates or oligosaccharide mimetics, respectively [37–41]. These molecules have been named "glycoclusters" or "cluster glycosides". While a number of glycoclusters have the size of small oligosaccharides, others are rather complex and multiply branched and, therefore, remind one of dendrimers or dendrons, respectively. It can be observed in the literature, that this structural relationship has often tempted chemists to rather apply the designation glycodendrimer and glycodendron laxly. In this paper the discussion of glycodendrimers is mainly restricted to those carbohydrate-containing molecules which allow a generationwise growth. The various branched glycoclusters which do not obey the principles of dendrimer chemistry are largely omitted in this chapter. Nevertheless, they can be of great value for glycobiological studies. Furthermore, branched or hyperbranched carbohydrate-containing polymers, called "glycopolymers" or "neo-glycopolymers" [42–53], respectively, have not been included into the discussion, even though they can help a great deal in the unraveling of the secrets of multivalency [54, 55]. Also carbohydrate-containing dendrimers in which glycoproteins have been multiplied on dendritic scaffolds in order to prepare multiple glycopeptide antigens [56] are not topic of this contribution.

Four major architectures of dendrimers containing carbohydrate derivatives will be distinguished here and introduced in the following section.

2
Architectures of Glycodendrimers

Four different classes of glycodendrimers can be distinguished according to the principal design of their molecular architectures.

(i) The first examples of glycodendrimers were molecules in which a non-carbohydrate dendritic core, e.g., a hyperbranched oligolysine dendrimer, a polypropyleneamine (POPAM) or a polyamidoamine (PAMAM) dendrimer, was built up first and was then functionalized with a coat of carbohydrate moieties in the periphery (Fig. 1, type A, cf. Sect. 3).
(ii) Alternatively the convergent approach is followed, in which glyco-coated dendrons are synthesized first and eventually assembled on an oligofunctional core molecule. The glycodendrons can either be synthesized in a divergent mode and functionalized with carbohydrates in the last step; or, vice versa, a glycocluster is synthesized first and then multiplied by a convergent approach, also leading to a glyco-coated dendron (Fig. 1, type B, cf. Sect. 4).
(iii) The complexity of natural glycoconjugates is most closely resembled by glycodendrimers which are built from carbohydrate-derived building blocks as the only "ingredient". By such an approach, so far only glycodendrons have been synthesized, which is possible by using either a divergent or a convergent strategy (Fig. 1, type C, cf. Sect. 5).

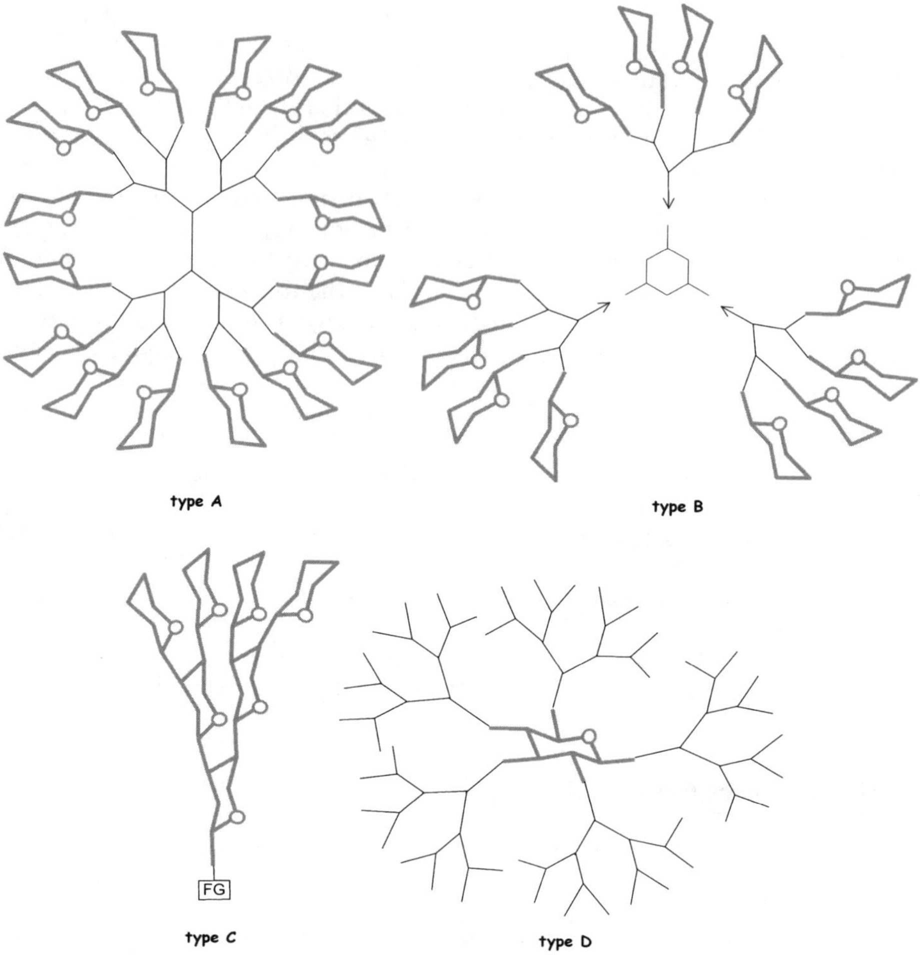

Fig. 1. Classification of glycodendimers according to their molecular architectures

(iv) Finally, by reversing the architecture of glyco-coated dendrimers, carbohydrates can be used as the dendrimer core and this has been realized by the synthesis of carbohydrate-centered glycodendrimers (Fig. 1, type D, cf. Sect. 6).

This classification of course does not intend to be the only possible one, yet it provides a useful subdivision and has been the basis for the content of this contribution. Many excellent reviews have been published surveying the chemistry and biology of glycodendrimers and glycoclusters and are recommended for reading [34, 58–63].

3
Sugar-Coated Non-Carbohydrate Dendrimers

Sugar-coated non-carbohydrate dendrimers comprise the first examples of this class of molecules at all. The basic idea, which prompted carbohydrate chemists to use this approach, was to substitute the complex carbohydrate interior of an oligoantennary glycoconjugate by a branched non-carbohydrate molecule which would basically only serve as a scaffold for the multiple presentation of sugar moieties, which are known as the principal carbohydrate epitopes in particular glycobiological system under investigation (Fig. 2). Based on their typical structural characteristics, dendrimers appeared as ideal candidates for scaffolding in this regard.

This relatively radical abbreviation of the natural example structures is affiliated with many practical advantages but also with a number of uncertainties which are traded in at the same time. As a prerequisite for this concept, it has been assumed that rather simple saccharides can serve as ligands for the carbohydrate-recognition domains (CRDs) of lectins. Lectins are a class of more or less specific proteins, specialized for the molecular recognition of carbohydrates [5, 64]. Indeed, monosaccharides are often sufficient to form the non-covalent carbohydrate-lectin complexes which are necessary to trigger the next event in a cascade of biological processes [65]. For other lectins, disaccharides are required for binding, and for a special class of "selective" lectins, the "selectins" [66], a complex tetrasaccharide, called sialyl-Lewis-x (sLe^x) [27] has been identified as the minimum carbohydrate structure for recognition. From this knowledge, it appears reasonable to test small oligosaccharides and even monosaccharides as lectin ligands in order to compete for their CRDs with the natural ligands and thus to assist in probing the biological role of carbohydrate-protein interactions. However, such interactions between simple sugars and lectin CRDs are characterized by weak affinities with typical binding constants in the millimolar or high micromolar range [67]. Apparently, the phenomenon of multivalency adds to the strength of carbohydrate-protein interactions that is needed for a significant biological effect [8]. Thus, the weak affinity of singular interactions is multiplied to an overall avidity, e.g., with binding constants in the nanomolar range. At first glance, understanding of this phenomenon is easy, as multiple interactions can obviously be reasoned on the basis of the branched, so-to-say multiple oligosaccharide structures found in nature on one hand, and on the other hand on the basis of multiple CRDs. Multiple CRDs (CRD clusters) are provided on one peptide strain of a lectin [64] or by membrane clustering of monovalent lectins, such as in the case of the selectins. Multivalency of carbohydrate-protein interactions implies a number of advantages such as an option for fine-tuning of biological response as well as an increase in specificity.

Multivalency of carbohydrate-protein interactions was proved impressively more than two decades ago, showing that binding affinity of the asialo glycoprotein receptor to a carbohydrate ligand is logarithmically improved with linear increase of the carbohydrate lectin ligand. This was demonstrated with a simple TRIS-derived synthetic glycoconjugate (Fig. 3). Molecules of this type

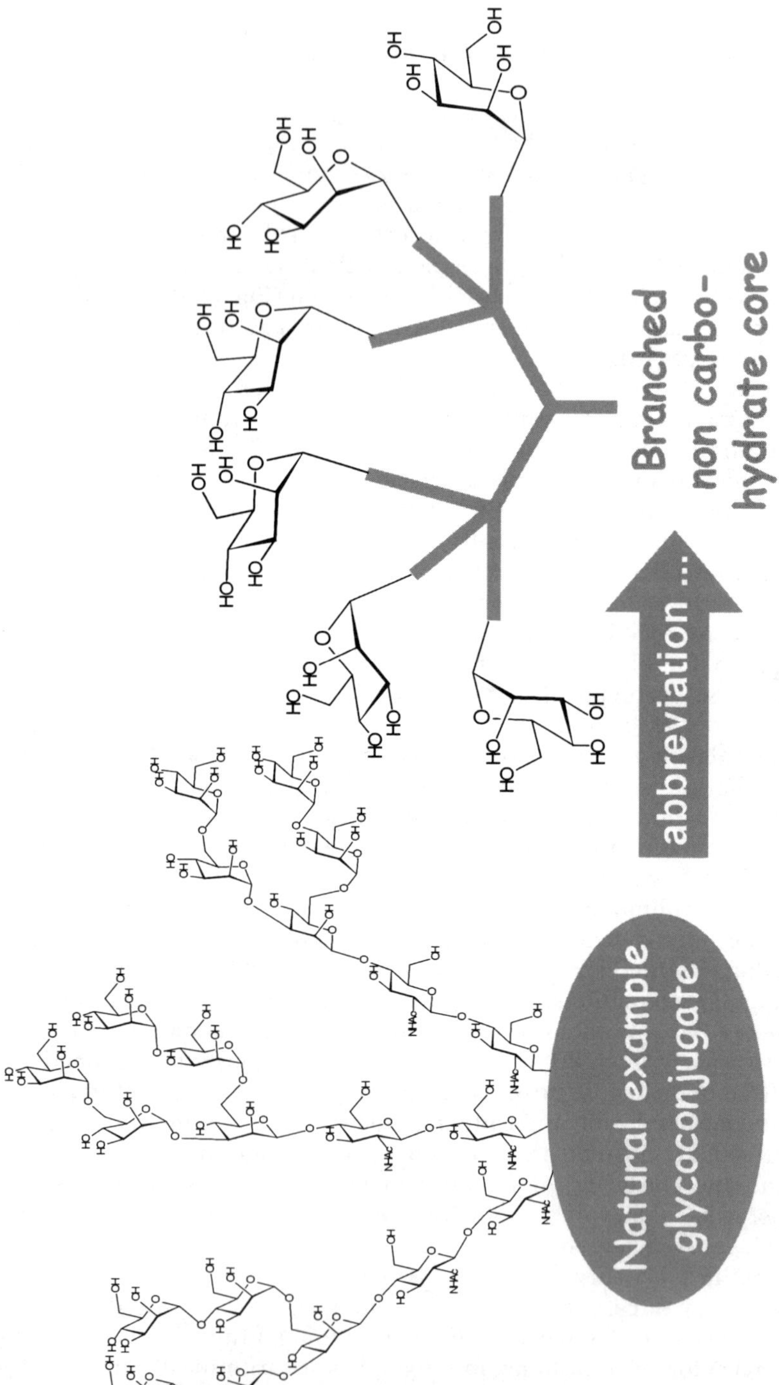

Fig. 2. Glycodendrimers can be considered as 'abbreviated' oligoantennary glycoconjugates

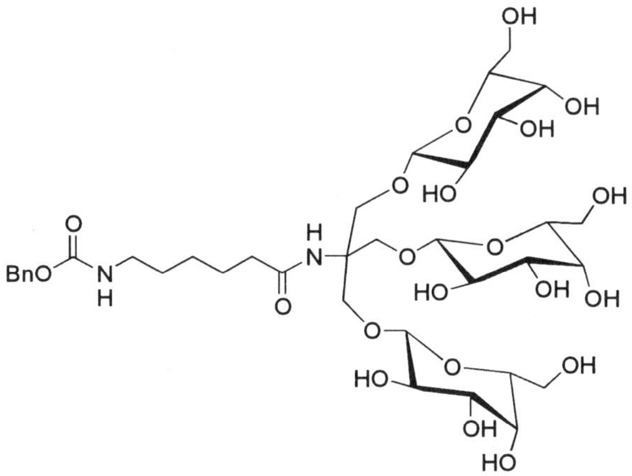

Fig. 3. Example of the first class of cluster glycosides, here a trivalent cluster galactoside

have been named cluster glycosides. For the observed binding effect, the term "cluster effect" or "glycoside cluster effect", respectively, has been coined [10].

Thus, the intriguing idea was born to interfere with carbohydrate-protein interactions effectively with the help of glycomimetics, which are designed as multivalent molecules, neo-glycopolymers or monodisperse neo-glycoconjugates, such as the glycodendrimers (Fig. 2). As protein-carbohydrate complexation is important in a wide range of medically significant interactions including signal transduction, inflammation and microbiological pathogenesis, this approach may also contribute to the development of a new class of carbohydrate-based therapeutics [68]. Furthermore, molecules are needed which help us to unravel the secrets of multivalent recognition. The physical chemistry of the effects observed are far from being fully understood and even many of the results which have been obtained in various biological assays may have to be interpreted with caution [69]. The queries asked about the role of multivalency in carbohydrate-protein interactions are being continually dealt with in current research and are being discussed elsewhere. Herein, the biology of the discussed glycodendrimers is only touched on.

Dendrimers offer some excellent options for addressing the role of multivalency in carbohydrate-protein interactions. As with smaller glycoclusters [70], the valency of the carbohydrate epitopes exposed can be exactly adjusted and even the distance and density, respectively, can be tuned. This promises a relatively systematic investigation of carbohydrate-protein interactions with regard to the issue of multivalency. Nevertheless, it remains mostly uncertain as to what extent the dendritic core of a sugar-coated dendrimer contributes to binding and how the molecular dynamics of a glycodendrimer can be designed so that it adds favorably to lectin binding. These problems are largely unsolved at present, whereas the synthesis of the desired molecules can be well controlled nowadays.

For the first glycodendrimer synthesis [36], a multibranched L-lysine core was utilized as a dendritic scaffold. This was elaborated in a solid phase peptide syn-

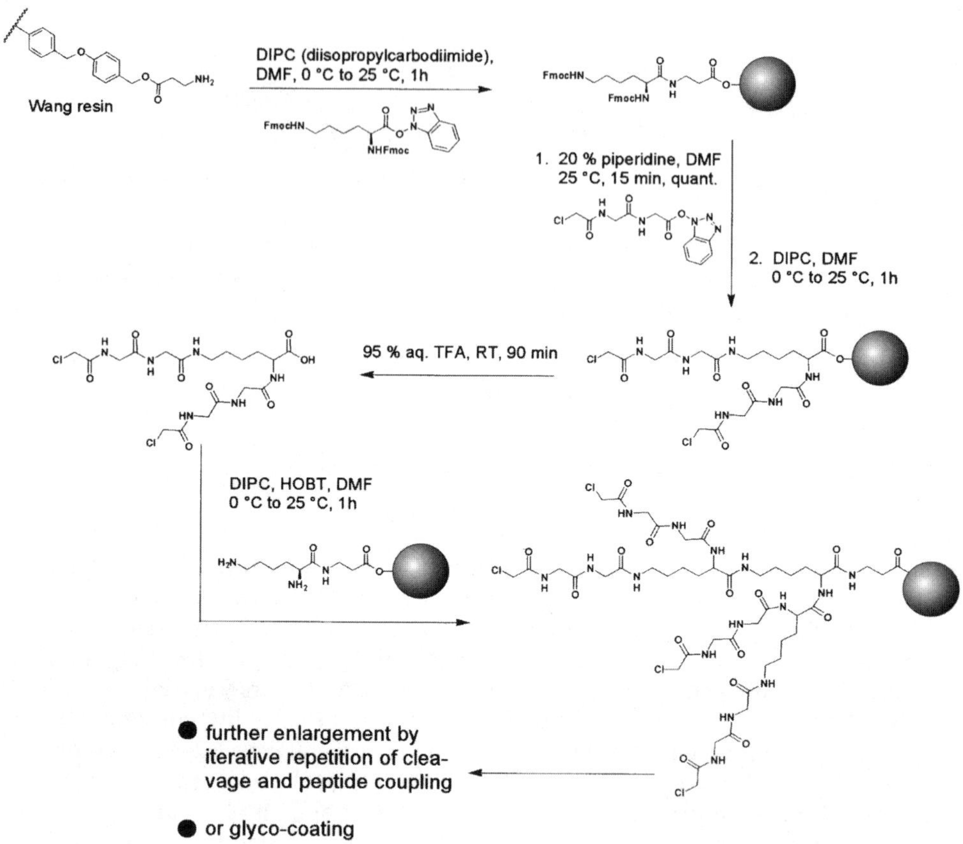

Scheme 1. Solid phase synthesis of L-lysine dendrimers

thesis on a *p*-benzyloxybenzyl alcohol (Wang) resin. Coupling of L-lysine was performed via an β-alanyl spacer which was anchored to the resin first, using the N^{α}-N^{ε}-di-Fmoc-L-lysine-HOBT ester (Scheme 1). Chloroacetyl-terminated diglycinyl spacers were used in the periphery of the hyperbranched molecules and thus, up to 16 *N*-chloroacetylated amino groups became available for sugar coating by thioether ligation in a nucleophilic displacement reaction (Scheme 2). After sequential hydrolysis and deprotection of the sugar hydroxyl groups, e.g., α-thiosialoside-bearing polylysine dendrimers were obtained [71].

Thiosialoside glycodendrimers were tested as inhibitors of influenza A virus. Influenza virus carries two proteins on its surface which are conserved in all phenotypes. The first protein is hemagglutinin, a lectin which binds to neuraminic acid residues, whereas the second protein is an enzyme which cleaves the glycosidic linkages of neuraminic acid. For the synthesis of the glycodendrimers, to probe influenza virus adhesion to host cells, thioglycosidic linkages were chosen in order to provide combined hemagglutinin and sialidase inhibition properties.

Various thioaldosides and thioketosides as well as a large number of glycosides bearing an ω-thiol as the aglycon have been used with polylysine scaffolds

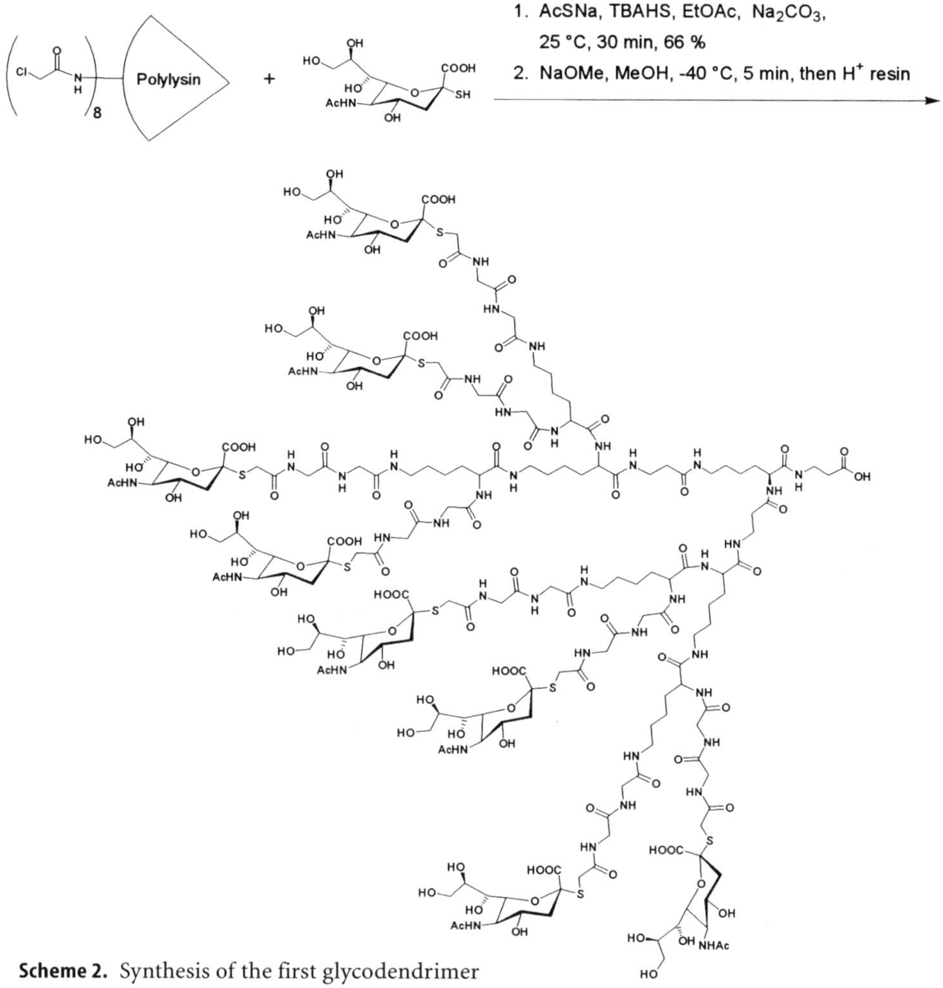

Scheme 2. Synthesis of the first glycodendrimer

(Fig. 4) [72, 73]. Even complex oligosaccharides, such as the biologically important T-antigen (Galβ1,3GalNAc) [74], the tetrasaccharide sLe^x and its trisaccharide analog 3′-sulfo-sLe^x [75] (Fig. 5) have been used for glyco-coating of dendrimers. Interestingly, sLe^x-coated oligolysine dendrimers could be prepared by a chemoenzymatic approach using a sialyl and a fucosyl transferase [76]. A chemoenzymatic approach was also employed when *N*-acetyllactosamine(LacNAc)-coated glycodendrimers were the target molecules [77]. First, *N*-acetylglucosamine(GlcNAc)-functionalized dendrimers with valencies of up to eight were prepared on the basis of L-lysine dendrons. These dendrimers were then further transformed enzymatically into dendritic *N*-acetyllactosamine (LacNAc) derivatives using a mix of UDP-glucose, UDP-glucose-4′-epimerase to turn UDP-glucose into the respective galactose derivative and GlcNAc-β-1,4-galactosyltransferase to catalyze the glycosyl transfer reaction.

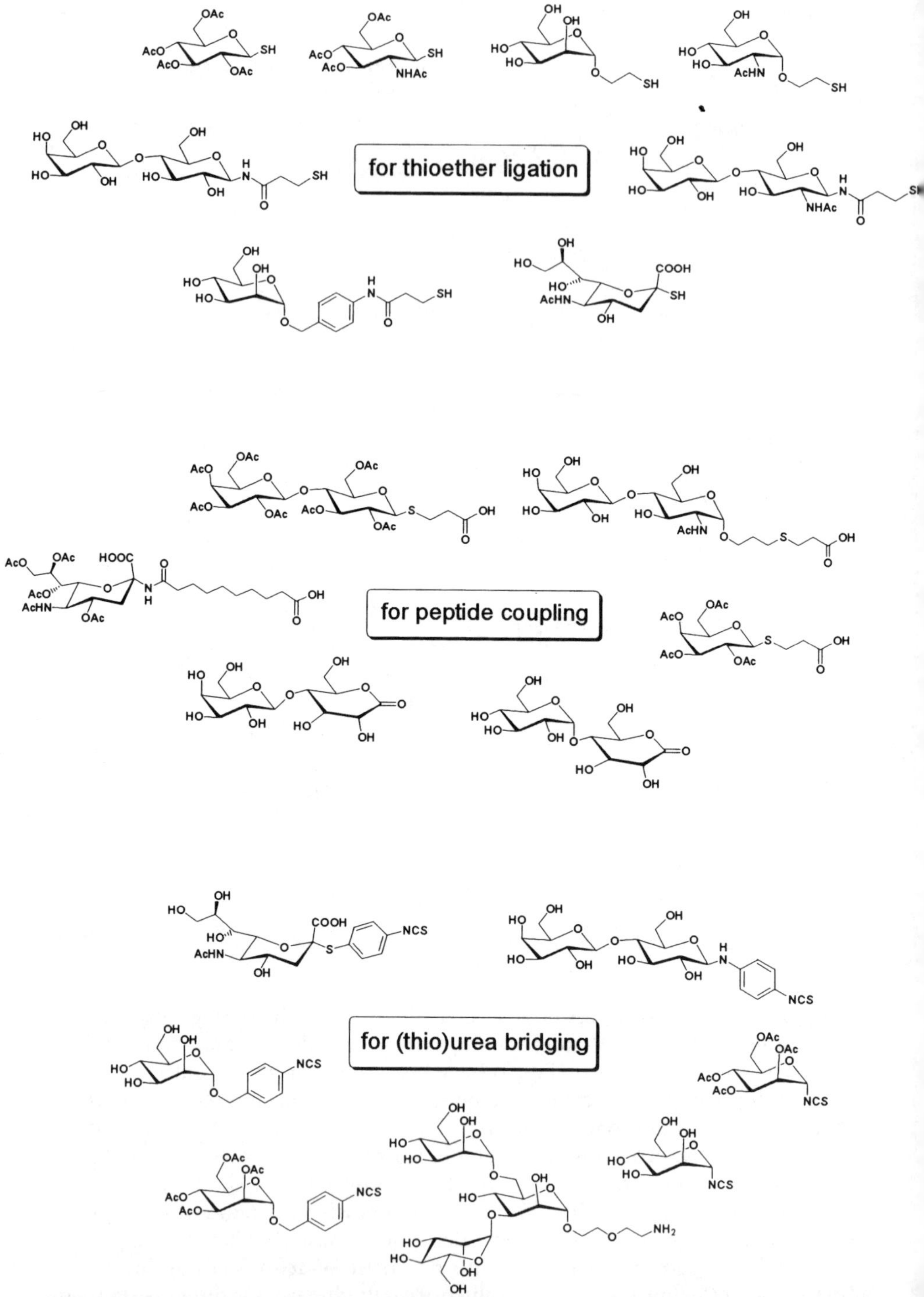

Fig. 4. A selection of sugar epitopes which have been used for peripheral functionalization of non-carbohydrate dendrimers using different ligation chemistries

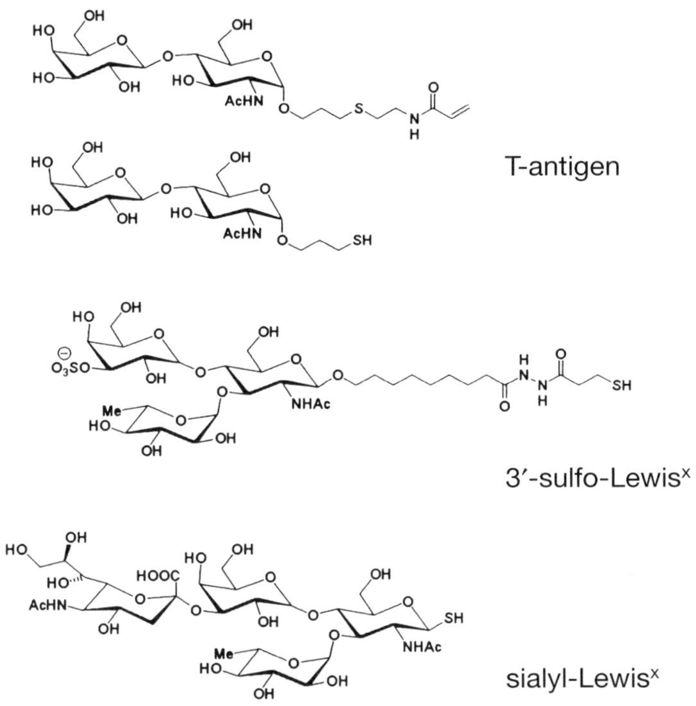

Fig. 5. Complex saccharides used as peripheral units in glycodendrimers

The effect of multivalency in the inhibition of binding of yeast mannan to the lectin concanavalin A and pea lectins has been evaluated by mannosylated lysine-based dendrimers [78]. A *p*-(3-thioacetyl)-propionamidophenyl α-D-mannopyranoside was employed in the nucleophilic displacement reaction with *N*-chloroacetylated oligolysine scaffolds to give glycodendrimers up to a valency of 16 (Fig. 6). A series of six such glycodendrimers differing in their sugar valencies were tested in the inhibition of binding of yeast mannan to concanavalin A and pea lectins in solid-phase enzyme linked lectin essays (ELLA) using *p*-nitrophenyl α-D-mannopyranoside as standard. The 16-mer was found to be 66- and 1383-fold more potent than *p*-nitrophenyl α-D-mannopyranoside, with con A and pea lectin respectively.

Oligolysine dendrimers were also favorable for selective functionalization [79]. Thus, fluorescein-labelled *N*-chloroacetylated L-lysine dendrons were prepared, followed by a chemoselective thioether ligation with fully deprotected glycoside derivatives. This approach was eventually used to synthesize antigen-bearing cluster mannosides [80], which are of biological relevance as highly mannosylated antigens lead to effective targeting to dendritic cells which may improve the efficacy of vaccines. With a *N*-chloroacetylated L-oligolysine core, which had been modified with a glyoxylyl function, two orthogonal chemoselective ligation reactions were applied, introducing 2-thioethyl α-D-mannopyranoside moieties by thioetherification and a hydrazine-modified peptide antigen

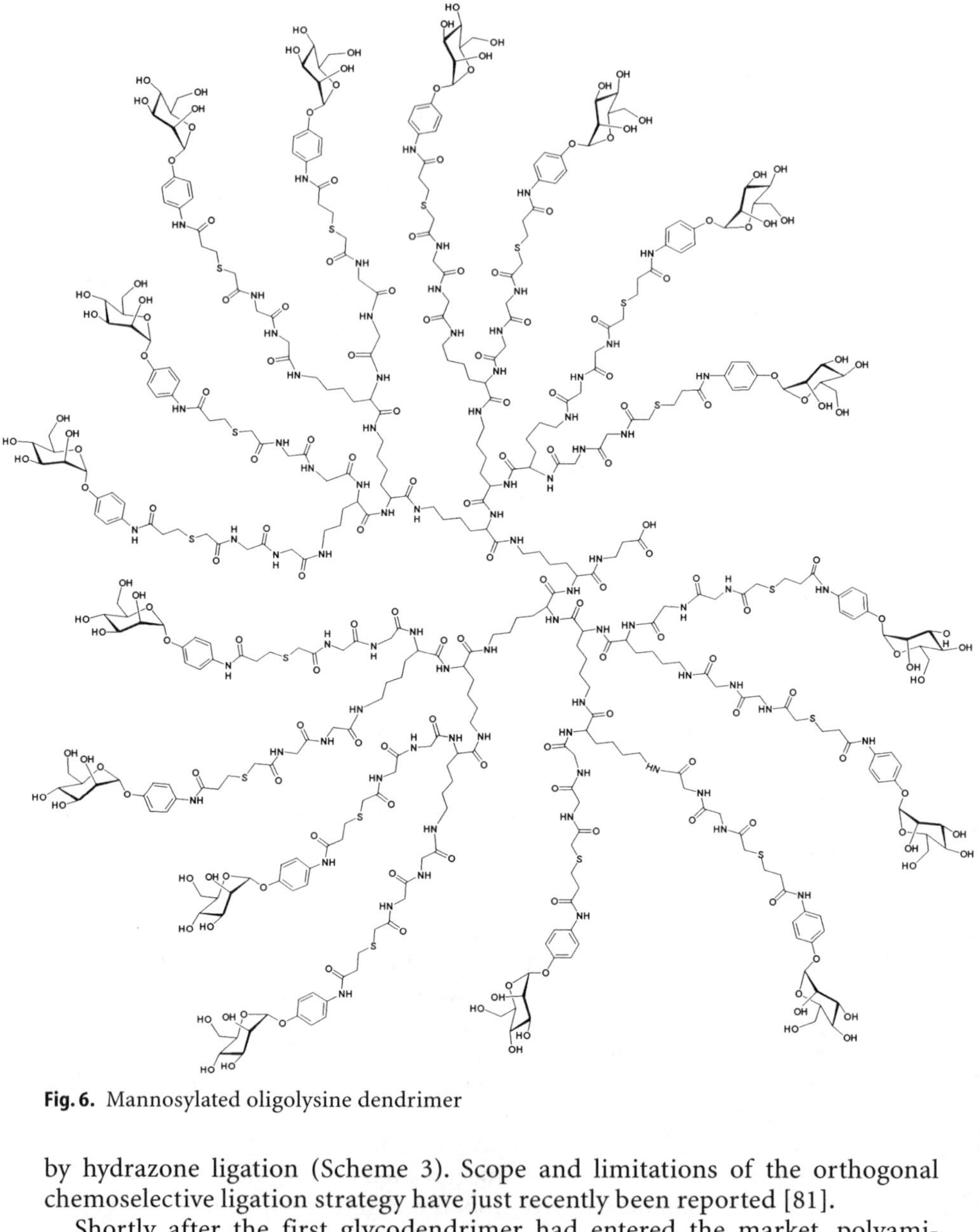

Fig. 6. Mannosylated oligolysine dendrimer

by hydrazone ligation (Scheme 3). Scope and limitations of the orthogonal chemoselective ligation strategy have just recently been reported [81].

Shortly after the first glycodendrimer had entered the market, polyamidoamine (PAMAM) and polypropyleneimine (POPAM) dendrimers were also introduced as core molecules for the synthesis of glycodendrimers (Fig. 7).

A large variety of carbohydrate derivatives (Fig. 4) have been employed in order to form dendrimers with 12, 24, 48 or 64 terminal glycan residues. For the first glycocoating of a PAMAM dendrimer, disaccharide lactones, *O*-*β*-D-glucopyranosyl-(1→4)-D-glucono-1,5-lactone and *O*-*α*-D-galactopyranosyl-(1→4)-D-glucono-1,5-lactone, were used as the acid components in a peptide coupling

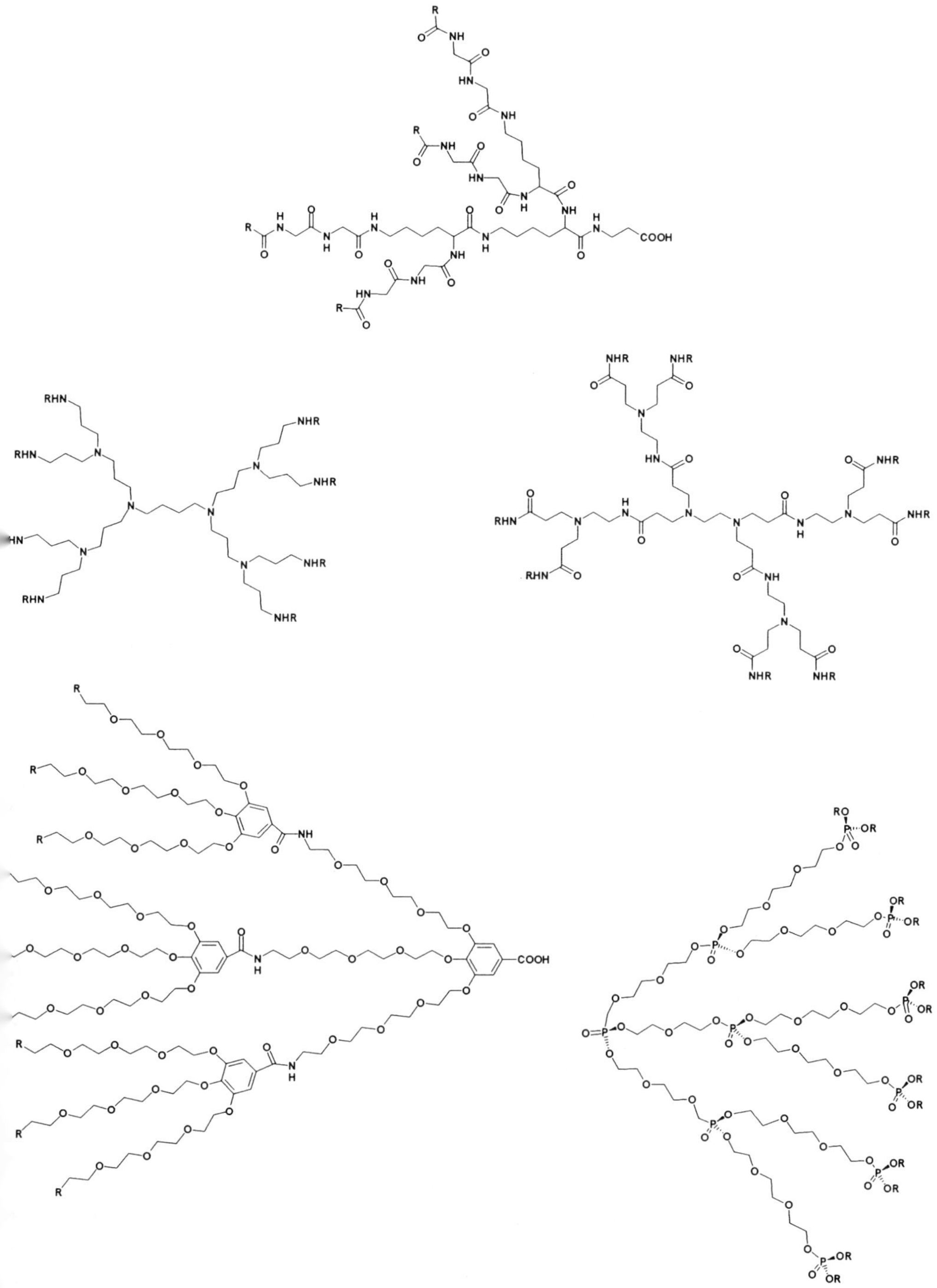

Fig. 7. A selection of dendritic scaffolds used for glycodendrimer synthesis; R = sugar residue

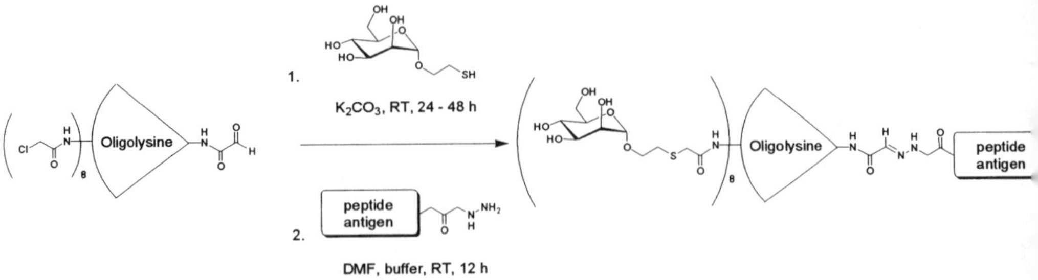

Scheme 3. An orthogonal chemoselective ligation strategy leads to antigen-bearing cluster mannosides

reaction, during which the terminal sugar ring is forced into the open-chain form (Scheme 4) [82].

Also the reaction of isothiocyanato-functionalized carbohydrate derivatives (Fig. 4) with branched oligoamines such as PAMAM dendrimers proved to be a successful ligation technique for the synthesis of glycodendrimers [30]. Many similar avenues to thiourea-bridged glycodendrimers have been elaborated during recent years [70] and extensively reviewed [42, 44, 58, 60, 61]. Thiourea-bridging has even been possible reacting PAMAM dendrimers with unprotected *p*-isothiocyanatophenyl α-D-mannopyranoside in water (Scheme 5) [83].

Thiourea-ligated glycodendrimers have also been used to investigate bacterial adhesion which is often dependent on carbohydrate-protein interactions.

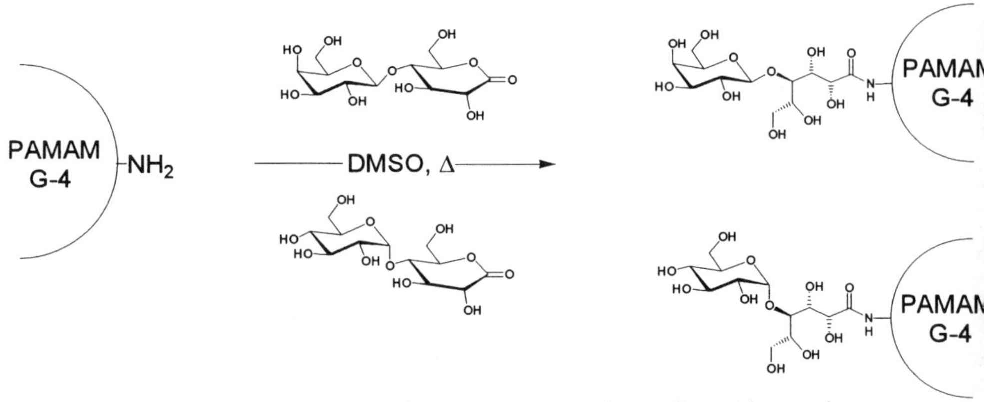

Scheme 4. Glycocoating of PAMAM dendrimers by peptide coupling with sugar lactones

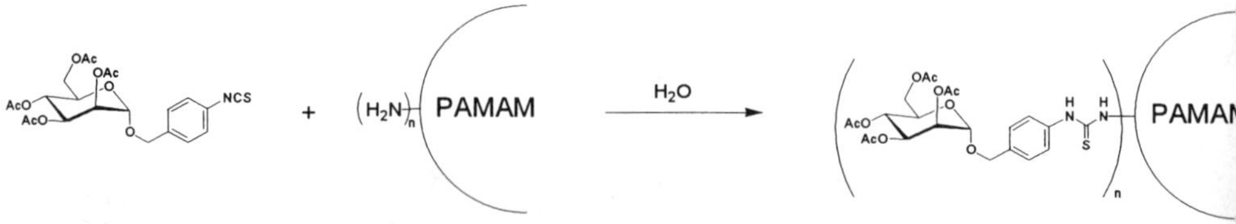

Scheme 5. Thiourea-bridging in water

Table 1. Inhibition of the hemagglutination of guinea pig erythrocytes by type 1 fimbriated *E. coli*

Inhibitor	1	2	3	4	5	6	7
Relative inhibitory potency	35	10	10	10	18	14	1

Bacteria use own lectins, which are expressed on protein appendages on their surfaces, called fimbriae or pili, to adhere to the glycocalyx of their potential host cells [84]. Mannose-specific bacterial adhesion, which is mediated by so-called type 1 fimbriae, can be inhibited by α-mannosyl-functionalized PAMAM dendrimers to a different extent. This has been tested in hemagglutination inhibition assays using a type 1 fimbriated *Escherichia coli* strain [85]. The inhibitory potencies obtained in this system with a series of systematically varied glycoclusters and glycodendrimers (**1**–**7**, Fig. 8) gave some interesting hints about the structural preferences required for effective inhibition of bacterial adhesion in this system. The measured values were valency-corrected and based on mannose derivative **7** as the monovalent reference compound (Table 1).

Thiourea-bridged phenyl β-D-lactoside-coated starburst dendrimers were used to investigate different types of galactoside binding proteins and to evaluate their potential to serve as high-affinity ligands for clinically relevant sugar receptors [86]. PAMAM dendrimers were also used for scaffolding of more complex oligosaccharides. Glycodendrimers bearing tumor related T-antigen were shown to strongly bind to mouse monoclonal IgG antibodies and to be suitable coating antigens in microtiter plates [87]. Other PAMAM-based oligosaccharide dendrimers were evaluated as ligands for cholera toxin [88].

Carbohydrate-coated glycodendrimers were also realized using preformed polypropyleneimine cores. Spacer-modified D-galactose and D-lactose derivatives were converted into their active *N*-hydroxysuccinimidyl esters, which could be coupled to the terminal amino groups of POPAM dendrimers. In addition, the attachment of trivalent cluster galactosides to dendritic cores with up to eight primary amino groups was demonstrated [89, 90]. The hydrodynamic properties of such POPAM glycodendrimers up to the fifth generation were investigated by velocity sedimentation, translation diffusion and viscosity measurements in 0.195% NaCl aqueous solution [91]. Thus, the complete functionalization of the initial dendrimers with sugar residues could be shown. The glycodendrimers were found to resemble spherical molecules with an inhomogeneous distribution of density.

In order to get access to symmetrical dendrimers which can readily be characterized by standard NMR techniques, new dendritic scaffolds were employed in glycodendrimer synthesis using dendrimers based on a 3,3-iminobis (propylamin) core [92–94]. Dendrimers with 2, 4, 8, 16 valencies in the first, second, third and fourth generation were transformed into the *N*-chloroacetylated analogs, followed by reaction with 2-thiosialic acid derivatives to give the respective sialodendrimers in 47–58% yield. These compounds are of interest as inhibitors of in human erythrocyte hemagglutination by influenza viruses and regarding their inhibition-properties in binding of human α_1-acid glycoprotein to slug lectin from *Limax flavus*.

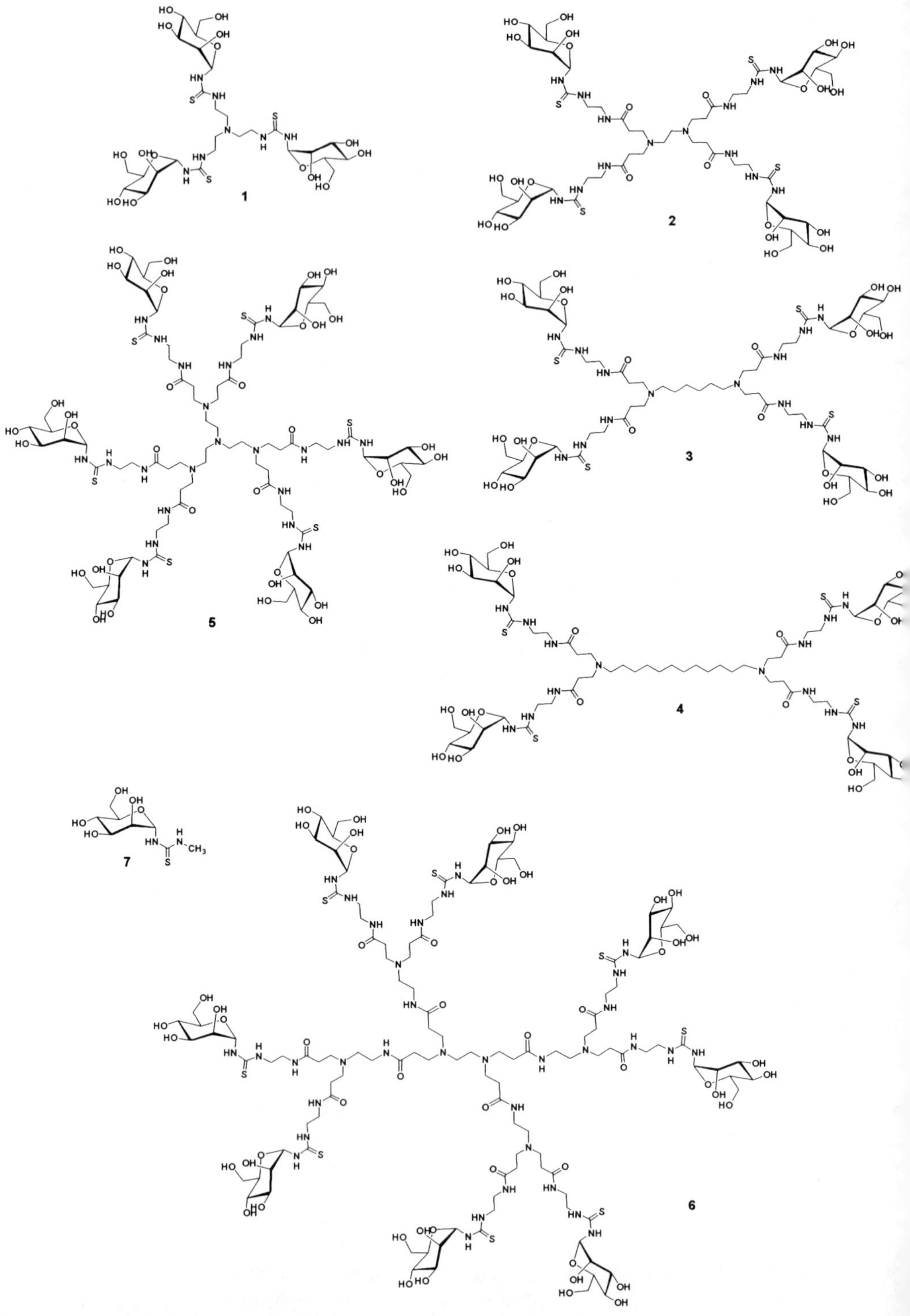

Fig. 8. Thiourea-bridged glycoclusters and glycodendrimers used for the inhibition of mannose-specific bacterial adhesion

Furthermore, more unusual hyperbranched molecules have been designed to serve as core molecules for the synthesis of glycodendrimers (Fig. 7). Based on a gallic acid core, dendritic scaffolds were obtained in combination with oligoethylene glycol spacers for the synthesis of glycodendrimers to aid further understanding of the multivalency effect. Lactosyl [95] and sialic acid residues [96] were scaffolded using the gallic acid-derived hyperbranched cores. The core units were synthesized starting from gallic acid methyl ester, which was equipped with tetraethylene glycol spacers having terminal azide groups. In the resulting molecule either the methyl ester was saponified or the azide function was reduced to the amine, to be ready for further branching.

Phosphotriester building blocks were also used as dendritic scaffolds (Scheme 6) [35, 60] which could be sugar-coated by nucleophilic displacement reactions.

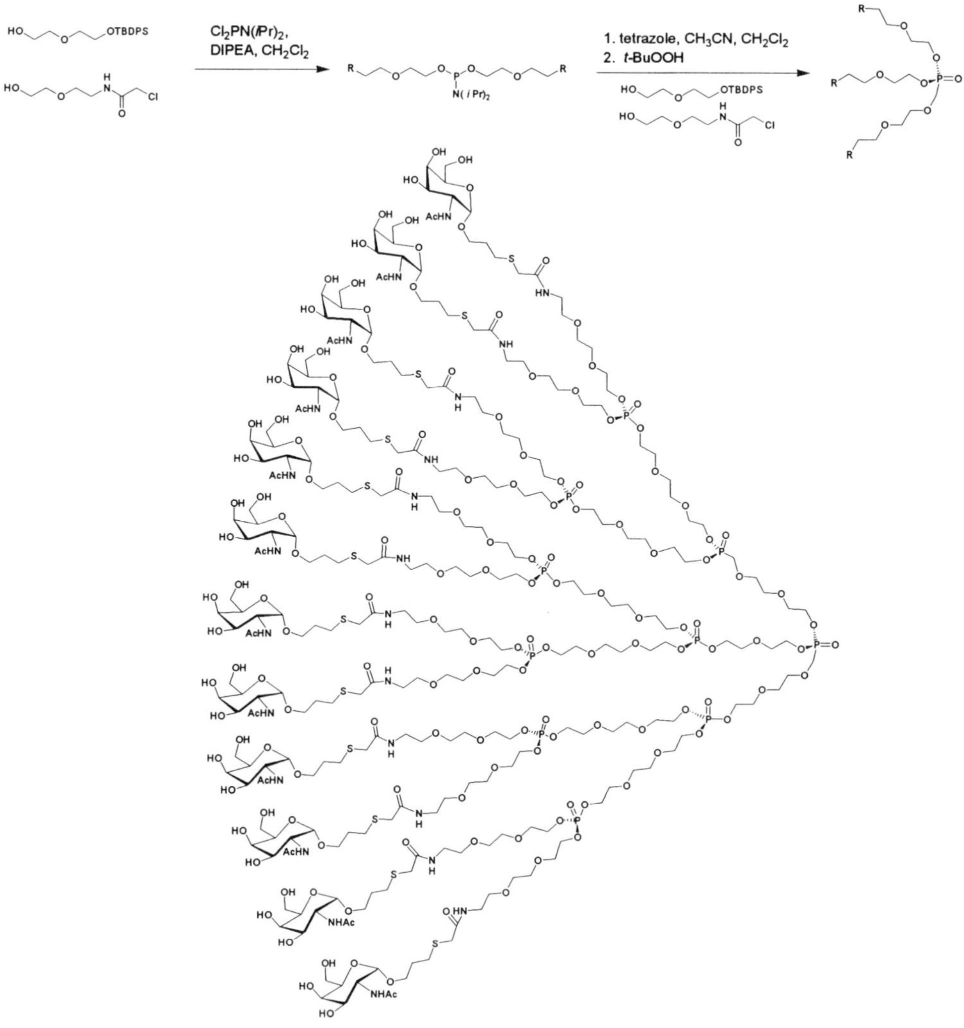

Scheme 6. Phosphotriester-based glycodendrimers

In conclusion, glyco-coated non-carbohydrate dendrimers have been synthesized in large variety since the first example has been published in 1993 [36]. Standard dendrimers, as well as more tailor-made, unusual dendritic cores were used and combined with a variety of carbohydrate epitopes of differing complexity, employing a number of high-yielding ligation chemistries. It has been the intention of this section to give a flavor of the principal approaches which are being pursued rather than to provide a comprehensive survey of all examples published so far.

4
Convergent Multiplication of Carbohydrate Wedges

The convergent alternative for the synthesis of dendrimers is favorable over the divergent preparation, regarding the ease with which the desired monodisperse target molecules can be separated from structurally imperfect impurities. The convergent methodology is, therefore, also an attractive approach for the synthesis of monodisperse glyco-coated dendrimers. Thus, relatively small non-carbohydrate dendrons have been built up and were in turn functionalized with carbohydrate epitopes in the periphery [97]. The resulting glyco-coated molecular wedges were then clustered on, mostly, trifunctional benzeneoid core molecules such as trimesic acid derivatives. The first example, which paved the way for the takeoff of this kind of chemistry was realized with a literature-known [cf. Fig. 3] TRIS-based cluster glucoside **8** which was peptide-coupled with the trimesic acid derivative **9** under standard conditions to yield the respective 9-mer (Scheme 7) [98]. To reduce sterical hindrance in the coupling step, a glycinyl-modified core molecule had been applied. Glycocluster **10** could be obtained in unprotected form after removal of the sugar protective groups.

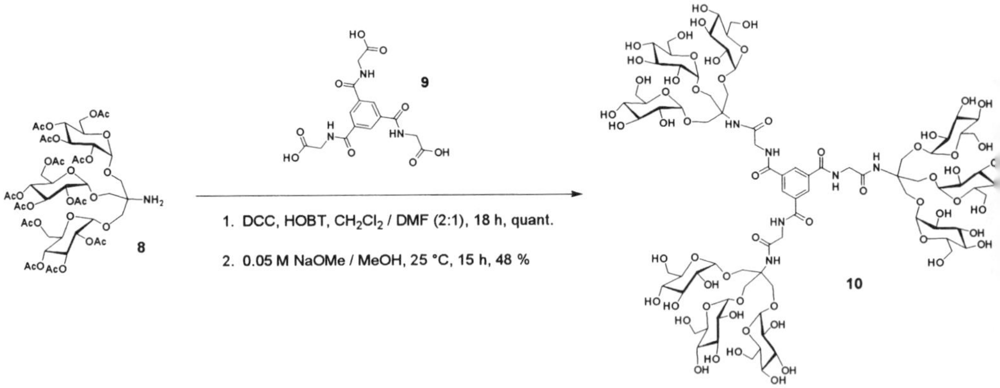

Scheme 7. Starting of with the convergent synthesis of glycodendrimers

By extension of this concept, using a 3,3-iminodipropionic acid derivative **11** as a molecular interface, the hexavalent glycodendron **12** was able to be synthesized (Scheme 8) and eventually be grown to the respective glycodendrimer with eighteen peripheral saccharide units in a DCC/HOBT-assisted peptide-

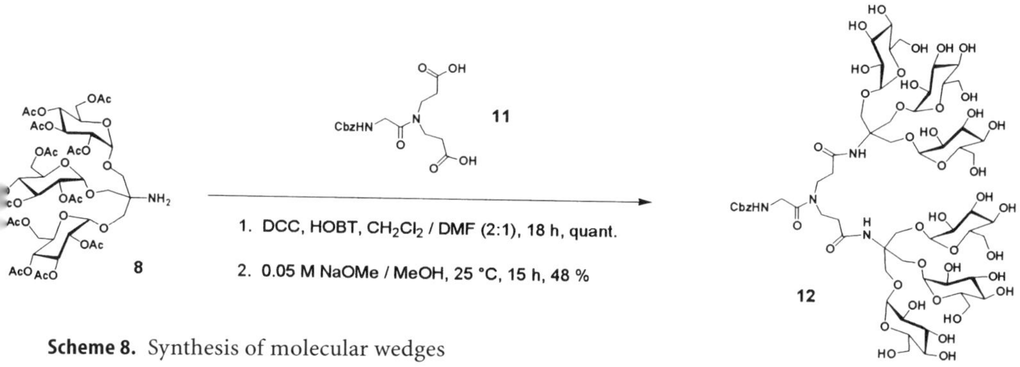

Scheme 8. Synthesis of molecular wedges

coupling reaction. The characterization of the products as monodisperse molecules by high-sophisticated NMR and MALDI-TOF-MS analysis is documented in the literature [98].

Intriguingly, the same chemistry can be utilized employing other, glycobiologically more interesting sugar residues than glucose. Thus, α-mannoside clusters were chosen to investigate analogous glycodendrimers as inhibitors of binding of the lectin concanavalin A to yeast mannan. Cluster **13** (Scheme 9) served as analog of **8** and furnished 9-mer **14** in a peptide coupling reaction, followed by deprotection of the hydroxyl protective groups [99]. In an analogous reaction sequence, glycodendrimer **16** was obtained from **9** and the glycodendron **15**.

The same chemistry, starting from the tetravalent core molecule **17** (Scheme 10), gave the glycodendron **18** [100], which was convergently assembled to glycodendrimer **19**, carrying 16 trivalent cluster mannoside units such as **13**.

When the 3-mer wedge **13** together with the glycodendrimers, **14**, **16**, and **19** were tested in an ELLA (enzyme linked lectin assay) setup, the 9-mer glycodendrimer proved to be the most active inhibitor with an IC_{50} value of 0.65 mM on a molar basis, compared to an inhibitory potency of 2.5 mM for methyl α-D-mannoside. How the, -relatively poor-, clustering effect, which was observed in this system, has to be interpreted has not yet been conclusively discussed.

Obviously, carbohydrate-functionalized dendritic wedges carrying one extra (potentially) reactive group can be used for the functionalization of molecules of many other classes such as the fullerenes [101]. Furthermore, spacer-equipped hexaalkoxytriphenylene core molecules were functionalized with clustered carbohydrate residues to yield discotic liquid crystals (Fig. 9) [102]. The presence of the flat aromatic core and the flexible alkyl side chains are crucial for the formation of the observed columnar mesophases. As glycolipids are naturally occurring examples of mesogenic systems, carbohydrate wedges have been used here as the hydrophilic terminal substructures.

The glycodendrons shown, may also be further employed as hyperbranched building blocks in a context beyond dendrimer chemistry. They have been attached to carrier molecules which offer a cavity, such as calixarenes (Fig. 10) and cyclodextrins (Fig. 11), respectively [103, 104]. These molecules are current-

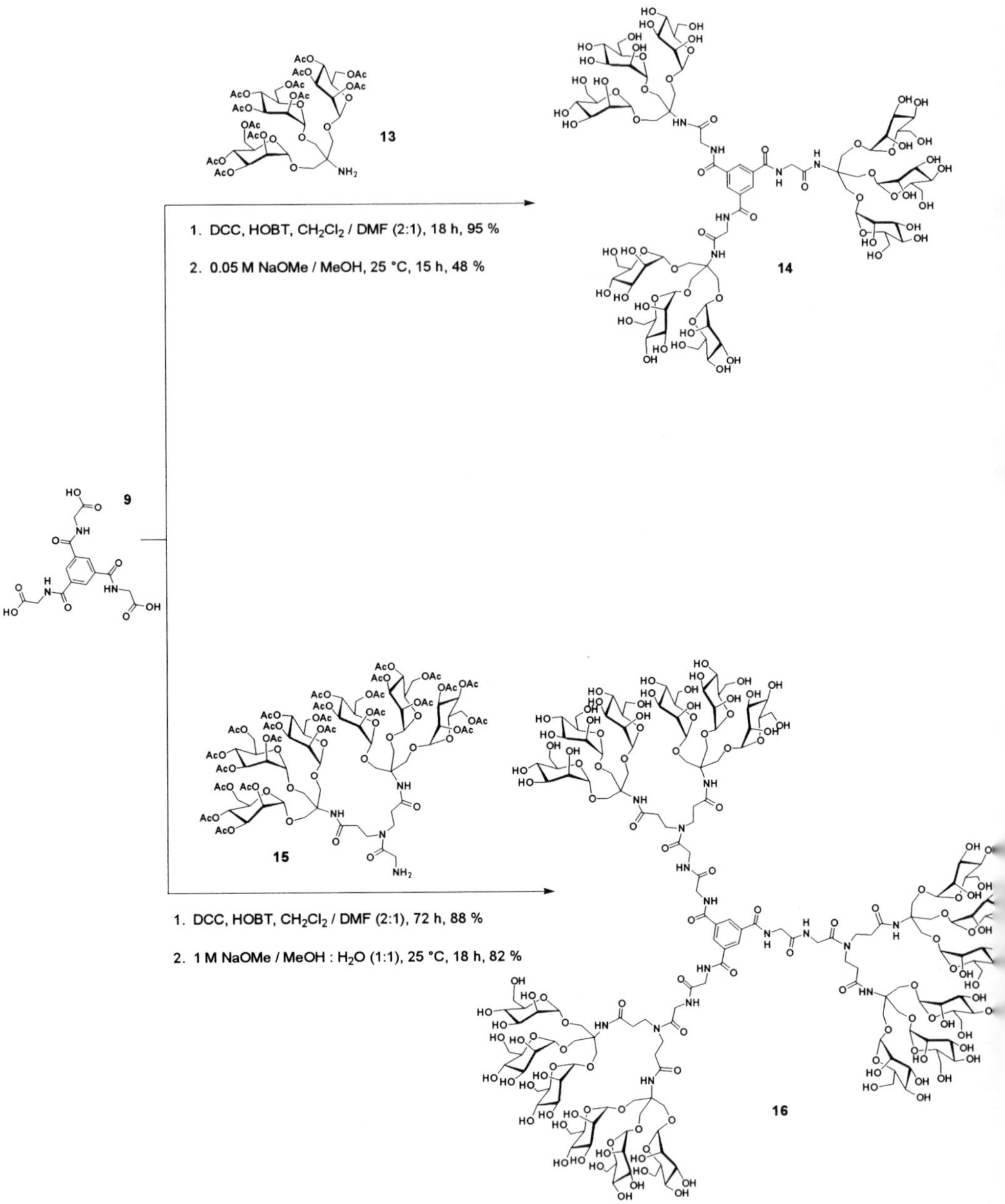

Scheme 9. Convergent synthesis of glycodendrimers

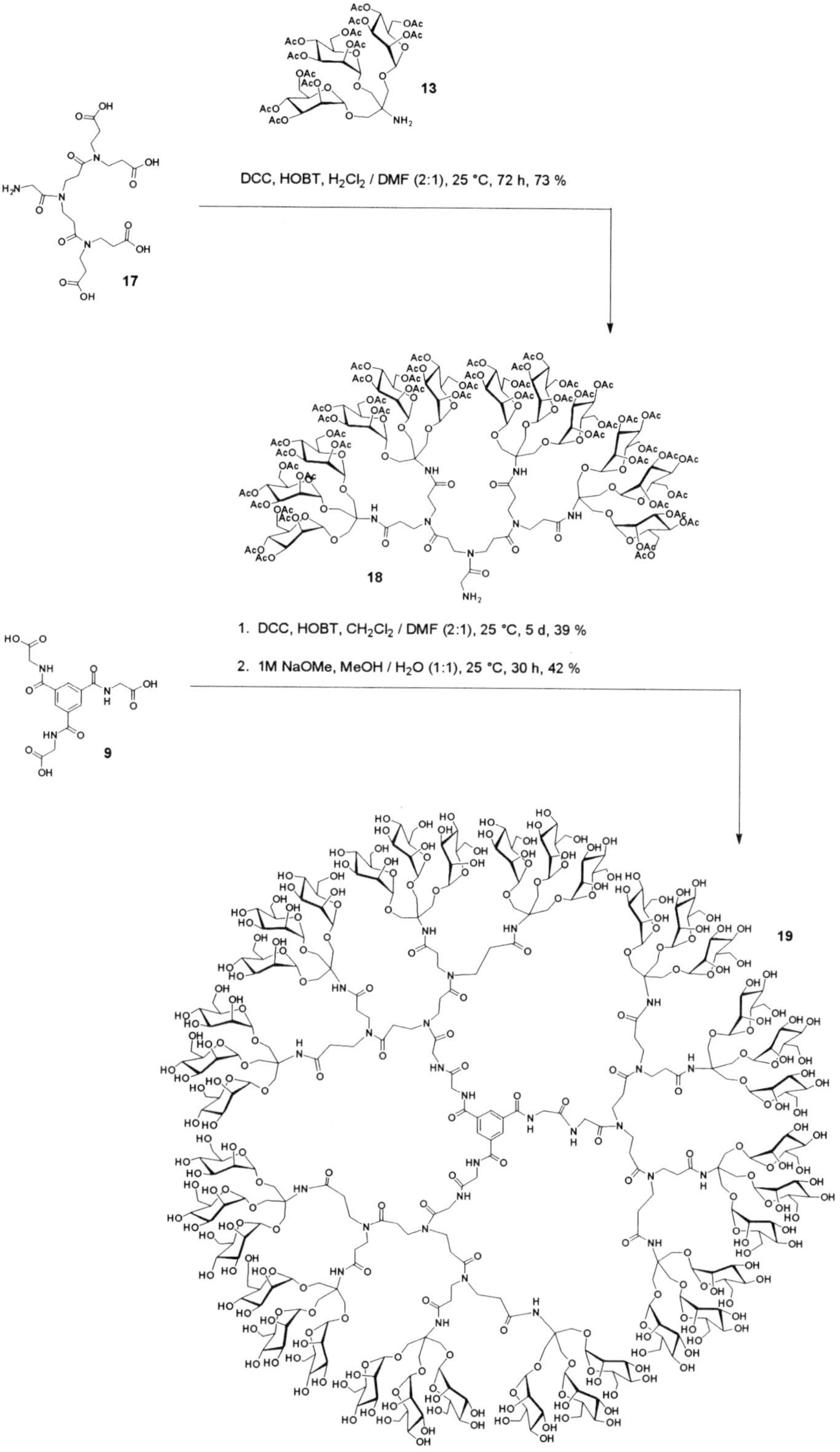

Scheme 10. Convergent synthesis of glycodendrimers

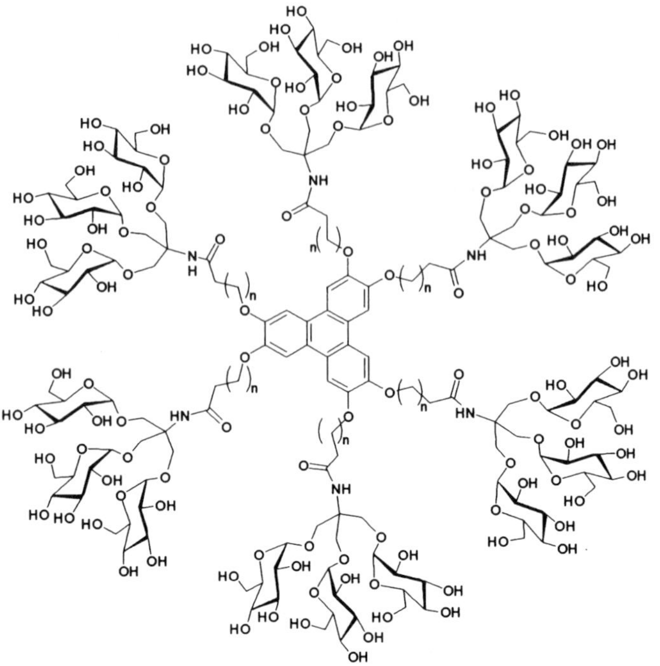

Fig. 9. Sugar-coated discotic liquid crystals

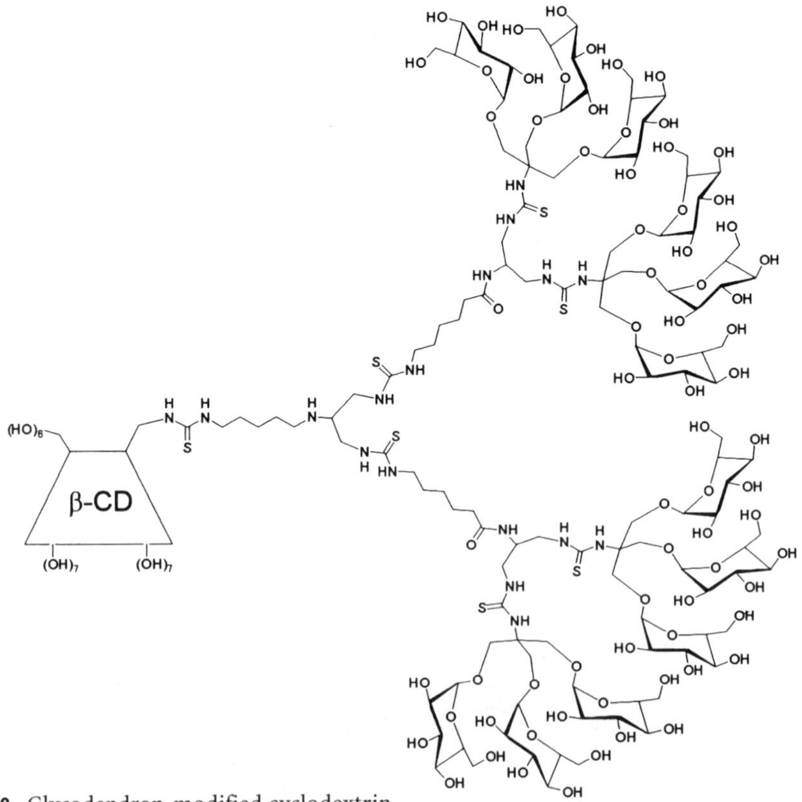

Fig. 10. Glycodendron-modified cyclodextrin

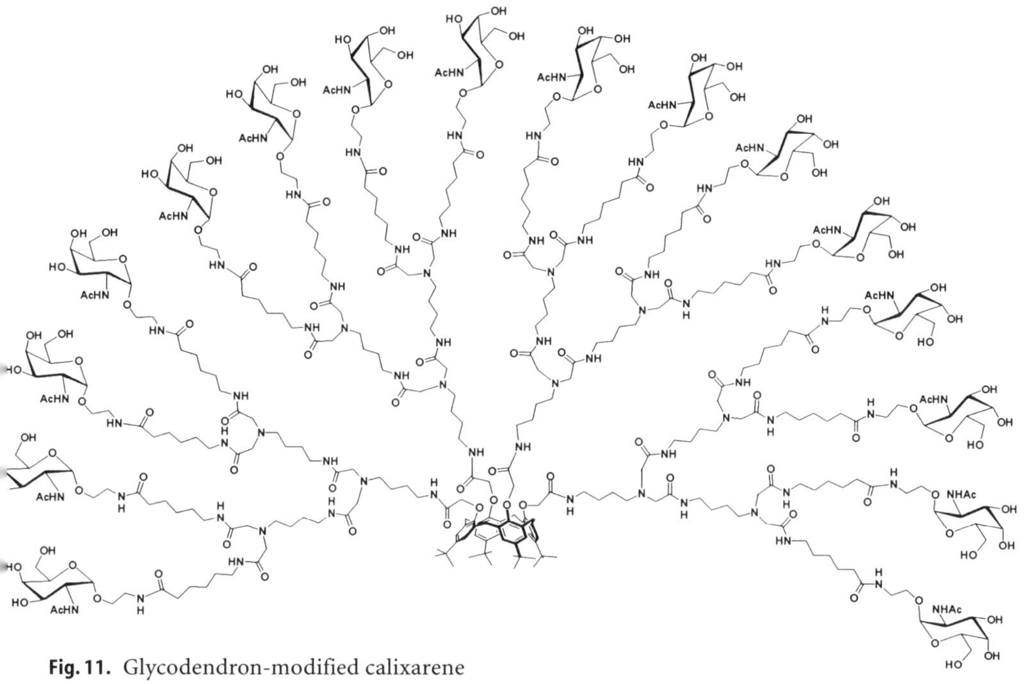

Fig. 11. Glycodendron-modified calixarene

ly being investigated as ligands for various lectins and for their potential as drug carriers.

Finally, the molecular wedge approach can also be combined with classical oligosaccharide synthesis. Oligosacharides themselves, as mentioned earlier, provide the structural requirements for branching and hyper-branching, respectively, and this is utilized by nature for the multiple presentation of carbohydrate epitopes. A careful investigation of the differences which arises from the variation of glycosidic connectivities in oligosaccharides regarding the structural presentation of carbohydrate epitopes in space as well as the resulting differences of biological properties has been carried out with cluster sialosides as inhibitors of influenza virus [105].

When the reducing end of an oligosaccharide is equipped with a functionalized spacer, the resulting molecules can once again be treated as molecular wedges and be clustered on an oligofunctional core molecule. In keeping with this idea, an aglycon-functionalized heptasaccharide has been synthesized according to classical carbohydrate chemistry, in which glucosyl units are branching out from glucosyl units [106]. The spacer-modified tetrasaccharide **22** was synthesized by glycosylation of **21** with **20** (Scheme 11). Acidic cleavage of the 4,6-*O*-benzylidene protecting group in **22** furnished **23**, which could be regioselectively glycosylated, once again using the trisaccharide donor **20**. Thus, the hyperbranched oligosaccharide wedge **24** was obtained. Deprotection at the "focal point" gave amine **25** which was employed in a peptide coupling reaction with a trimesic acid-derived core (Scheme 12) to yield the oligosaccharide glycodendrimer **24**.

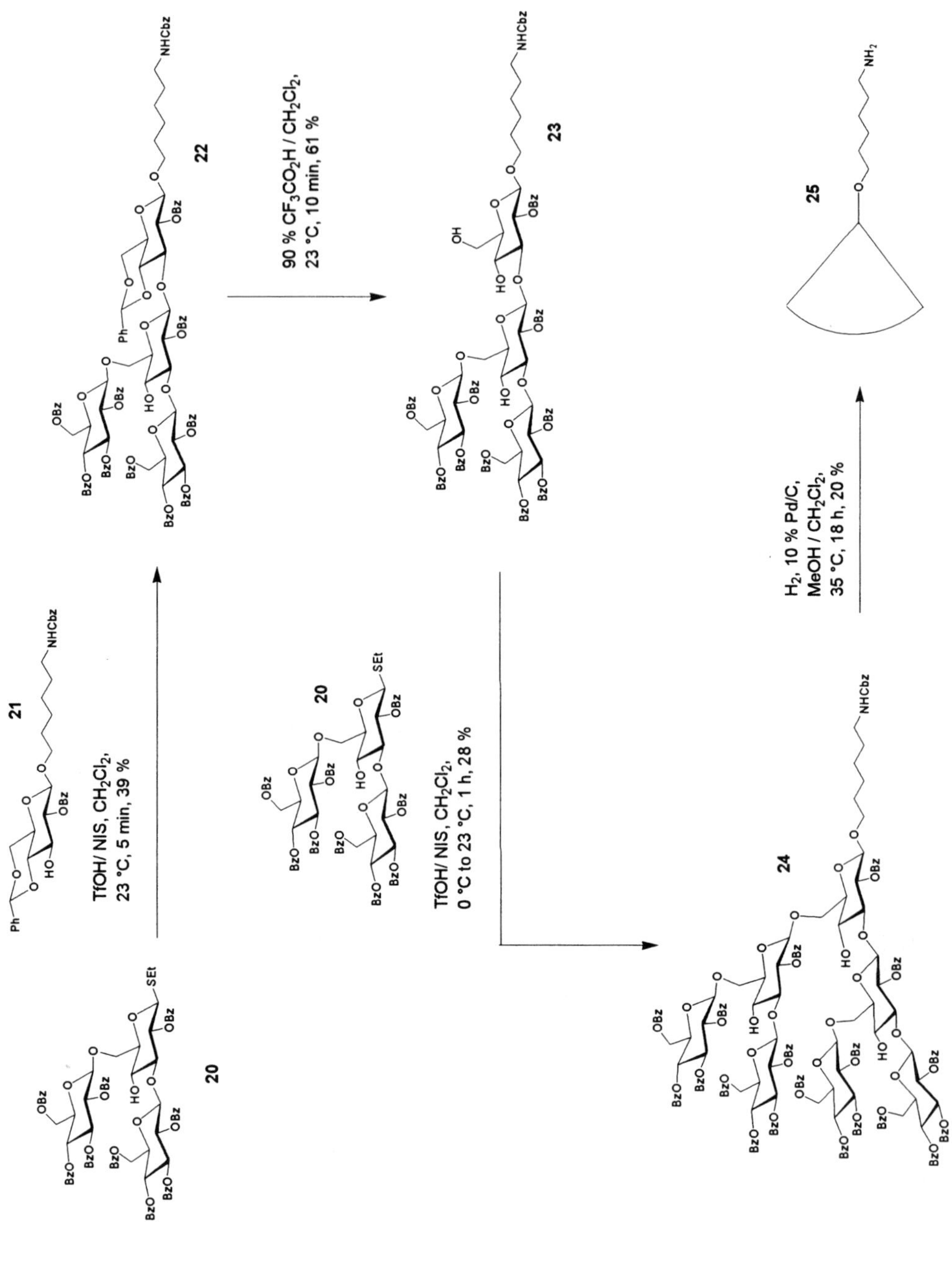

20
21
TfOH/ NIS, CH2Cl2, 23 °C, 5 min, 39 %
22
90 % CF3CO2H / CH2Cl2, 23 °C, 10 min, 61 %
23
20
TfOH/ NIS, CH2Cl2, 0 °C to 23 °C, 1 h, 28 %
24
H2, 10 % Pd/C, MeOH / CH2Cl2, 35 °C, 18 h, 20 %
25

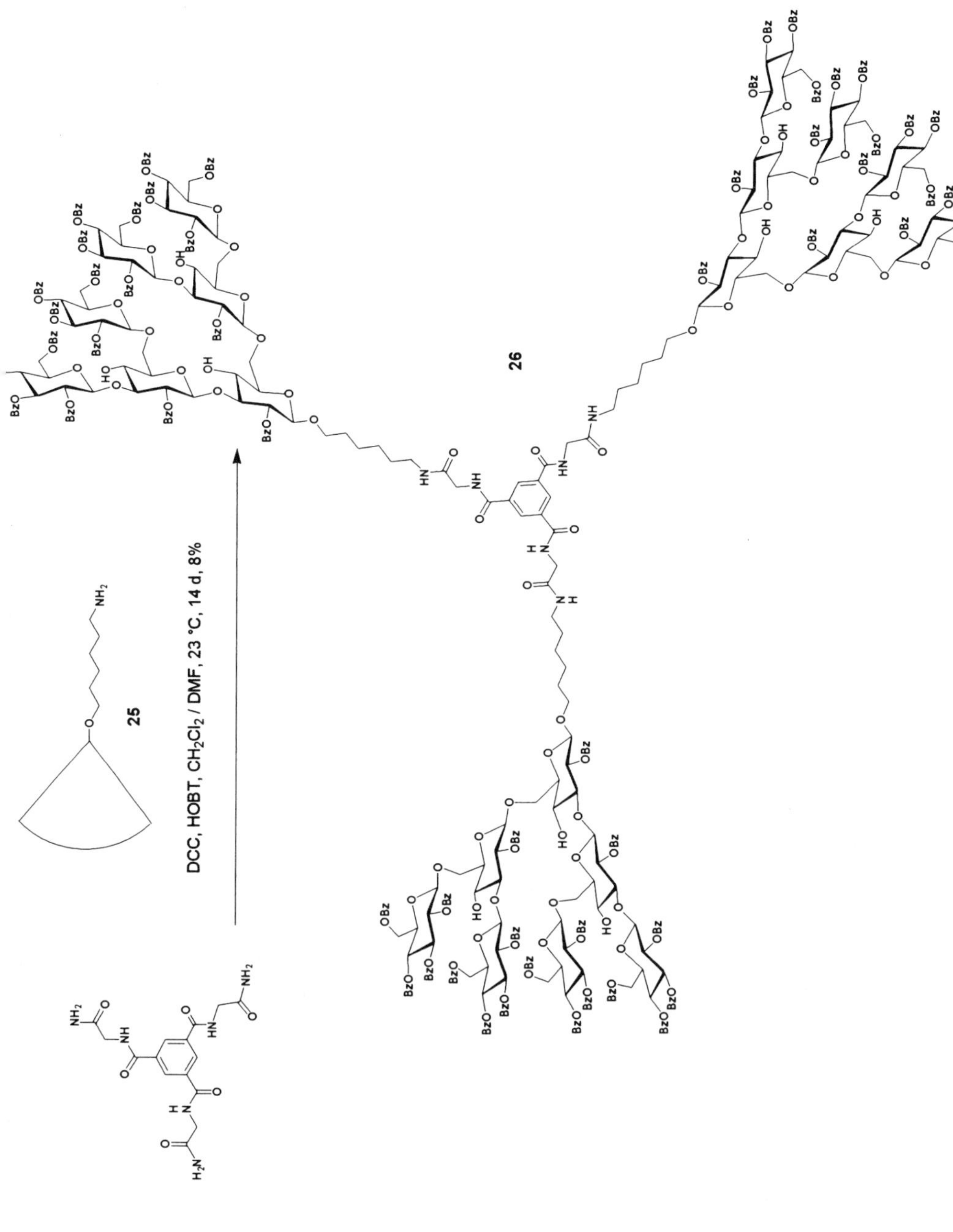

Scheme 12. Convergent synthesis of an oligosaccharide glycodendrimer

5
Dendrimers from Carbohydrate Building Blocks

What one might call the ultimate combination of carbohydrate and dendrimer chemistry is the generationwise construction of glycodendrimers or glycodendrons, respectively, from carbohydrate-derived building blocks only. This may lead to very complex carbohydrate-containing molecules with a strong structural relation to cell-surface carbohydrates. Earlier generations of such glycodendrimers can apparently be designed as monodisperse molecules, whereas higher glycodendrimer generations will rather consist of a mixture of more or less structurally imperfect molecules, due to increase of sterical hindrance. This, however, should not hamper their use in glycobiology as potential nano-scale neoglycoconjugates, which might mimic substructures of the glycocalyx. Also from the point of view of material sciences the issue of dendritic purity does not have to be a critical one in this case. Moreover, the use of carbohydrate building blocks for dendrimer synthesis, is advantageous because it starts from renewable resources and offers possible advantages which can less easily be achieved with other dendrimers, such as biocompatibility, tuning of hydrophobicity and design of hydrophilic cavities, for example.

Growing dendrimers with carbohydrate derivatives is possible using AB_2-type building blocks (Fig. 12). Starting from one and the same AB_2 molecule in which functional groups A and B are orthogonally protected, respective deprotection steps lead to building blocks of type I and of type II. Functions A and B have been chosen so, that in the next step a high-yielding chemical reaction can lead to the protected glycodendrimer of the first generation (G-1, Fig. 12). In an iterative reaction sequence, higher generations will emerge by either a divergent or a convergent approach.

Two quite different AB_2-type saccharides have been designed to meet the requirements of such a methodology (Fig. 13). The first molecules which was used, has been designed for enlargement by peptide coupling. It carries a Boc-protected amino function at the aglycon terminus and two methyl ester groups branching out from the 6-position of the sugar ring. This AB_2 building block was synthesized by exhaustive Michael-type addition of a 6-amino-6-deoxy glucoside to methyl acrylate [107]. Starting from this building block, a saponification

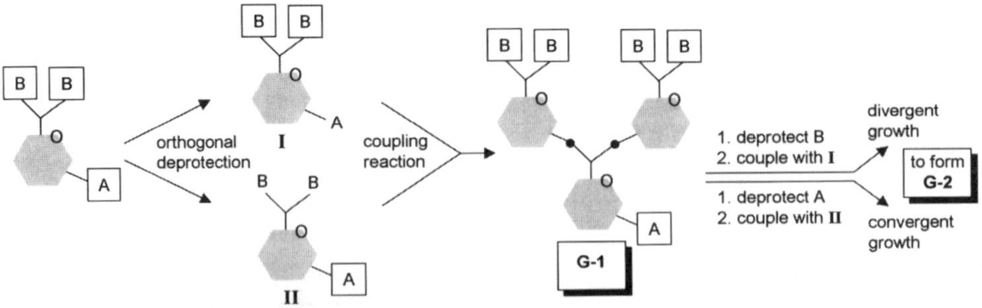

Fig. 12. Dendritic growth with carbohydrate-derived AB_2 building blocks

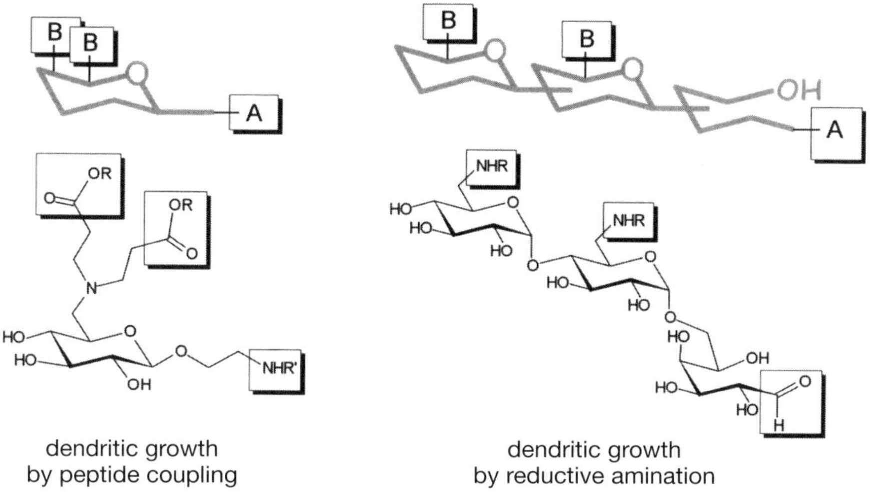

Fig. 13. Two different AB_2-type saccharides for the synthesis of glycodendrimers

step furnished the diacid **27**, while acid-catalyzed deprotection of the amino function led to **28** (Scheme 13). These two components could then be peptide-coupled to give the G-1 glycopeptide dendron **29** after deprotection [108]. Glycodendron **29** may also be regarded as glycopeptide mimetic. It resembles structural elements of naturally occurring glycopeptides and may be an interesting ligand for carbohydrate-binding proteins such as lectins. Within the concepts of dendrimer chemistry it can be grown generationwise to form higher dendron generations, in a convergent or divergent manner. Here, convergent growth proved to be advantageous over a divergent approach and led to the G-2 glycopeptide dendron **30** after removal of the Boc-protective group in **29** and reaction with **27**. Repetition of the reaction sequence consisting of NHBoc-deprotection and peptide coupling led to the G-3 glycodendron **31**, which was able to be obtained without structural defects.

This approach to glycopeptide dendrons implies a number of additional possibilities which have not yet been evaluated. Thus, the synthesis may be carried out on solid phase, possibly even in an automated fashion. Furthermore, the Boc-protected amino group at the focal point of the glycodendron might certainly be employed in a final convergent step leading to more spherical glycodendrimers. Finally this concept also offers the option to employ carbohydrate wedges such as **34** (Scheme 14) in which the peripheral carboxyl groups have been functionalized with carbohydrates [109]. This strategy allows to adjust the glycopeptide dendrons to the chemical nature of specific glycoconjugate oligosaccharides by choice of the respective terminal glycosyl units, e.g. *N*-acetyl-glycosamine, galactose, lactosamine, or sialic acid residues, to name but a few relevant ones.

A different approach to dendrimers from carbohydrate building blocks has utilized a trisaccharide AB_2-type building block (Fig. 13) which was designed for

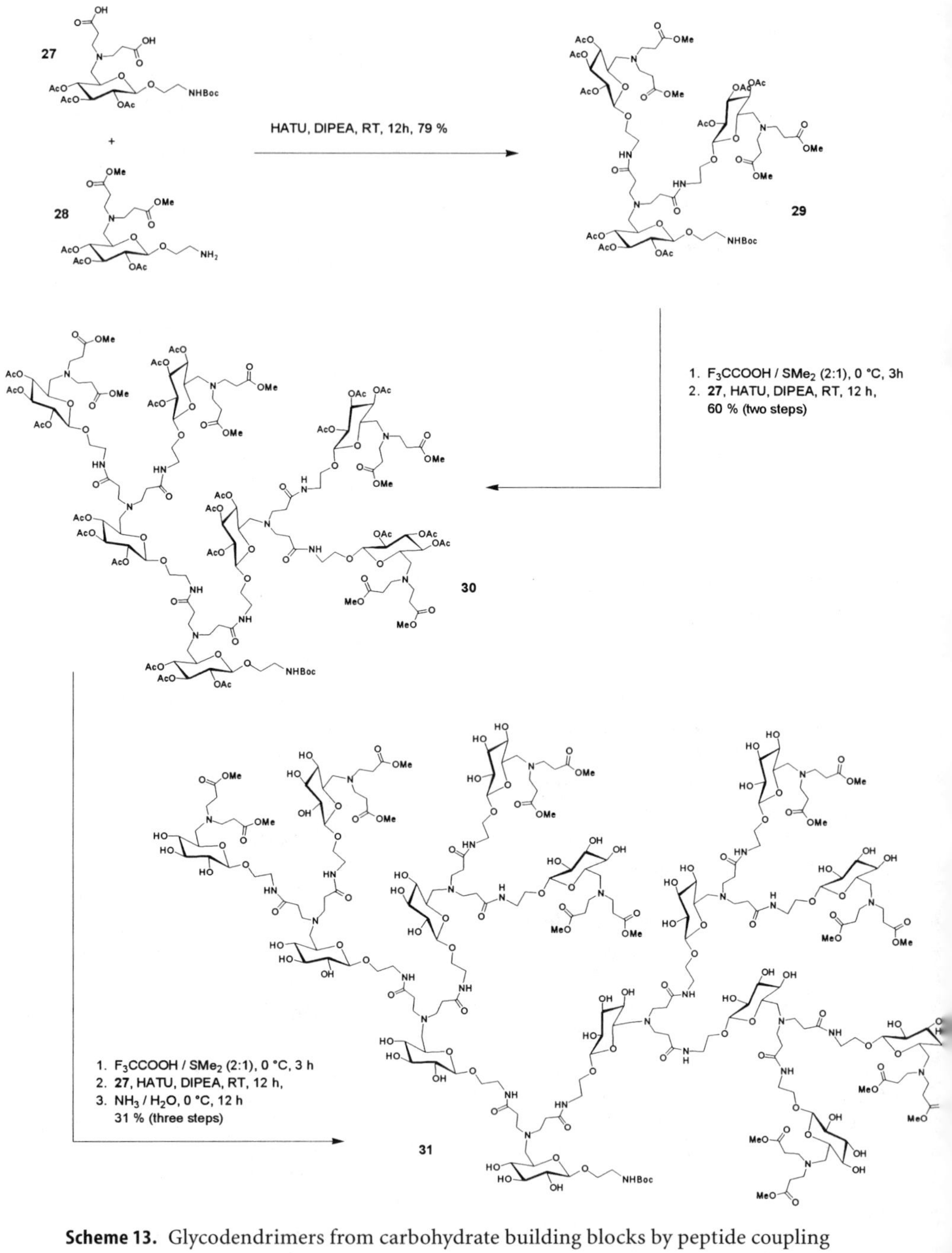

Scheme 13. Glycodendrimers from carbohydrate building blocks by peptide coupling

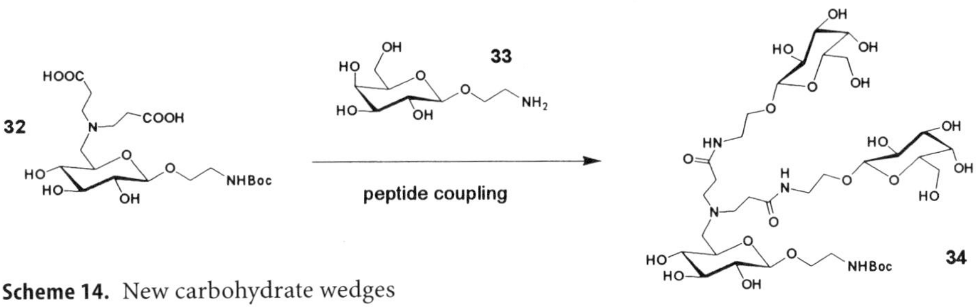

Scheme 14. New carbohydrate wedges

reductive amination as the connecting chemistry [110]. Here, secondary amine functions at the 6′- and 6″-postions resemble the two reactive sites A, whereas the non-reducing end of the trisaccharide with its carbonyl reactivity formed function B. The reactivity of the anomeric functionality in this system was blocked by a 1,3-dioxolane ring resulting from isopropylidenation of the 1- and 2-positions of the terminal sugar ring.

The required trisaccharide building block was obtained from a glycosylation reaction of diisopropylidene galactose (**36**) and the maltosyl trichloroacetimidate **35**, leading to trisaccharide derivative **37**, in which the 6′- and 6″-positions had to be converted into 6-amino-6-deoxy functions regioselectively (Scheme 15). This was possible via a series of standard reactions, giving rise to the dibromide **38** as the key intermediate of the synthesis. It formed the starting material for both the complementary trisaccharide building blocks **39** and **40**, allowing the synthesis of G-1 glycodendron **41** by reductive amination. The trisaccharide diamine **39** was synthesized as bis-secondary amine to avoid overalkylation which was observed with the respective primary amines [111]. The carbonyl component **40**, required for reductive amination, was obtained after acid-catalysed removal of the isopropylidene protective groups of the trisaccharide's galactose moiety. Thereby, the reducing trisaccharide is obtained, which is in equilibrium with its open-chain form (Scheme 15). The open-chain trisaccharide is removed from the equilibrium in the reductive amination reaction with **39**, leading to G-1 glycodendron **41**. This complex molecule now awaits further conjugation and enlargement by either convergent or divergent strategies.

6
Carbohydrate-Centered Dendrimers

The structural and chemical features of the core molecule from which the dendritic shells are emerging, is of importance for the overall shape of a dendrimer. Obviously the multiplicity of the chosen initiator core is a relevant feature in this regard. The most widely used cores for dendrimer synthesis are di- or trifunctional molecules. Core molecules with much higher multiplicity are rarely described or have only been recently introduced. Also the stereochemical properties of a dendrimer initiator core, besides those of the branching units, are important for the properties of the dendrimer. Asymmetric building blocks, e.g.

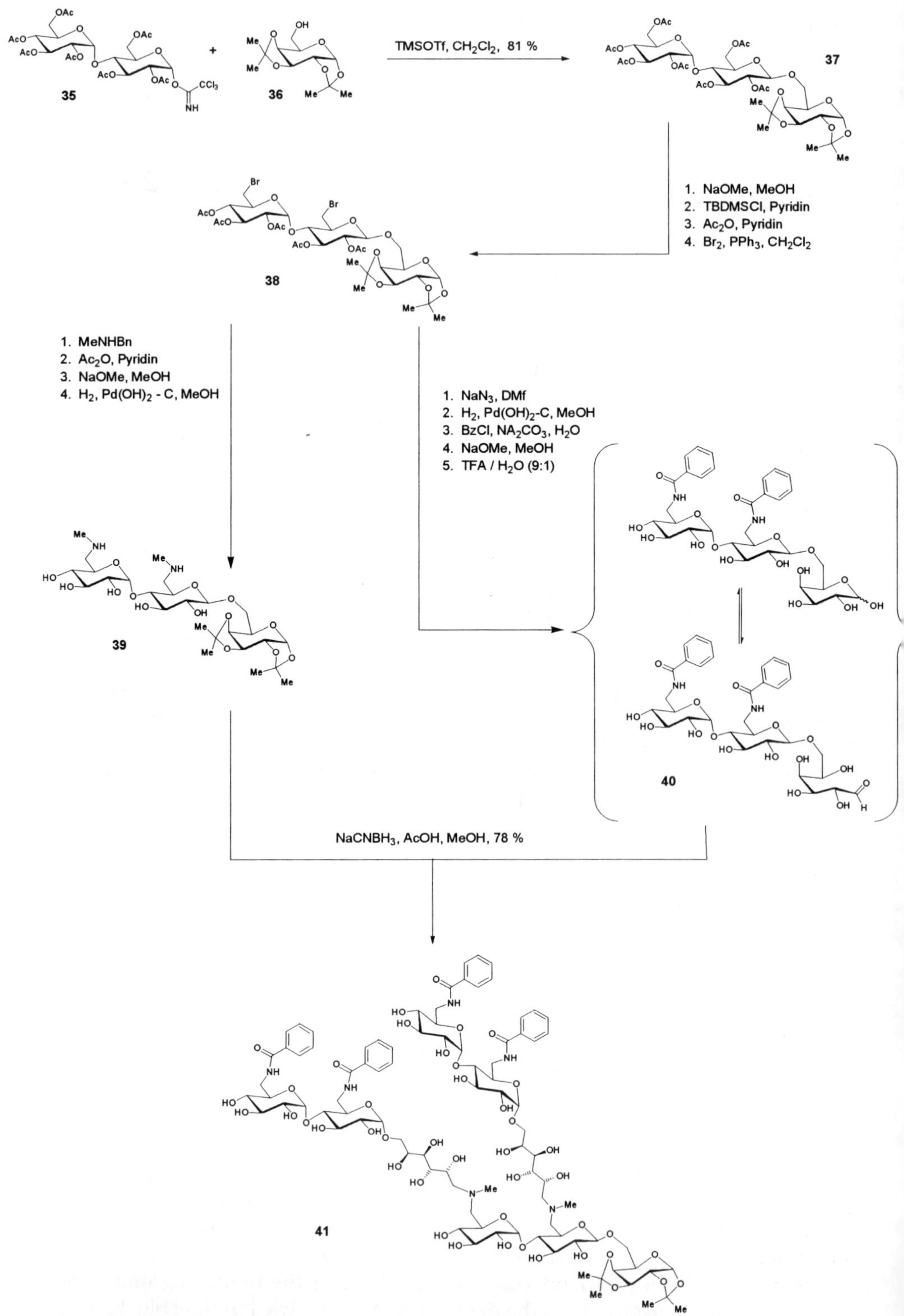

Scheme 15. Glycodendrimers from carbohydrate building block by reductive amination

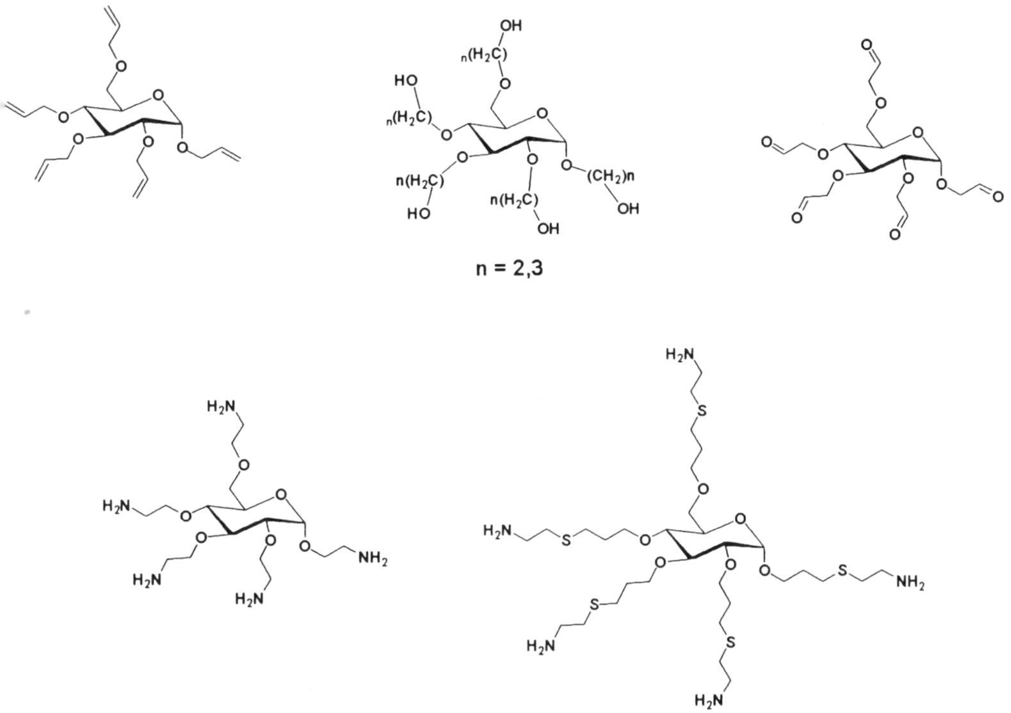

Fig. 14. "Octopus glycosides": glucose derivatives functionalized as pentavalent core molecules

from the chiral pool, have been employed to prepare chiral dendrimers [112] and to use dendrimers in stereoselective catalysis [113].

Carbohydrates are among the most commonly used chiral pool compounds and have just recently received attention as core molecules in dendrimer synthesis. Due to their large structural variety they offer a great potential for the manipulation of the three-dimensional shape of dendrimers. Thus, monosaccharides have been functionalized as pentavalent initiator cores, as shown with D-glucose (Fig. 14), which has been converted into a series of derivatives, which have been named "octopus glycosides" [114]. Variable functional groups in the "outer shell" of these molecules allow to utilize different connectivities for the construction of the first dendrimer generation.

The first carbohydrate-centered glycodendrimer has initiated from the octopus glucoside 2-aminoethyl 2,3,4,6-tetra-*O*-(2-aminoethyl)-α-D-glucoside (**42**), as PAMAM (polyamidoamine) dendrimer (Scheme 16) [115]. Carbohydrate-centered PAMAM (half)generations, **43**, **44**, and **45**, up to G-2 were synthesized in excellent yields and could be analyzed as monodisperse molecules. Interestingly, the specific rotation values of the carbohydrate-centered dendrimers obtained decreased dramatically with increasing molecular weight, according to the "dilution effect" which is known for chiral dendrimers [116]. The molar rotation values, however, remained in the same range around 400 (Table 2).

It will be of interest to investigate, if alteration of the sugar core configuration, e.g. when D-glucose is substituted by D-galactose or D-mannose. Furthermore,

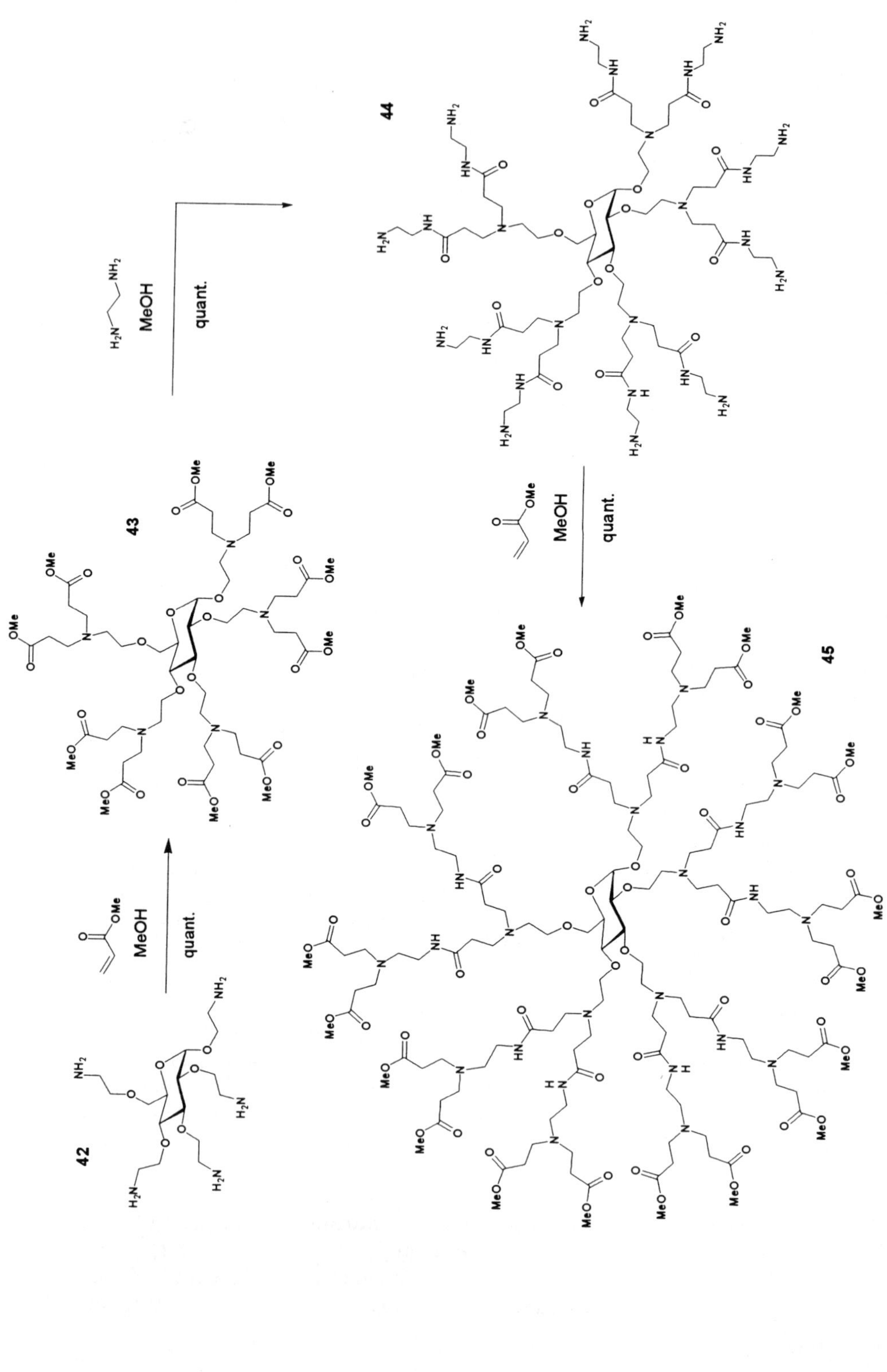
42
MeOH
quant.
43
MeOH
quant.
44
MeOH
quant.
45

Table 2. Specific and molar rotation values of glucose-centered PAMAM dendrimers

Compound	42	43	44	45
Specific rotation values	$[a]_D^{28} = +76.8°$ (c = 0.60, MeOH)	$[a]_D^{27} = +34.6°$ (c = 0.96, MeOH)	$[a]_D^{25} = +32.9°$ (c = 0.26, MeOH)	$[a]_D^{25} = +13.1°$ (c = 0.38, MeOH)
Molar rotation values	+304	+434	+505	+427

instead of monosaccharides, the use of di- or other oligosaccharides, according to the same synthetic strategies might be highly attractive. This allows to easily raise the multiplicity of a dendrimer core, and this will result in a given number of peripheral functions in much earlier generations as when the same dendrimer is constructed from classical, e.g. difunctional initiator cores. New PAMAM dendrimers may be obtained this way which may be better suited for transfection, e.g., as the "activated" PAMAM dendrimers used so far [117]. Toxicity and biocompatibility problems might be concurrently improved.

When the non-reducing disaccharide trehalose was used as core molecule instead of glucose in the same sequence as mentioned above (Scheme 16), structural defects were observed in the first PAMAM half-generation (**47**, Scheme 17) [118]. This problem may be solved by using longer spacers to reduce steric hindrance or, in the contrary, utilized when structurally imperfect PAMAM dendrimers are required [117].

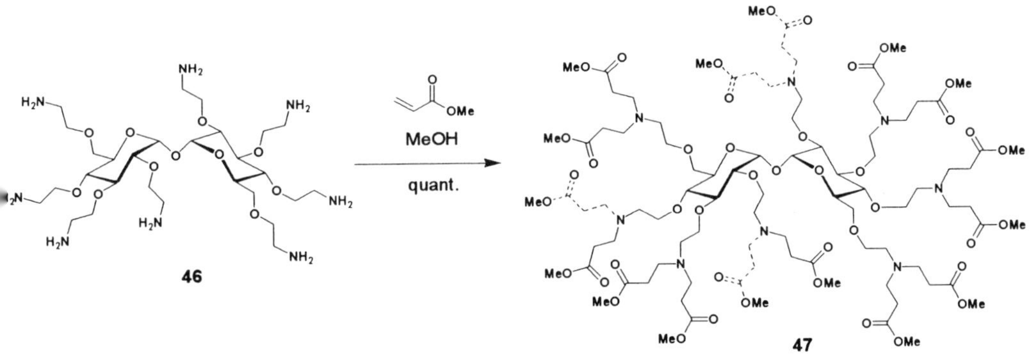

Scheme 17. Structurally defect trehalose-centered PAMAM

Most of the possibilities of carbohydrate-centered dendrimers still await exploration. This is a promising field as carbohydrate chemistry offers much more than synthesis of octopus glycosides. The special reactivity of the anomeric center, e.g., can be used to provide selectively functionalized core molecules for dendrimer synthesis, leading to the respective selectively functionalized carbohydrate-centered dendrimers which are otherwise difficult to obtain (Fig. 15).

A first entry into this kind of chemistry has been made with carbohydrate-centered carbosilane dendrimers in which the aglycon moiety of the core (48) is not involved in the iterative reaction sequence leading to dendritic inflation

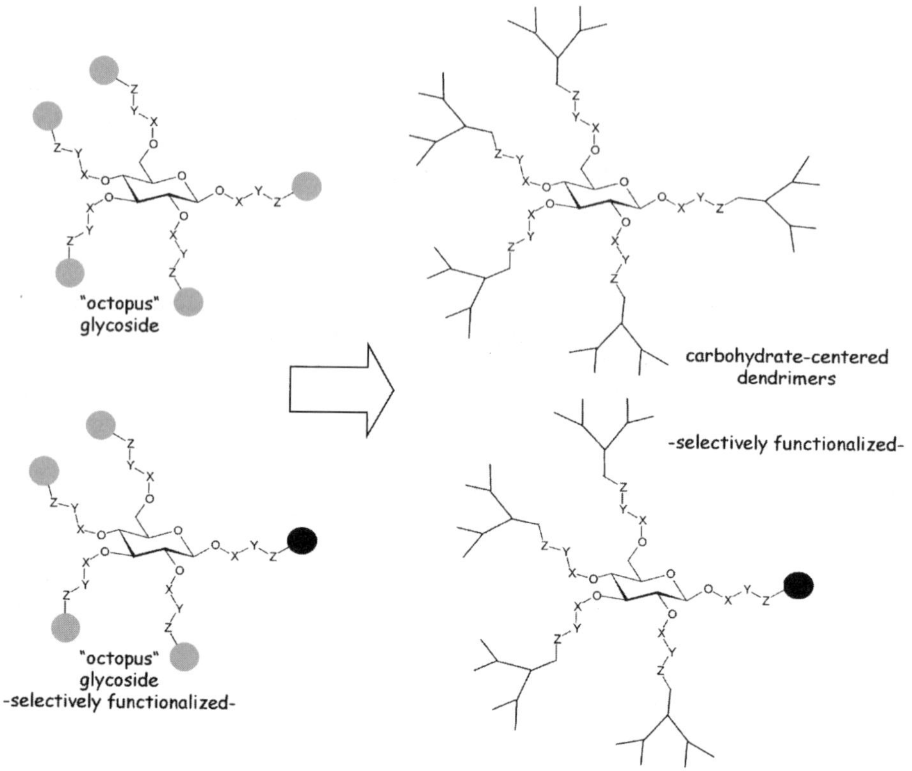

Fig. 15. Carbohydrate cores can also be employed for the construction of selectively functionalized dendrimers

(Scheme 18) [119]. It will be also interesting to see, how this kind of carbohydrate-centered dendrimer compares with the much more polar carbohydrate-based PAMAM dendrimers, introduced earlier. Especially in carbohydrate-protein interaction studies, the effects should be different, due to the variable chemical character of the dendrimer generations used.

Furthermore, carbohydrate moieties have been used at the focal point of Fréchet dendrons in order to allow self-assembly of the so-obtained dendrons

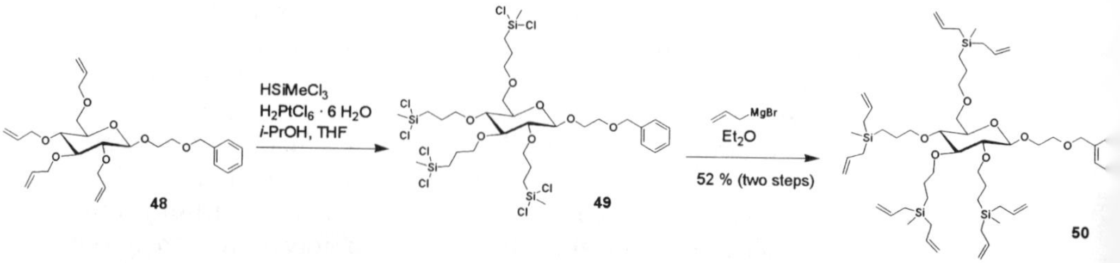

Scheme 18. Selectively functionalized carbohydrate-centered hyperbranched carbosilanes

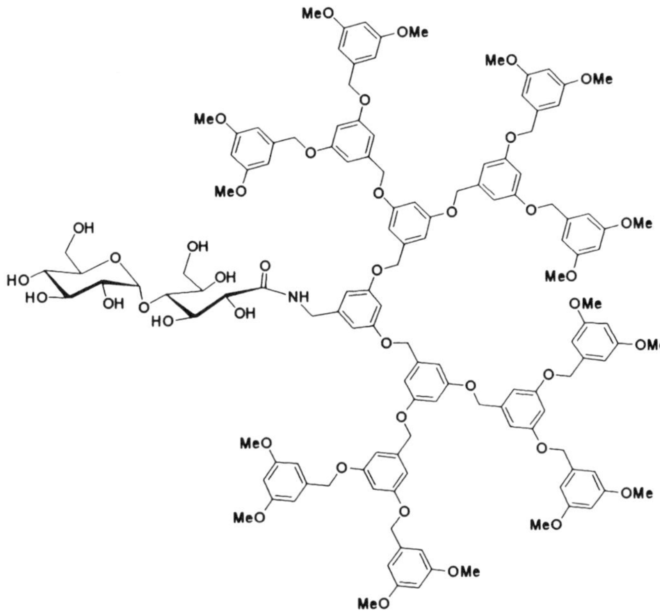

Fig. 16. Glyco-centered dendrons for supramolecular chemistry

(Fig. 16) in organic solvents due to hydrogen-bonding interactions [120]. It has been observed in this system that the size of the formed supramolecular particles decreased with the increase of dendron generations. As these aggregates were rather labile entities, attempts to cross-link them in situ were made to obtain spherical dendritic structures [121].

7
Perspectives

Combining carbohydrate with dendrimer chemistry has provided multivalent neo-glycoconjugates resembling attractive structures, which are useful molecular tools in glycobiology by mimicking the complexity of the naturally occurring multibranched glycoconjugate oligosaccharides. The valency and shape of such molecules can be favorably controlled in order to tackle the questions connected with multivalency in carbohydrate-protein interactions. Moreover, it can be assumed that the potential of carbohydrate-containing dendrimers has by far not been fully exploited as with regard to their possible usefulness as selectively functionalized carrier molecules and their value in supramolecular chemistry and in material sciences. From the point of view of molecular design, many options for glycodendrimer architectures are still awaiting "materialization". It is not difficult to predict that the glycodendrimers will further expand and conquer new areas of natural sciences beyond those which are occupied so far.

8 References

1. Varki A (1993) Glycobiology 3:97
2. Kobata A (1993) Acc Chem Res 26:319
3. Dwek RA (1996) Chem Rev 96:663
4. Varki A, Cummings R, Esko J, Freeze H, Hart G, Marth J (eds) (1999) Essentials of Glycobiology. Cold Spring Harbor Laboratory Press, La Jolla
5. Lis H, Sharon N (1998) Chem Rev 98:637
6. Aebi M, Helenius A (2001) Science 291:2364
7. Kiessling LL, Pohl NL (1996) Chem Biol 3:71
8. Mammen M, Choi SK, Whitesides GM (1998) Angew Chem 110:2908; Angew Chem Int Ed 37:2754
9. Bertozzi CR, Kiessling LL (2001) Science 291:2357
10. Lee YC, Lee RT (1995) Acc Chem Res 28:321
11. Vliegenthart JFG, Montreuil J (1995) In: Montreuil J, Vliegenthart JFG, Schachter H (eds) Glycoproteins, New Compr Biochem, 29a: p 13
12. Boons G-J, Demchenko AV (2000) Chem Rev 100:4539
13. Jung K-H, Müller M, Schmidt RR (2000) Chem Rev 100:4423
14. Gridley JJ, Osborn HMI (2000) J Chem Soc, Perkin Trans I: 1471
15. Křen V, Thiem J (1997) Chem Soc Rev 26:463
16. Takayama S, McGarvey GJ, Wong C-H (1997) Chem Soc Rev 26:407
17. Crout DHG, Vic G (1998) Curr Opin Chem Biol 2:98
18. Schmidt D, Sauerbrei B, Thiem J, (2000) J Org Chem 65:8518
19. Koeller KM, Wong C-H (2000) Chem Rev 100:4465
20. Sears P, Wong C-H (2001) Science 291:2344
21. Herzner H, Reipen T, Schultz M, Kunz H (2000) Chem Rev 100:4495
22. Kahne D (1997) Curr Opin Chem Biol 1:130
23. Hilaire PMS, Meldal M (2000) Angew Chem 112:1211; Angew Chem Int Ed 39:1162
24. Seeberger PH, Haase W-C (2000) Chem Rev 100:4349
25. Ye X-S, Wong C-H (2000) J Org Chem 65:2410
26. Sears P, Wong C-H (1999) Angew Chem 111:2446; Angew Chem Int Ed 38:2300
27. Roy R (1997) In: Witczak ZJ, Nieforth KA (eds) Carbohydrates in Drug Design. Marcel Dekker Inc, New York, p 83
28. Patel A, Lindhorst TK (2000) J Org Chem 66:2674
29. Kötter S, Krallmann-Wenzel U, Ehlers S, Lindhorst TK (1998) J Chem Soc, Perkin Trans I: 2193
30. Lindhorst TK, Kieburg C (1996) Angew Chem 108:2083; Angew Chem Int Ed Engl 35:1953
31. Biessen EAL, Noorman F, van Teijlingen ME, Kuiper J, Barrett-Bergshoeff M, Bijsterbosch MK, Rijken DC, van Berkel TJC (1996) J Biol Chem 271:28024
32. Dell A, Morris HR (2001) Science 291:2351
33. Unverzagt C (1997) Angew Chem 109:2078; Angew Chem Int Ed Engl 36:1989
34. Lindhorst TK (1996) Nachr Chem Tech Lab 44:1073
35. Roy R (1996) Polymer News 21:226
36. Roy R, Zanini D, Meunier SJ, Romanowska A (1993) J Chem Soc, Chem Commun:1896
37. Kitov PI, Sadowska JM, Milvey G, Armstrong GD, Ling H, Pannu NS, Read RJ, Bundle DR (2000) Nature 403:669
38. Dubber M, Lindhorst TK (2000) J Org Chem 65:5275
39. Hansen HC, Haatja S, Finne J, Magnusson G (1997) J Am Chem Soc 119:6974
40. Das SK, Roy R (1999) Tetrahedron Lett 40:4015
41. Dubber M, Lindhorst TK (2001) Synthesis:327
42. Roy R (1996) In: Khan SH, O'Neill RA (eds) Modern Methods in Carbohydrate Synthesis. Harwood academic publishers, Amsterdam, p 378
43. Bovin NV, Gabius JH (1995) J Chem Soc: 413

44. Roy R (1996) TIGG 8:79
45. Furuike T, Aiba S, Suzuki T, Tabahashi T, Suzuki, Y, Yamada K, Nishimura, S-I (2000) J Chem Soc, Perkin Trans I:3000
46. Reuter JD, Myc A, Hayes MM, Gan Z, Roy R, Qin D, Yin R, Piehler LT, Esfand R, Tomalia DA, Baker JJr (1999) Bioconj Chem 10:271
47. Liang R, Loebach J, Horan N, Ge M, Thompson C, Yan L, Kahne D (1997) Proc Natl Acad Sci USA 94:10554
48. Pohl NL, Kiessling LL (1999) Synthesis: 1515
49. Strong LE, Kiessling LL (1999) J Am Chem Soc 121:6193
50. Zistler A, Koch S, Schlüter AD (1999) J Chem Soc, Perkin Trans I: 501
51. Nomura K, Schrock RR (1996) Macromolecules 29:540
52. Lees WJ, Spaltenstein A, Kingery-Wood JE, Whitesides GM (1994) J Med Chem 37:3419
53. Sigal GB, Mammen M, Dahmann G, Whitesides GM (1996) J Am Chem Soc 118:3789
54. Fan E, Zhang Z, Minke WE, Hou Z, Verlinde CLMJ, Hol WGJ (2000) J Am Chem Soc 122:2663
55. Lundquist JJ, Debenham SD, Toone EJ (2000) J Org Chem 65:8245
56. Vepřek P, Ježek J (1999) J Peptide Sci 5:5
57. Ježek J, Velek J, Vepřek P, Velková V, Trnka T, Pecka J, Ledvina M, Vondrašek J, Písačka M (1999) J Peptide Sci 5:46
58. Zanini D, Roy R (1998) Architectonic Neoglycoconjugates: Effects of Shapes and Valencies in Multiple Carbohydrate-Protein Interactions. In: Chapleur Y (ed) Carbohydrate Mimics, Concepts and Methods. Wiley-VCH, Weinheim, p 385
59. Jayaraman N, Nepogodiev SA, Stoddart JF (1997) Chem Eur J 3:1193
60. Roy R (1997) In: Driguez H, Thiem J (eds) Top Curr Chem 187: Glycoscience, Synthesis of Substrate Analogs and Mimetics, p 241
61. Roy R (1996) Curr Opin Struct Biol 6:692
62. Kiessling LL, Gestwicki JE, Strong LE (2000) Curr Opin Chem Biol 4:696
63. Roy R (1999) Carbohydr Eur 27:34
64. Weis WI, Drickamer K (1996) Annu Rev Biochem 65:441
65. Weis WI, Taylor ME, Drickamer K (1998) Immunol Rev 163:19
66. Lasky LA (1995) Annu Rev Biochem 64:113
67. Toone E (1994) Curr Opin Struct Biol 4:719
68. Yarema KJ, Bertozzi CR (1998) Curr Opin Chem Biol 2:49
69. Dimick SM, Powell SC, McMahon SA, Moothoo DN, Naismith JH, Toone EJ (1999) J Am Chem Soc 121:10286
70. Roy R, Pagé D, Perez SF, Bencomo V (1998) Glycoconjugate J 15:251
71. Roy R, Zanini D, Meunier SJ, Romanowska A (1994) In: ACS Symposium Series 560:104
72. Page D, Zanini D, Roy R (1996) Bioorg Medicin Chem 4:1949
73. Zanini D, Park WKC, Roy R (1995) Tetrahedron Lett 36:7383
74. Roy R, Baek MG, Xia Z, Rittenhouse-Diakun K (1999) Glycoconjugate J 16: S53
75. Roy R, Park WKC, Zanini D, Foxall C, Srivastava OP (1997) Carbohydr Lett 2:259
76. Palcic MM, Li H, Zanini D, Bhella RS, Roy R (1998) Carbohydr Res 305:433
77. Zanini D, Roy R (1997) Bioconj Chem 8:187
78. Page D, Zanini D, Roy R (1996) Bioorganic & Medicinal Chem 4:1949
79. Grandjean C, Rommens C, Gras-Masse H, Melnyk O (1999) Tetrahedron Lett 40:7235
80. Grandjean C, Rommens C, Gras-Masse H, Melnyk O (2000) Angew Chem 112:1110; Angew Chem Int Ed 39:1068
81. Granjean C, Gras-Masse H, Melnyk O (2001) Chem Eur J 7:230
82. Aoi K, Itoh K, Okada M (1995) Macromolecules 28:5391
83. Kieburg C, Lindhorst TK (1997) Tetrahedron Lett 38:3885
84. Sharon N (1987) FEBS Lett 217:145
85. Lindhorst TK, Kieburg C, Krallmann-Wenzel U (1998) Glycoconjugate J 15:605
86. Andre S, Cejas Ortega PJ, Perez MA, Roy R, Gabius HJ (1999) Glycobiology 9:1253
87. Baek M-G, Rittenhouse-Olson K, Roy R (2001) Chem Commun:257
88. Thompson JP, Schengrund C-L (1997) Glycoconjugate J 14:837

89. Peerlings HWI, Nepogodiev SA, Stoddart JF, Meijer EW (1998) Eur J Org Chem: 1879
90. Ashton PR, Boyd SE, Brown CL, Nepogodiev SA, Meijer EW, Peerlings HWI, Stoddart JF (1997) Chem Eur J 3:974
91. Pavlov GM, Korneeva EV, Jumel K, Harding SE, Meijer EW, Peerlings HWI, Stoddart JF, Nepogodiev SA (1999) Carbohydrate Polymers 38:195
92. Zanini D, Roy R (1997) J Am Chem Soc 119: 2088
93. Zanini D, Roy R (1998) J Org Chem 63:3486
94. Llinares M, Roy R (1997) Chem Commun: 2119
95. Roy R, Park WKC, Wu Q, Wang SN (1995) Tetrahedron Lett 25:4377
96. Meunier SJ, Wu Q, Wang SN, Roy R (1997) Can J Chem 75:1472
97. Jayaraman N, Stoddart JF (1997) Tetrahedron Lett 38:6767
98. Ashton PR, Boyd SE, Brown CL, Jayaraman N, Nepogodiev SA, Stoddart FJ (1996) Chem Eur J 2:1115
99. Ashton PR, Boyd SE, Brown CL, Jayaraman N, Stoddart JF (1997) Angew Chem 109:756; Angew Chem Int Ed Engl 36:732
100. Ashton PR, Hounsell EF, Jayaraman N, Nilsen TM, Spencer N, Stoddart JF (1998) J Org Chem 63:3429
101. Caedullo F, Diederich F, Echegoyen L, Habicher T, Jayaraman N, Leblanc RM, Stoddart JF, Wang S (1998) Langmuir 14:1955
102. Barbera J, Garces AC, Jayaraman N, Omenat A, Serrano JL, Stoddart JF (1999) Am Chem Soc, Polym Preprints 40:488
103. Roy R, Kim JM (1999) Angew Chem 111: 380; Angew Chem Int Ed 38:369
104. Baussanne I, Benito, JM, Ortitz Mellet C, García Fernández JM, Law H, Defaye J, Chem. Commun 2000:1489
105. Sabesan S, Duus, JØ, Neira S, Domaille P, Kelm S, Paulson JC, Bock K, J Am Chem Soc 1992, 114:8363
106. Colonna B, Harding VD, Nepogodiev SA, Raymo FM, Spencer N, Stoddart JF (1998) Chem Eur J 4:1244
107. Kieburg C, Sadalapure K, Lindhorst TK (2000) Eur J Org Chem: 2035
108. Sadalapure K, Lindhorst TK (2000) Angew Chem 112: 2066; Angew Chem Int Ed 39:2010
109. Sadalapure K, Lindhorst TK, unpublished results
110. Turnbull WB, Pease AR, Stoddart JF (2000) Chembiochem, 1:70
111. Dubber M, Lindhorst TK (2001) Synthesis: 327
112. Seebach D, Rheiner PB, Greiveldinger G, Butz T, Sellner H (1998) Top Curr Chem 197:125
113. Bolm C, Derrien N, Seger A (1999) Chem Commun: 2087
114. Dubber M, Lindhorst TK (1998) Carbohydr Res 310:35
115. Dubber M, Lindhorst TK (1998) Chem Commun: 1265
116. Seebach D, Lapieree J-M, Greiveldinger G, Skobridis K (1994) Helv Chem Acta 77:1673
117. Weber M, Nachr Chem (2000) 48:18
118. Dubber M, Dissertation (2001) Universität Hamburg
119. Boysen M, Lindhorst TK (1999) Org Lett 1:1925
120. Ikeda A, Numata M, Shinkai S (1999) Chem Lett: 929
121. Numata M, Ikeda A, Shinkai S (2000) Chem Lett: 370

Author Index Volume 201–217

Author Index Vols. 26–50 see Vol. 50
Author Index Vols. 51–100 see Vol. 100
Author Index Vols. 101–150 see Vol. 150
Author Index Vols. 151–200 see Vol. 200

The volume numbers are printed in italics

Printing (Computer to Film): Saladruck Berlin
Binding: Stürtz AG, Würzburg